# What's Goin' On?

## Trends and Issues in American Education

### Edited by
### Howard E. Taylor

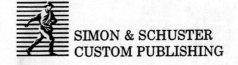

Copyright © 1998 by Simon & Schuster Custom Publishing.
All rights reserved.

This copyright covers material written expressly for this volume by the editor/s as well as the compilation itself. It does not cover the individual selections herein that first appeared elsewhere. Permission to reprint these has been obtained by Simon & Schuster Custom Publishing for this edition only. Further reproduction by any means, electronic or mechanical, including photocopying and recording, or by any information storage or retrieval system, must be arranged with the individual copyright holders noted.

Printed in the United States of America

10 9 8 7 6 5 4 3 2 1

*Please visit our website at www.sscp.com*

ISBN 0-536-01104-4
BA 98062

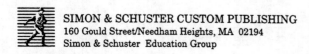

SIMON & SCHUSTER CUSTOM PUBLISHING
160 Gould Street/Needham Heights, MA 02194
Simon & Schuster Education Group

# Contents

**Section I: Urban Education and Environments**

Contexts That Promote Success For Inner-City Students . . . . . . . . . . . . 2
    Norman A. Newberg
    *University of Pennsylvania*
    Randall B. Sims
    *Senior Project Coordinator, Say Yes to Education*

Next Steps in Inner-City Education . . . . . . . . . . . . . . . . . . . . . . . 18
    Margaret C. Wang
    *Temple University Center for Research in Human Development and Education*

Life in the Bricks . . . . . . . . . . . . . . . . . . . . . . . . . . . . . . . . . 33
    Charles Bruckerhoff
    *University of Connecticut*

Child Rearing and Education in Urban Environments . . . . . . . . . . . . 45
    Josephine A. Bright
    *Wheelock College*
    Christopher Williams
    *California State University, Northridge*

Safe Schools: Policies and Practices That Impact Teachers and
Teacher Training Programs . . . . . . . . . . . . . . . . . . . . . . . . . . . 56
    Dawn Kum-Walks, Ph.D.
    *Assistant Professor, Old Dominion University*

**Section II: Vital Signs of American Schools**

International Comparisons . . . . . . . . . . . . . . . . . . . . . . . . . . . . 64
    Kevin Bushweller
    *Associate Editor, The American School Board Journal*

**Section III: Addressing Developmental, Sexual, Cultural, Cognitive, and Ableness Diversity**

Meeting the Developmental and Instructional Needs of Middle
School Students: What Do Teachers Need to Know? . . . . . . . . . . . . 98
    M. Lee Manning
    *Old Dominion University*

Creating Safe and Open Schools for Gay, Lesbian, and
Questioning Youth: A Rationale and Practical Strategies . . . . . . . . . 110
    Howard E. Taylor
    *Old Dominion University*

Supporting the Invisible Minority ........................ 117
   John D. Anderson

Let's Stop Ignoring Our Gay and Lesbian Youth ............ 122
   Ann T. Edwards

What Does It Mean? ..................................... 125
   Natalie G. Adams
    *Georgia Southern University*

Multicultural Education ................................. 133
   S. Rex Morrow, Ed.D.
    *Associate Professor of Education, Old Dominion University*

MI and Curriculum Development ........................... 148
   Thomas Armstrong

How Our Brain Learns, Remembers, and Forgets ............ 163
   Robert Sylwester

Promising Practices That Foster Inclusive Education ..... 175
   Alice Udvari-Solner and Jacqueline S. Thousand

## Section IV: Promising Practices in Contemporary American Education

Technological Literacy for Educators .................... 190
   Richard C. Overbaugh
    *Old Dominion University*

Integrating Children's/Young Adult Literature into the Social Studies .. 207
   Katherine T. Bucher
    *Old Dominion University*

Characteristics, Responsibilities, and Qualities of Urban School Mentors ........................ 215
   Edith Guyton
    *Georgia State University*
   Francisco Hidalgo
    *Texas A&M University, Kingsville*

Modeling and Mentoring in Urban Teacher Preparation ..... 221
   Geneva Gay
    *University of Washington, Seattle*

Teachers' Perspectives on School/University Collaboration in Global Education ..................................... 232
   Timothy Dove, James Norris, and Dawn Shinew

School Uniforms and Safety .............................. 245
   M. Sue Stanley
    *California State University, Long Beach*

Service Learning ....................................... 253
   Richard J. Kraft
    *University of Colorado at Boulder*

The Teacher's Place in the Formation of Students' Character . . . . . . . 272
    Carolyn Gecan
    *Thomas Jefferson High School*
    Bernadette Mulholland-Glaze
    *Mt. Vernon High School*

Family Group Conferencing . . . . . . . . . . . . . . . . . . . . . . . . . 281
    Bruce R. Taylor and Glenn Kummery

# Section I

## *Urban Education and Environments*

# Contexts That Promote Success For Inner-City Students

NORMAN A. NEWBERG
*University of Pennsylvania*

RANDALL B. SIMS
*Senior Project Coordinator, Say Yes to Education*

This article examines the interventions associated with the Say Yes to Education program, a college tuition-guarantee program that was promised to graduating sixth graders from low-income families in the School District of Philadelphia in 1987. The program attempted to change the context of school and the odds that defeat inner-city students. The program made a critical point: Without transforming relationships and widening the sense of possibility, students would not be able to take advantage of the opportunity. The article presents research on student outcomes and analyzes two stories of students' lives as examples of how particular contexts can produce success.

On June 19, 1987, 112 sixth-grade students from Belmont Elementary School received a historic offer from George Weiss and his former wife Diane Weiss of Hartford, Connecticut. The terms of the offer were as follows: Upon graduation from high school the students' college or vocational training tuition would be paid in full by the Say Yes to Education Foundation (SYTE) funded by the Weisses. The model was similar to the I Have A Dream program, a relatively new type of third-party educational intervention for disadvantaged youth. Most of the Belmont students came from low-income families, and all were African American. Fifty-three attended special education classes, having been classified as learning disabled (Mezzacappa, 1987a).

SYTE set up a program site at the University of Pennsylvania's Graduate School of Education. Program staff consisted of an executive director and two project coordinators. The Weisses and Executive Director Norman Newberg established collaborative relationships with policy-level decision makers at the School District of Philadelphia, with the University of Pennsylvania, and with a variety of human services providers. As students worked their way toward high school graduation, SYTE coordinators provided such services as tutoring, counsel-

---

Norman A. Newburg and Randall B. Sims, "Contexts That Promote Success for Inner City Students," *Urban Education*, May 1996, Vol. 31, No. 2, pp. 149–176. Reprinted by permission from Sage Publications, Inc. (US), Corwin Press.

AUTHORS' NOTE: The authors would like to thank Michelle Fine for her helpful comments and suggestions on various drafts. They would also like to thank their research assistant, Leanne Gorfinkle. Funding for this research was provided by the Lilly Endowment and the Say Yes to Education Foundation.

ing, regularly scheduled home visits, advocacy with schools, mentoring, college visitations, internships, and summer school enrichment programs (Mezzacappa, 1987b). The cost of this program from grades seven through twelve averaged $1,100 per year for each student.

This article examines the interventions associated with the SYTE program and explores the potential implications this work may have for improving outcomes for inner-city youth. The article is organized into four sections. First, we present an overview of the program and its vision for students, their families, and their schools. Second, we discuss the quantitative and qualitative research that has been conducted on the SYTE program. Third, we describe two SYTE students who, though burdened by severe obstacles, were able to build successful futures with the program's help. Finally, we examine these individual lives as successes that were not idiosyncratic but rather illustrative of how positive contexts make success possible.

## The Program

The SYTE program was structured to be more than a vision of the future that makes no sense within the psychologies of adolescents. Few students could imagine in 1987 that they might be able to take advantage of the gift in 1993. Rather, SYTE staff consciously built contexts for meaningful, reliable relationships with students, teachers, and students' families, fostered a capacity to trust, and provided settings in which positive "possible selves" could be invented and explored (Markus & Nurius, 1986). The staff discovered in their first meeting with the students that they were reluctant or unable to express the kind of person they wished to become. How did the staff discover this?

Initially, two African American project coordinators, a male and a female, visited each student's home to discuss the gift with their families and to interview the students about their strengths, weaknesses, talents, and interests. The coordinators picked up a wealth of information about available family supports and also assessed the gaps in a family's ability to provide for their children. Over half of the mothers and fathers of these children did not complete high school. And only 12 parents out of 211 had any college experience, generally 1 or 2 years. Therefore, the program was asking a large proportion of parents to help their children achieve educational goals they had not attained themselves.

Each initial student interview concluded with the following question: What would you like to be when you finish school? Fewer than 25% of all students made a response. They could not imagine a possible future. They did not see their parents as role models of school success. And consequently, they did not connect school with the ability to construct a viable future. Further, the school system they attended communicated, through its tracking and grade retention policies, that success in school was not for all children, and more precisely, not for them (Oakes, 1985). Minority students from low-income families were particularly at risk for subject failure, retentions in grade, and finally dropping out. These policies depress students' interest in school and academic pursuits. They feel devalued and as a result become alienated from school. Using Steele's (1992) term, African American students from low-income families "disidentify" with school as the place that can help them. The program set out to create an alternative vision of the future for SYTE students, their families, and the schools they attend.

On the day the gift was announced, there were shouts of joy and tears of happiness. However, the terms of the gift would prove to be quite challenging. The ensuing years of careful support provided by staff required students to work hard, be persistently motivated to succeed, and believe that education could make a positive difference in their life chances. The SYTE staff was responsible for communicating the vision and translating its meaning into the incremental steps students had to take to achieve their goals.

A deep relationship developed between staff and students. A female student captured the level of mutual trust this way:

> You all [staff] are not just people who are like our friends. We can talk to you. We go out to lunch and dinner with you. It's like you are older friends, more than counselors and coordinators. It's a different kind of feeling. You all listen. Most people don't listen. You all listen.

In addition to ongoing advocacy and counseling of students, the program sponsored extensive tutoring opportunities. Students recruited from the University of Pennsylvania and other area colleges tutored SYTE students during and after school hours. When students were attending high school and needed to work after school, staff made agreements with the Penn Bartram School for Human Services, where over half the student were enrolled, for tutors to be available all day. As a demonstration to the rest of the school population that their success in school also was valued by the program, SYTE installed 100 tutors, available to all students. A SYTE student recalls how this service operated at Human Services:

> They had tutors for us every day, any time you need them. And they [SYTE staff] gave the tutors not just to Say Yes, but the whole school. Because if they gave it only to Say Yes, people might become jealous. And those people might need help too. So it was good of them to give it to everybody, whoever needs it.

That comment expresses a tension that existed between the SYTE students and the rest of the school's population. In schools, essential services like tutoring can be scarce or limited. Although it was not always possible to offer SYTE's resources to all students, the program made a conscious effort to extend beyond itself, to promote help as an entitlement for all.

The SYTE program tried to model the kinds of activities that need to happen if more inner-city students are to succeed. Therefore, in addition to providing tutoring for all students, SYTE upgraded the work-study component of the school by recruiting a full-time volunteer to develop higher quality job placements for all students. Similarly, prior to SYTE's involvement, Human Services only offered its students a brief SAT preparation course, eight 45-minute sessions over 1 month. SYTE, in contrast, sponsored an intensive program over 3 months, a 45-minute session twice each week. The program brought in an outside specialist experienced in training students for the test. This program disclosed some of the weaknesses in the academic preparation of students, especially in vocabulary development. The principal, observing the SAT class, felt defensive at first, when she saw students' scores on practice tests. She claimed, "Our students do not do well in this kind of test"—an allusion to the alleged bias in the vocabulary students are expected to know. Midway into the course, the principal instituted a school-wide campaign to strengthen vocabulary acquisition. Teachers introduced new words and reinforced their usage in all basic subject classes. The SAT course was one more example to the faculty and administration that SYTE had high expectations for its students. If SYTE students were experiencing more pressure to succeed, teachers, too, were experiencing pressure to improve the quality of education so that more students might be successful in their academic subjects.

Staff carefully monitored students' grade performance. Regularly scheduled meetings with the students' classroom teachers, counselors, and administrators were held to assess student progress and to explore alternative methods of instruction when that seemed indicated. If SYTE was going to change the odds for student success, the staff believed that a holistic approach was essential: one that combined an array of social supports, academic monitoring, advocacy for school reform, and deep parental involvement.

The staff met monthly with the parents to discuss their children's development and other topics of interest to them, such as discipline, talking to school officials about the progress of one's child, teenage sexuality, and setting and reaching goals for the future. With the parents of children in special education, the project coordinators organized a series of workshops conducted by parents who had faced problems in accessing appropriate services for their special-needs children. Parents were also involved in planning and cooking for holiday celebrations, implementing outreach to disaffected families, chaperoning student trips and events, and representing the program in public forums.

By high school, the meetings with parents in group settings were less frequent, as parents and their children began to separate from each other. Over the 6 years, whenever a student experienced some kind of difficulty in school, the first intervention was a home visit to assess how the family was coping with the problem. If parents were not able to help their child, staff took on surrogate parental responsibility by advocating for better placements in school, accessing needed social, medical, or academic services, and mediating family conflicts.

The transitions from elementary to middle and from middle to senior high school are often fraught with danger. For students with academic difficulties, the transition to an unfamiliar environment and the different expectations of a new school for which they have not been prepared aggravates already existing problems (Eccles & Midgley, 1989; Newberg, 1991). Each of these school-level transitions for SYTE students was a disastrous experience: 23% of the students were retained in seventh grade; 30% in ninth grade. Both statistics were lower than citywide averages. For example, at Human Services and University City, two of the high schools SYTE students attended, 35% and 52%, respectively, of all ninth graders were retained (School District of Philadelphia, 1990b). There is evidence suggesting that the cause of these precipitous drops in achievement may be associated with a mismatch between the structure and organization of school levels, which put large numbers of students at risk (Eccles et al., 1993; Fine, 1994).

Generally, students disperse to a variety of settings for high school. The SYTE leadership felt that it would be helpful to concentrate a substantial proportion of its students in a small, caring school so that academic and social development could be supported. Newberg created an alternative to University City's large comprehensive high school in 1992, Penn Bartram School for Human Services, that had a reputation for knowing how to shepherd a class of students through high school. Human Services included students from grades 10 through 12. The program's sponsors and Newberg convinced the school district to expand the school to include ninth grade and invited SYTE students to form the first freshman class.

Not all of the interventions SYTE advocated were as successful as those just described. This was particularly true with regard to the special education students. Approximately half of the SYTE population were enrolled in special education classes, having been classified as learning disabled. These students attended University City High School, a large comprehensive high school. SYTE sponsors and staff were concerned that the prognosis for these students would not be good if they took the standard special education curriculum. The dominant instructional strategy was to ask students to complete worksheets. There was little active learning, no hands-on experience applying academic learning to work settings, and little, if any, evaluation of the effectiveness of these strategies in producing positive student outcomes.

SYTE's executive director discussed these conditions with the superintendent of schools, who recommended that he start planning with the special education faculty to see if a better program could be designed for all ninth-grade students. Intensive conversations ensued involving all of the relevant faculty, the principal, several parents, and supervisors and directors from the central office of special education. After 5 months of planning, a proposal for funding was written that substantially restructured the special education program so that work experience was at the center of the curriculum with academics taught in relation to specific skills required by the workplace. The proposal emphasized work ethics, problem solving, conflict resolution with coworkers, and close monitoring of the curriculum every 6 weeks to determine students' mastery of skills and concepts.

The president of the University of Pennsylvania was able to get a promise of a $1 million grant to field-test the special education program at University City High School. However, the clearer the dimensions of this proposal became, the more resistant the faculty became, because the proposal required basic shifts in personnel, relocation of activities, and changes in instructional strategies. The faculty finally rejected the proposal they helped shape. A compromise solution was devised by the faculty in which the least successful students would be released to attend a vocational school. This solution left their program intact and provided a weak support

## Table 1

### Comparison and SYTE Students Six Years After Leaving Elementary School

| Group | Graduates | | Dropouts[a] | | Still Enrolled | | Total[b] | |
|---|---|---|---|---|---|---|---|---|
| | N | % | N | % | N | % | N | % |
| All students[c] | | | | | | | | |
| Comparison | 22 | 28 | 43 | 54 | 15 | 19 | 80 | 100 |
| SYTE | 44 | 44 | 36 | 36 | 20 | 20 | 100 | 100 |
| Regular education students[d] | | | | | | | | |
| Comparison | 18 | 32 | 28 | 50 | 10 | 18 | 56 | 100 |
| SYTE | 32 | 60 | 11 | 21 | 10 | 19 | 53 | 100 |
| Social education students[e] | | | | | | | | |
| Comparison | 4 | 17 | 15 | 63 | 5 | 21 | 24 | 100 |
| SYTE | 12 | 26 | 25 | 53 | 10 | 21 | 47 | 100 |

SOURCE: Addendum to Sclessinger's (1993) study presented to the Board of Directors of the SYTE Foundation in August, 1993.

a. Reasons for being classified as a dropout include being a nonattender, i.e., a student over the compulsory age of 17 who has been absent (unexcused) for 30 consecutive days, being committed to a correctional institution, and joining the job corps before graduation. Most of the students classified as dropouts were nonattenders.

b. Twelve SYTE students and five comparison group students were excluded from this analysis. Students were excluded because they moved from Philadelphia, transferred to a parochial or private school, had incomplete records, or were deceased.

c. ($x^2 = 6.53$, $df = 2$, $p < .05$).
d. ($x^2 = 11.25$, $df = 2$, $p < .01$).
e. ($x^2 = .80$, $df = 2$, not significant).

system for the students slated for the vocational school. Predictably, these students did not do well in this setting; neither did the students who remained in the comprehensive high school. Only 12 out of 53 special education students graduated. In the comparison group, the results were 4 out of 24 (see Table 1).

Thus, SYTE became a program that attempted to change the context of school and the odds that defeat inner-city students' success. In a sense, the program was much more than a gift that changed the opportunity structure for students from low-income families. The program was making a critical point: Without transforming relationships and widening the sense of possibility, students would not be able to take advantage of a better opportunity.

## Research Data

Little research has been reported on tuition-guarantee programs and the effects on those who participate (U.S. General Accounting Office, 1990). Of the data that have been reported, none, to our knowledge, have shown their results in relation to a comparison group. We present the following data to redress that omission.

The graduating Belmont sixth-grade class of 1986 was compared with the graduating Belmont sixth-grade class of 1987, the SYTE students. Four criteria were used to determine the similarity of the comparison group:

1. The comparison and SYTE groups were similar in the proportion of male and female students.
2. The mean ages of the comparison group and SYTE students were similar.
3. Regular education students in both groups performed similarly on standardized tests in reading and mathematics administered in the sixth grade.
4. Students in both groups attended schools that served families with similar socioeconomic characteristics, and were significantly poorer than their counterparts in West Philadelphia (Schlesinger, 1993).

By August 1993, 44 SYTE students graduated from School District of Philadelphia highschools. An additional five students received high school diplomas or graduate equivalency diplomas from private or out-of-state schools. Three students were deceased. One hundred eligible SYTE students were compared with a group of 80 students similar in background. The results indicate that 44% (44 students) of the SYTE group and 28% (22 students) of the comparison group completed high school. The difference is significant at the $p < .05$ level.

Comparing the percentage of male students in SYTE who graduated with a comparison group, the results were 46% and 22%, respectively ($p < .05$); for females, it was 43% compared to 34%. These data show that SYTE male students did significantly better than the comparison group, but there was no significant difference in the female groups. The low graduation rates for females may be attributed to the fact that 50% of the SYTE females delivered babies by age 16. No comparable statistics exist for the comparison group.

If we disaggregate the regular education students from the special education population, we find that 61% of SYTE and 32% of the comparison group graduated ($p < .01$). The dropout rates for the two groups also differ: 36% for SYTE and 54% for the comparison group (see Table 1). For regular education students, the dropout rate was 20% for SYTE and 50% for the comparison group ($p < .01$).

The SYTE dropout rates can also be compared to those rates for the city and two particular high schools. The pattern of results in urban schools is disturbingly regular. In West Philadelphia, where SYTE students attended school, students leave elementary school 1 to 2 years behind grade level. By ninth grade, they are 2 to 3 years behind, which explains why 50% of all Philadelphia high school students do not earn enough credits to move on to tenth grade in 1 year (McMullan, Snipe, & Wolfe, 1994). And the 50% retention in ninth grade is predictive of the 30% to 80% high school dropout rates 1 to 2 years later. School districts use a variety of methods to compute dropout rates. The most reliable method identifies a cohort of incoming ninth-grade students and follows their progress for 4 years. The School District of Philadelphia initiated such a study citywide for the ninth-grade class of 1984–1985. Over the course of 4 years 31.6% were identified as having dropped out of school (School District of Philadelphia, 1990a). Extending this study through June 1989, an analysis of the data reveals that 38% of the class were identified as having dropped out. Although a 32% dropout rate is the approximate average for the city, the dropout rate for the high school that was in the feeder pattern for SYTE students in 1992, University City High School, was 73.6%. At Kensington High School, of those ninth-grade students starting high school in 1988, 81.5% had dropped out by June 1992 (Mezzacappa, 1992). These data provide additional points of comparison for the SYTE students, illustrating the program's capacity to reduce dropout rates significantly and increase graduation rates. Forty-seven SYTE graduates are currently enrolled. In 2-year or 4-year colleges; however, postsecondary data about the comparison group are not available.

A qualitative study consisting of interviews of 85 SYTE students, 15 selected parents, and 15 teachers, was conducted by program staff in the spring, summer, and fall of 1993. Twenty-six students were excluded from the interview sample because they moved from Philadelphia, transferred to a parochial school, or were deceased. Program staff was trained to use a 24-question protocol; each interview lasted between 1 and 2 hours. Statements were verified by cross-records. In the analysis, the sample was stratified not only by student characteristics but by the social contexts of their lives (continuity in the home and school; access to positive, caring adults; exposure to abuse; and selling and/or using drugs).

## Two Stories

In this section, we examine two students out of the total sample, Michelle and Roger.[1] Roger has been retained in grade once, and Michelle twice. Both come from severely stressed homes where they experienced neglect and abuse. Yet both graduated from high school and were accepted to 4-year colleges. We explore what made a difference in their lives and extrapolate the range of conditions, circumstances, opportunities, and so forth that move a population in a positive direction. Are there particular sets of things that appear to predict success? How do students operate in certain circumstances? How does contact with affirming or disconfirming people (parents, friends, teachers, or SYTE staff) make a difference? Why do some people give up in difficult circumstances, whereas others do not?

The contrasting stories illustrate how lives are mutually produced and, consequently, move beyond a record of individual defeat or success. In using a social context and an institutional approach, we hope to contribute to the recent thinking posited by Rutter (1987), Spencer (1995), Spencer, Swanson, and Cunningham (1991), Swanson and Spencer (1991), and Winfield (1991) about how resilience operates in the lives of inner-city students. We argue that resilience must be more than an idiosyncratic quality of individuals. Rather, resilience is a function of the interaction of individuals and their social contexts such as home, school, and community institutions.

### Michelle

Michelle is the second of seven children. Her mother was 19 when she was born. Neither parent finished high school, but her father did complete a General Education Degree. Her parents separated when she was 7 years old. The family income is below the poverty line and the mother receives welfare assistance. Michelle tells an apocryphal story that at her birth, her dad said to her mother that she was no longer his "first girl." That remark seems to have engendered jealousy and hostility between Michelle and her mother that persisted through her adolescence. They fight frequently. She believes her mother hates her and wishes that she had never been born. Throughout her years growing up, until she turned 18, her mother beat her "with ironing cords or whatever she could pick up." Sometimes she hit her with her fists. Her mother has a long history of drug dependence.

Michelle's family moved to several different locations in her early childhood, causing her to miss school. She claims that she was retained in first grade "because my mom and dad were never around." From June 1987 to June 1993, she lived with three different family units: her mother for 2 years, her father for 1 year, and her grandmother for 3 years. For short periods of time she has lived with an uncle and an aunt. Each of these households was crowded with younger cousins and various members of an extended family.

For the first three years of the SYTE Program, staff made frequent attempts to visit Michelle's mother at her home. Each time staff were told that she was unavailable. Contact with Michelle was limited to school or on the street. Although she often promised to attend program activities, she rarely did.

In reviewing her school records, staff learned that she had performed below grade level in reading in fourth grade and was recommended for a special reading class that used programmed instruction on a computer. As a reward for completing work successfully, she was allowed to play computer games for the remainder of the class period. Within 1 month, her reading teacher noticed that she was reading at grade level; therefore, she was not allowed to work on the computer any longer. She reports that she "cried hard" when she lost access to computers.

Throughout middle school, Michelle's grades were below average, but passing. It was apparent to teachers and program staff that she had the ability to do better but lacked the motivation to do so. In June 1990, she announced that she was not going to attend a regular high school; she wanted to attend a vocational school to major in cosmetology. The staff tried repeatedly to convince her to attend a comprehensive high school so that she could take college preparatory subjects. She refused to listen. By October of that year, she regretted her decision and transferred to her neighborhood high school, where she took an academic track curriculum. None of her vocational-education courses were transferable. She spent the entire year catching up. When it became apparent that she would fail three major subjects, she stopped working. Ironically, she passed both Algebra I and Physical Science in summer school with ease.

In tenth grade, Michelle's attendance was sporadic. Even though she failed one subject that year, she was recommended for a special program for students who have the potential to attend college but are underachievers. The Motivation Program, as it was called, was a small school within University City High School, with a faculty that worked with a cohort of students for 3 years. In such an intimate environment, students were seen and heard, and the anonymity of the larger school was significantly reduced. The smaller organization also made advocacy for Michelle by SYTE staff simpler. If we requested a roster change, it did not require a flurry of paper marched through the bureaucracy to effect the change; one conversation sufficed. That summer, SYTE staff found her a paid internship in an office using a computer. Michelle was elated. Her excellent performance earned her the opportunity to continue this work part-time during the school year. However, without much explanation, she did not return to work that fall.

Staff notes "Michelle got into frequent fist fights with peers and seemed unperturbed if she was suspended. She looked angry. Her home situation continued to frustrate her, especially when her mother moved back into her grandmother's home." Michelle was one of several SYTE students at this high school who seemed unmotivated and/or often got into fights. The staff decided to hire a therapist who initially worked with a group of students including Michelle and eventually scheduled individual sessions with those students in greatest need. By the winter of 1992, when she was in eleventh grade, Michelle accepted that she did have problems. She wanted help. She also stated that she wanted to go to college. She said of the counseling, "It seemed like I was talking to my best friend. I could tell him anything, and he would help me out." Concurrently, her grades improved. Previously, she had achieved mostly Ds and Cs, and at least one failure. Now she was consistently making As and Bs. In the summer of 1993, she took a college-level course in data processing and achieved a B as a final grade.

Michelle made some basic changes in her life. She became trusting of caring adults, such as her supervisor. Therapy helped change Michelle's attitude toward her brothers and sisters. Previously she used them as a target for her anger against her mother. She said,

> It wasn't that I was angry at them. It's just that they was there at the time when I was angry, so I took it out on them. But now my attitude has changed towards them. I don't holler at them at all like I used to. But if I do, I always end up apologizing to them.

She no longer resisted academic learning. Her English teacher, Mr. Muller, got her interested in books and reading. She admired him because he made her think. She appreciated the reach for excellence he demanded of students. "He pushes students to do the best that they can do," she remarked. "Then he pushes them to do better." She continued,

> Most teachers are different: If you don't want to work, they leave you alone. But Mr. Muller, if he see that you don't want to work, he will find out what the problem is, why you don't want to work, and will try to help you out.

Slowly, Michelle understood the value of supportive guidance. She also appreciated the pressure and expectations that motivated her to excel. She used writing as a way to make sense out of her experience. Her English teacher encouraged her to write an essay for a contest on the importance of getting an education. She described how she came to believe that she could strive to compete:

> And we had to write a paper about education and I just wrote how I felt because my teacher has taught me a lot, when my counselor wasn't around, I needed somebody to talk to. So I talked to my teacher . . . and he helped me out a lot. He was like a friend to me too. And he always pushed me to do better and kept telling me don't let this get you and don't let that get you. And I started taking his advice. And after that, my counselor's advice. And it changed the way I started thinking about things and that's what I wrote about in my paper.

Michelle won first place in the contest and received a plaque to commemorate the occasion.

Several times in high school, Michelle had been on the verge of dropping out. Those were also times when she resisted being helped. SYTE staff understood that students move in and out of connection and the program did not give up on her. At some level, she had to learn the value of help and to trust the helper before she could accept it. She is now experiencing success and sees the consequences for some of her classmates who have dropped out and shakes her head.

We asked Michelle to reflect on why some SYTE students dropped out of school. Again, she raised the theme of pressure to excel. By this time, she understood that pressure must be internal:

> Maybe they [the dropouts] thought that it was too much of a struggle for them. . . . Maybe they just used to doing things like if they think it's okay, then that's just good enough. They not used to pushing themselves, or somebody pushing them to do things. At first I didn't like being pushed cause my mom never pushed me or nothing. . . . And then I started getting used to it. Now it's helping a lot.

Michelle felt that being in the program "had a major effect on [her] life." She explained, "Now I know if I really put my mind to it I can do something":

> 'Cause I just want to further my education and I really want to go to college and learn about computers. . . . If I went to a trade school, they won't teach you everything. They will probably try to teach you the basics but there's a lot of little things that you got to understand, too.

Michelle's comment may refer to her experience at the vocational high school. The focus was narrow in that school and, in retrospect, too limiting for her goals. She has learned that education is a necessary requirement for getting a good job, but she was also not sanguine that going to college guarantees employment:

> It [college] would help me do a better job than I would have if I didn't go to college. Though a lot of people who went to college doesn't have a good paying job. But if I go to college, it's only what I make of myself. So in computers and stuff, the education I get from college, I can use that to help me get a job.

Michelle is the first one of her family who will attend college. Aunts and uncles in her family "graduated from high school but they went into the service or a trade school." She is looking for role models that will inspire her to attain a college degree. She will not find them in her family.

But SYTE bridges this gap in Michelle's experience through visits to many college campuses and by frequent contact with college tutors.

Michelle is now goal oriented. "Getting a high school diploma," she says, "means that I have reached my goal, done what I had to do and that's just my reward for finishing what I had to do." She is able to take her goal-directedness and project it into the future. She imagines that 5 years from now she will have her "own office": "[I will be] working in a big, big business building. Maybe a manager. 'Cause I'm into computers." She allows that she might have a child and marry. When we asked if anything could interfere with her attaining her goal, she said emphatically, "No, nothing at all. I want to live. See, I'm not a party person, so I don't be out that much, so I just stay in the house and do whatever I have to do."

## Roger

At the writing of this article, Roger was 18; he has an older brother, a younger brother 13, and a sister. His father graduated from high school and served in the armed services. His mother did not complete high school and for most of his school-age years was unemployed and received welfare assistance. His father died when Roger was 4 years old. When Roger was in sixth grade, his mother developed a crack dependency. He remembers that "things in the house started disappearing and living conditions were horrible." His mother abdicated her authority to set and enforce limits. "She started to become less of a parent," he recalls, "and more of a sister, like a friend or something. Because she wasn't like forcing us to come to school; it was optional." Discipline was lax or nonexistent; bedtimes and wake-up times were his decision. If he got into trouble at school, his mother would not hit him, which for him would have been an expression of concern. By seventh grade, he was 5 feet 6 inches and growing, whereas his mother was 5 feet 5 inches. Perhaps, he conjectured, she feared he might return the slap. He maintains that he would not have.

Roger was quick-tempered and resorted to fist-fighting with little provocation. Mostly he fought peers, but he had several altercations with teachers and administrators and was capable of more extreme violence. In a fight over a girl with a rival male, he cut the other boy and the girl with a knife. That incident took him to court several times; eventually the charges were dropped. When asked why he carried a knife, he told us that his mother advised him to carry one and be quick to use it if someone jumped him. He acted on that advice.

During the first marking period of seventh grade, Roger attended SYTE's after-school tutoring classes with some regularity. His tutor reported that he was bright, was eager to learn, and demonstrated fairly good reading skills. However, by the start of the second marking period, his attendance at school and at tutoring class dropped abruptly. When program staff visited his home, his mother said he often played hooky from school.

In seventh grade, Roger moved at least four times and attended four different schools. Family members and friends offered housing to him and his family, but invariably a conflict developed, and they were forced to move after a short period. At one point they lived in a public shelter. Each of these moves was punctuated by a violent episode at school, which necessitated a transfer to a different school. Roger described seventh grade as "hard":

> I was into fighting everybody for no good reason. And my life changed big, big during seventh grade. That was when I left home for a month and hung out with some friends. And then when I came back to school I got left back. I had to take seventh grade over again.

In an end-of-the-year report, staff wrote,

> We are trying to find Roger's address so that we can get him into a stable tutoring program and offer other support. He needs social and academic

services interventions very badly. It's obvious that his mother cannot manage him. She appears to have personal problems [an allusion to her addiction] which makes it impossible for her to supervise his school work.

From the shelter, his mother reconnected with her closest childhood friends, Joan and Cherrell Ryan, who lived in a house with their mother, Miss Lillian Ryan. She became Roger's adopted grandmother; her daughters, each of whom had a son Roger's age who was also part of SYTE, became his adopted aunts, and their sons became his cousins. Gradually, Roger increased his stays at Grandmother Lillian's house and finally moved in and became a part of the extended family. In some ways, this was a positive step, because he did not have to worry about finding food or a place to sleep. But because his mother was not living in the house, he was not treated like the other children. He complained that his adopted aunts treated "their kids better" than they treated him. He felt "lower than these people in this house in a way," but inside he felt he was not. "So in a way," he realized, "that feeling inside me was what kept me feeling up."

One of Miss Ryan's daughters, whom Roger calls Aunt Joan, was a teacher's aide at a local school. She offered to use her influence to get him into her school where she could watch over his behavior. She made it clear to him that he "wasn't coming there to mess up her reputation." She also told him that if he showed off, "she would be the one to handle it." She reinforced her promise when Roger misbehaved in school by hitting him with a ruler while other students watched. By this time he had also been told by the school district that the next time he was expelled for fighting he would be placed in a disciplinary boarding school.

The principal of this latest school noticed Roger's potential to do academic work, but also understood his need for close supervision. The next time Roger got into a fight with a teacher, he did not make an official report of the incident. Rather, he put him in a special education class. He got his assignments from his regular teacher daily but was required to do the work in the special education class and have it checked by his regular teacher at the end of the day. He thrived in the small class of 10 students and soon was helping the teacher by tutoring the other students. The special education teacher got him involved in after-school swimming and skating classes. Roger appreciated these activities as strategies for keeping him out of trouble.

For a period of two years SYTE staff could not ascertain Roger's whereabouts. They could not keep up with where he was living, and Roger rejected the program's offers of help. "I wasn't aware of what the program offered me and I didn't take it seriously," he said. "I was like, I got better things to do than to go to these tutoring sessions and talk to these people." Once he started to respond positively to offers of support in school, Roger's Aunt Joan decided to expand his network. She contacted a SYTE program coordinator and told him that Roger was attending school regularly and she hoped the program would get involved with him. Roger graduated from eighth grade as class valedictorian with a straight B average. Program staff renewed contact with him and convinced him to attend the annual summer program that presented a preview of courses he would encounter in ninth grade.

The transitional summer program Roger attended was not only a preview of ninth-grade courses, but also an introduction to the Human Services faculty who would teach SYTE students in the fall. Roger explained his experience of this transition:

> I got to know a lot of teachers and when I started high school . . . they treated me different 'cause they already knew me. It wasn't like I was a new freshman, they knew what type of person I was, they knew what kind of work I could do.

Roger enjoyed the summer program and understood its purpose. But once he entered Human Services without his aunt's watchful eye and the highly personalized attention he got from the special education teacher, he reverted to some of his old behaviors. He did not attend consistently. When he did attend, occasionally he could be seen walking the halls or shooting baskets in the gym. If he was asked why he was not in class he would say, "I'm not here today." He often drank beer and sometimes hard liquor before coming to school. He did not adjust well to the relative freedom of even a small high school setting. His first report card recorded four failures and one passing grade. When he brought his report card home, no comments were made. He understood more clearly than ever before that he was on his own. He explained, "If I want to do better, I gotta make my own effort to do it." By year's end he made a slight improvement by passing some subjects with Ds and recording two failures.

Human Services, a school of 270 students, prides itself in paying close attention to students' needs. If a student fails two subjects, he or she is required to meet with the faculty to discuss whether the student can stay at the school and under what terms. The faculty agreed to let Roger stay, provided his grades and attendance improved. He recalled that they also asked "a biology teacher, Ms. Shakir, to represent [him]." Roger explained why:

> So that anything I did wrong in the school came back to her. Teachers sent daily reports to her about my work. And she started signing me up for stuff. She taught an after school science club. And I had to take that.

Ms. Shakir was no pushover: She demanded that Roger work up to capacity. He responded to her call for discipline and hard work:

> She's always been hard on me. If you try to get away with it real easy, she will not let it go. If she feels you can do the work, she'll make you do it. And a person who doesn't want to do anything, she'll talk to you and say you're not giving up. She'll tutor me next fall and give me an independent study on advanced biology.

In addition to demands for academic responsibility, the faculty figured out a way for Roger to be thoughtful about disruptive behavior in the school, and his own behavior began to change:

> I remember like in the beginning of the 10th grade, they had this thing called The Disciplinary Society—it was like all the kids that was in trouble in school had to go before this panel . . . the teachers elected a student to be the head of the panel. So they elected me and a couple days later I was chosen as sophomore class president and so I started to get active in like making the school better, trying to teach the Human Services spirit.

Roger had been virtually on his own, as he said, for years. He wanted to be someone's concern. His adopted grandmother and aunts provided shelter and some minimal guidance. The school's and SYTE's support compensated in some measure for the vast neglect and abuse Roger had experienced when he lived with his biological mother.

In October 1990, staff members took Roger and several other SYTE students to Hartford to spend the weekend with the Weisses, the program's sponsors. George Weiss introduced Roger to one of his business partners, Mike Christiani, who casually encouraged him "to call if he ever needed someone to talk to." Roger accepted Mike's offer. They struck up a friendship, discovering that they shared common interests in basketball, pizza, and hunting. Mike made an agreement with Roger: "If you make Honor Roll on the first marking period, you can come to Hartford and spend the weekend with me and my family." Roger was intrigued by the offer and the person who made it:

> Mike told me, he said, "You do your best and don't worry about what other people say." He said, "Sometimes it's hard, so you just take advantage of every opportunity you have and get the work done." I wanted to be the same type of person that Mike is. People make role models. And I wanted to be like him. He got a nice wife, and a nice house, and I want the same thing.

Roger said, "People *make* role models"—not *have* role models. Mike is a self-made success and was showing Roger how to do it for himself. The incentive produced the desired result. Roger made honor roll each marking period. He visited with Mike and his family several times that year: "hanging out, going for hikes, helping him in his office." Roger believed that Mike and he developed "a father-son relationship."

Roger demonstrated a knack for latching onto people who could help him. George Weiss, program sponsor, gave him his 800 number as he did with many SYTE students. Roger, however, was one of the few who called him on a regular basis, sometimes two or three times a week. He talked to him about his schoolwork or complained if program staff did not respond fast enough to a request he made for a tutor or some other need. Increasingly, SYTE became his surrogate family. Program coordinators served as father, mother, and sister: "Sims, I'll tell him everything; he knows me and what I've been through. I trust him." The older female coordinator, Robin Wall-Hill, he thought of as a mother:

> I never had that kind of relationship with my mom. Like places my mother is supposed to accompany me, Robin is always there. She's always taking me places and telling me right from wrong. And then Leanne, she's like a big sister—a little sister 'cause she's short. She helped me select different colleges, told me what to look for in a college. She helped me out with the applications.

Roger valued his adopted families, but he maintained contact with his biological family. His mother, a recovering addict, and Roger's younger siblings moved to another city. She is working and has repeatedly asked Roger to rejoin his family. Although he stays in touch by phone and occasional visits, he is clear that it would not be in either of their best interests for him to return. He has a sense that he is better off with his adopted family. Although he does not say it explicitly, he does not want to leave his SYTE contacts. Roger has mastered the ability to communicate with people, taking advantage of opportunities given to him. He is not ashamed to ask for help or to offer it, and he is developing a sense of reciprocity with fellow students in need:

> I don't want to regret not calling anybody, if somebody was in trouble or needed someone to talk to. I would call them because I don't want them to have to go through it alone. Try to be their friend.

We asked Roger to reflect on reasons some SYTE students have dropped out and to expand on how his analysis might be relevant to himself:

> Like one of the cases where a student dropped out of school, it was like a financial thing. He needed money and he needed it bad. . . . They say with a long-term education, it pays off in the end. . . . But the person I'm talking about didn't see it that way. He needed money now. So he dropped out of school and started selling drugs and he's still selling. That guy could have been me. . . . But, I mean, money isn't everything. Since I changed my life, I know I'll make honest money. There's a lot of people, Ms. Shakir, Mr. Sims, they all educated and make money. Mr. Weiss wasn't always rich. He went to college and now he's making lots of money.

Roger has made a connection between getting a college education and the ability to make money. He also is developing a sense of the need for delayed gratification—seeing education as a necessary step before he can build the kind of life he wants. He says, somewhat grandly, "I'll go to school the rest of my life." His plans have ranged from pursuing a career in medicine to becoming a physical therapist. The shift from medicine to physical therapy came during his internship in the summer of 1993 with a University of Pennsylvania scientist. He tried out his career goals with the scientist who gave him a clearer sense of how he might pursue a variety of career paths in allied medical fields. Roger adjusted his goal to one he thought was more realistic given his good, but not outstanding, school performance.

## Discussion and Conclusion

We selected these two students to present because their lives represent many of the stresses that destroy inner-city youth. They also represent hopeful stories of how resources and sustained support can change the odds from predictable failure into futures that may be satisfying and fulfilling. Both students experienced abandonment and deprivation in their biological families. They were raised by a single female parent or by a grandmother. In Roger's case, the grandmother was a surrogate. Addiction has taken its toll on both families, exacerbating the poverty they experienced and contributing to the instability of the home setting (Anderson, 1990). Michelle and Roger lived nomadic existences in first and seventh grades, respectively, necessitating frequent school changes. Chronic absence from school led to their retention in grade. The instability of the home setting and psychological and physical abuse by parents left these students with rage that manifested itself in frequent fights with peers and adults.

What seemed to make a positive difference in their lives was finding a home that was moderately stable. Michelle found a haven with her grandmother, although that relief was interrupted each time her mother returned to live in the house. For Roger, it was the serendipity of getting himself "adopted" by Miss Ryan. We are struck, in Roger's story, by the generosity of relative strangers. However, it is not that uncommon in African American communities for "grandmothers" to take homeless children in and provide modest shelter (Stack, 1974). Neither student's home setting was ideal or crisis free, but each setting provided some respite.

The psychological damage these students experienced scarred them in ways that further isolated them from people who might care for them. Slowly, these students learned to accept their problems and began to differentiate between those people who would harm them and those who could offer help. For example, Michelle's anger at her mother seems justifiable. The SYTE staff and her therapist helped change the context in which she experienced people, and her anger started to be more positively directed.

In the larger SYTE study we are conducting, and in our direct experience, we have seen repeatedly how students resist accepting help with academic, social, or personal problems. They have internalized a macho-cool attitude that makes help-seeking a sign of weakness (Nelson-LeGall & Jones, 1991). Students are afraid they will be held up to derision by their peers if they are seen as accepting help. Some of the reluctance to ask for help goes beyond peer culture, defining school as a hostile, intrusive, and unreliable institution, especially for low-income students who experience frequent failure. By contrast, Michelle and Roger learned to use help because the context in which they received it was perceived to be caring and trustworthy.

Michelle was willing to bond with a therapist who helped her cope with her anger at home and in school. Roger and Michelle were able to bond with caring teachers, SYTE staff, and mentors. In accepting help, they learned to value their abilities to learn and succeed academically, and they developed an appreciation for interdependence—not independence (without others) or dependence (on others). The pressure of higher expectations was motivating. They began to appreciate the value of standards and the connection between knowledge and their capacity to improve their life's circumstances. In the past, possible selves for Roger and Michelle were mostly negative. Now they can expand their self-concepts to include selves they would very much like to became (Markus & Nurius, 1986). We are not suggesting that their lives are problem free. They are not.

A failed test, a fight in their neighborhood, or a slight from a parent or guardian can set them back for a while. They walk an emotional tightrope between their negative reactions to stress and adversity and the more positive response of resilience (Rutter, 1987). What is different in their lives is that they possess strategies, and supports, ways of thinking and acting, that protect them from reverting back to their former behavior. They know their histories and are willing to talk about them reflectively. They value their goals and are willing to work to achieve them. They can reach for inner and external resources that will support them. As Rutter notes, it is the coping strategies individuals bring to a stressful situation that determines their capacity to be resilient.

Michelle's and Roger's stories are about individual resiliency. They do bounce back and, for them as individuals, that is important. But their stories may be more generalizable because they are stories about the contexts that create and sustain resiliency. Human Services and the Motivation Program are the kinds of school environments that grow resilient students by building trusting relationships (Noddings, 1992). These organizations back off from the traditional stance of a school organized to make success a scarce commodity. Instead, they present contexts that instill and engage commitment and challenge. Teachers like Shakir and Muller do not "dumb down" the curriculum. They will not accept passivity and sloppy performance as the norm. They demand good work and they get it. But their institutional context also supports that aim. Witness the mobilization of the faculty in Roger's behalf.

Institutional contexts make a difference. School size and faculty commitment to students and their learning are key factors in helping improve the performance of students from low-income families (Oxley, 1994). Researchers have been reporting that large, impersonal schools are inappropriate settings for children. Over the last 5 years, efforts have been made in several cities to restructure comprehensive high schools into smaller units called *schools within schools* or *charters*. In Philadelphia, over 100 charters have been established within existing high schools. Results over 4 years for students in charters indicate improvement in attendance and subjects passed and performance substantially better than in a control group (McMullan, Snipe, & Wolfe, 1994).

In a sense, Roger's and Michelle's stories demonstrate the power of surrogate families, temporary communities that are created for students who need support and advocacy. The natural grandmothers, the adopted ones, the therapists, the demanding and caring teachers, and the SYTE program coconstructed an ecological system for raising children our society has placed at risk. Key has been the continuity of caring provided by SYTE for more than 6 years. It acted as a third-party advocate and broker for students between these various systems. Its advocacy was goal-directed, improving the quality of life for each student through education.

In describing our initial interviews with SYTE students, we noted that less than 25% could describe what they wanted to become after finishing school. Children who live in poverty do not grow up imagining positive "possible selves." They are locked into limited, often negative, self-concepts. Programs like SYTE make a difference by expanding children's aspirations and by providing supports for their realization.

Racism, unemployment, and unresponsive schools are corrosive blights that rot and destroy human potential. These social ills are at the root of our society's perpetuation of cultures of poverty and must be addressed vigorously by citizens and government. However, as we organize to address these global problems, it remains possible to change schools and students' relational lives so that we can reduce the number of children who have to beat the odds we sought to change through SYTE.

## Note

1. The names of the students and parents mentioned in the article have been changed to protect anonymity.

# References

Anderson, E. (1990). *StreetWise.* Chicago, IL: University of Chicago Press.

Eccles, J. S., & Midgley, C. (1989). Stage-environment fit: Developmentally appropriate classrooms for young adolescents. In C. Ames & R. Amu (Eds.). *Research on motivation in education* (3rd ed.; pp. 139–186). San Deigo, CA: Academic Press.

Eccles, J., Midgley, C., Adler, T., Wigfield, A., Buchanan, C., Reuman, D., Flanagan, C., & McIver, D. (1993). Development during adolescence: The impact of stage-environment fit on adolescents' experiences in schools and families. *American Psychologist, 48,* 90–101.

Fine, M. (Ed.). (1994). *Chartering urban school reform.* New York: Teachers College Press.

Markus, H., & Nurius, P. (1986). Possible selves. *American Psychologist, 41*(9), 954–969.

McMullan, B., Snipe, C., & Wolfe, W. (1994). *Charters and student achievement: Early evidence from school restructuring in Philadelphia.* Bala Cynwyd, PA: W. C. Wolf.

Mezzacappa, D. (1987a, June 20). 6th-grade class is offered a gift of college tuition. *The Philadelphia Inquirer,* pp. 1A, 6A.

Mezzacappa, D. (1987b, November 1). Gift of schooling may be hard to accept. *The Philadelphia Inquirer,* pp. 1, 1-B.

Mezzacappa, D. (1992, June 18). Urban obstacles make diplomas precious at Kensington High: Their pomp defies the circumstances. *The Philadelphia Inquirer,* pp. B1, B4.

Nelson-LeGall, S., & Jones, E. (1991). Classroom help-seeking behavior of African-American children. *Education and Urban Society, 24,* 27–40.

Newberg, N. A. (1991). Bridging the gap, An organizational inquiry in an urban school system. In D. Schön (Ed.), *The reflective turn: Case studies in and on practice* (pp. 65–83). New York: Teachers College Press.

Noddings, N. (1992). *The challenge to care.* New York: Teachers College Press.

Oakes, J. (1985). *Keeping track.* New Haven, CT: Yale University Press.

Oxley, D. (1994, March). Organizing schools into smaller units: Alternatives to homogeneous groupings. *Phi Delta Kappan,* 521–526.

Rutter, M. (1987). Psychosocial resilience and protective mechanism. *American Journal of Orthopsychiatry, 37,* 317–331.

Schlesinger, M. (1993). *A Study of a tuition guarantee program.* Unpublished doctoral dissertation, Temple University, Philadelphia, PA.

School District of Philadelphia, Office of Accountability and Assessment. (1990a). *A study of the ninth-grade class of 1984–85* (No. 9024, pp. 1–2). Philadelphia, PA: Author.

School District of Philadelphia Office of Accountability and Assessment. (1990b). *Superintendent's Management Information Center, 1989–1990.* (No. 9021, pp. 98, 158). Philadelphia, PA: Author.

Spencer, M. B. (1995). Old issues and new theorizing about African-American youth: A phenomenological variant of ecological systems theory. In R. L. Taylor (Ed.), *Black youth: Perspectives on their status in the United States* (pp. 37–70). Westport, CT: Praeger.

Spencer, M. B., Swanson, D. P., and & Cunningham, M. (1991). Ethnicity, identity and competence formation: Adolescent transition and identity transformation. *Journal of Negro Education, 60*(3) 366–387.

Stack, C. (1974). *All our kin: Strategies for survival in a Black community.* New York: Harper & Row.

Steele, C. M. (1992, April). Race and the schooling of Black Americans. *Atlantic Monthly,* 74–75.

Swanson, D. P., & Spencer, M. B. (1991). Youth policy, poverty, and African American youths' identity and competency. *Education and Urban Society, 24*(1), 148–161.

U.S. General Accounting Office. (1990). *Promising practice: Private programs guaranteeing student aid for higher education* (GAO/PEMD-90-16). Washington, DC: Author.

Winfield, L. F. (1991). Resilience, schooling, and development in African-American youth. *Education and Urban Society, 24,* 5–14.

# Next Steps in Inner-City Education
## Focusing on Resilience Development and Learning Success

### MARGARET C. WANG
*Temple University Center for Research in Human Development and Education*

Nowhere are the problems and needs of children as great as in this nation's urban communities, where the lives of so many children are in disarray. Their families, neighborhoods, and community agencies—including schools—are desperately depleted of resources and spirit. Problems abound, including unemployment, crime, child abuse and neglect, and addiction to drugs and alcohol. This litany of troubles in urban America sometimes overshadows the problem of widespread academic failure in the schools, which could cripple the next generation.

Although there is much neglect and despair, there is every reason to hope and work for improvements. Cities contain many rich and promising resources for children and families. Despite the difficulties of urban life, many children and youths manage to rise above the problems of inner-city life and mature into healthy, competent, well-educated adults. Children are remarkably resilient; they respond readily to caring adults and a supportive community. If only we can find the means to magnify the "positives" in the lives of all urban children, we can rekindle hope for remaking urban education into a system that fosters resilience and educational success in inner-city communities.

This concern for improving the prospects for education and life circumstances of urban children and youths, particularly those living in inner-city communities, sparked the establishment of the National Center on Education in the Inner Cities (CEIC) at the Temple University Center for Research in Human Development and Education (CRHDE). CEIC was established in 1990 with initial funding support from the Office of Educational Research and Improvement of the U.S. Department of Education, as one of the 10 national R&D centers. The purpose of this article is to discuss the work of CEIC and its implications for charting the next-step efforts to significantly improve the capacity for education in urban America.

---

Margaret C. Wang, "Next Steps in Inner City Education," *Education and Urban Society,* May 1997, Vol. 29, No. 3, pp. 255–276. Reprinted by permission from Sage Publications, Inc. (US), Corwin Press.

AUTHOR'S NOTE: The research reported herein is supported, in part, by the Office of Educational Research and Improvement (OERI) of the U. S. Department of Education through a grant to the National Center on Education in the Inner Cities (CEIC) at the Temple University Center for Research in Human Development and Education (CRHDE). The opinions expressed do not necessarily reflect the position of the supporting agencies, and no official endorsement should be inferred.

# The National Center on Education in the Inner Cities: An Overview

The work of CEIC is organized under three R&D programs that address the fundamental question, "What conditions are required to bring about massive improvements in the development and learning of children and youth in this nation's inner cities?" The first focuses on the family as an agent in the educational process. Studies in this program aim to explore and enhance family life and its contribution to education from multicultural and multigenerational perspectives. A particular emphasis is placed on families with special needs, including those with problems related to substance abuse and academic underachievement. The second program concentrates on school factors that are effective in fostering educational resilience and learning success among inner-city children and youths. Among the expected outcomes of the projects included in this program is a detailed database on features and outcomes of school programs and practices that support a high standard of achievement among students, even among those with special needs or those who live in highly adverse circumstances. The third program addresses the community and its connections with, and capacity for, education. This program focuses on the analysis of relationships between schools and a wide variety of community resources such as government agencies, businesses, religious institutions, and social and medical services agencies. The central question of this program is how educators and other human and social services providers can enhance each other's efforts to improve the prospects of inner-city children and youths through cooperation, coordination, and mobilization of the latent energies and resources of the urban community.

## A Perspective on Education in the Inner Cities

Fundamental to the unique perspective of the work of CEIC is the concept of resilience. Grounded in developmental and ecological contexts, the resilience construct provides a broad cross-program and cross-project orientation for the work of CEIC (Wang & Gordon, 1994). A basic premise is the desirability of finding what is positive in the lives of children, schools, and communities even in the face of adversity. Although not forgetting for a moment the details, complexity, and history of the problems that cities face, CEIC researchers focus on the "positives" of urban life, the vast resources of the cities, and, most important, the resilience and potential for development and learning of children and youths.

## Design Principles

Four basic principles undergird the design of CEIC's programs of research and development: (a) focusing on solutions that promote development and educational success, (b) connecting research to problems faced by children and families in urban America, (c) forging a transdisciplinary approach to research and development, and (d) building on existing structures. These principles are briefly discussed below.

### *Focusing on Solutions*

It seems logical to expect that schools in urban communities would provide a place of refuge and hope for children. To an extent some do, and others valiantly try. Schools provide breakfast and lunch, and they play leadership roles in forging school-community connections among services for children and families in need. Sadly, though, in the critical matter of basic learning, many schools fail. This lack is particularly serious in schools with a high concentration of students from economically disadvantaged homes, which adds to the adversity facing children and families in urban communities, particularly those in the most inner of the inner cities. Surely other efforts will come to naught if we fail to offer powerful forms of instruction in this nation's inner-city schools.

Schools are our primary focus as we seek to improve the capacity for education in urban communities. However, school improvement efforts must also take into account that significant learning occurs outside the schools—in homes and the larger community setting. The capability of schools can be greatly enhanced through a better understanding and appreciation of community resources and influences, family educational goals, and factors that foster resilience and learning success. Thus the search for answers to CEIC's fundamental question must embrace families as well as all elements of the community.

The growing problems facing today's children, youths, and families stem from a variety of economic, political, and social pressures. The solutions are by nature complex; they require long-term programs of study that integrate knowledge and expertise from many disciplines and professions. One thing is certain: Human development and education must be key considerations in the rebuilding of our nation's inner cities. Solid research is essential to thoroughly understand the problems and resources of the inner city; to help raise consciousness about the opportunities in the community, especially among those in a position to shape policies and provide services; and, perhaps most important, to make improvements on the basis of what is known from research and school practices that promote student achievement. Creativity, clear vision, and realism are necessary as we seek to improve the depleted and unhealthy environments affecting the development and learning of children and youths in inner cities. There is much at stake for all of us.

## Connecting Research to Urban Problems

Research and researchers can contribute to making a difference in efforts to improve learning and life circumstances for children and youths in inner cities. This assumption has played a major role in shaping the research and development agenda at CEIC. It is widely acknowledged that few researchers are active in sustainable efforts to significantly improve our capacity for healthy development and educational success of children and youths in urban situations. Unfortunately, this neglect has been less than benign. Researchers, along with many others, have abandoned the heart of most cities; often there is an estrangement that compels researchers to "stay out." In many cases, researchers hold paternalistic attitudes or are only interested in testing their own ideas and theories. Frequently, they fail to join with practitioners in seeking concrete solutions to problems. The work of CEIC forges genuine partnerships that can make a difference in inner-city education.

Another assumption undergirding the work of CEIC is that cities are complex ecosystems. The economic, political, and ecological climate of the city bears on education in important ways. In particular, it influences the motivation of students and their beliefs about the future. Community agencies desperately need to coordinate their efforts in service to children and families; as they presently stand, agencies are fractionated, not coordinated, in their service efforts. In a broader context, the disciplines that might help to study and provide an understanding of life and learning in the inner city are not coordinated either. Unfortunately, schools are part of this disconnected nonsystem.

Perhaps the assumption stated above might be more properly framed as a challenge to researchers to prove the value of their work by applying their theories and research results in the difficult circumstances of the inner city. Researchers must become genuine partners with parents, teachers, and other community residents and help devise practical ways in which their research can be made useful. The needs in our cities are fundamental and, thus, require fundamental insights and tools. Although mindful of the need to avoid overly high expectations for improving these conditions through research, center researchers widely believe in the importance of making optimal use of what research *can* contribute to enhancing the capacity for education and quality of life. Research is needed at every stage: to delineate problems, to communicate effectively about them, to design and test interventions, to foster improvements, and to disseminate findings to the people of the inner cities.

## A Transdisciplinary Approach to Educational Research and Development

A knowledge and understanding of city life is central to improving cities' capacity for education. Significant advances can be made through building on what is known from research that works and through establishing communication channels and interactive connections across disciplines, professional services, and theories and practices. A key strength of CEIC is the expertise of a multidisciplinary team of researchers and collaborating practitioner colleagues who make use of diverse "ways of knowing" as they engage in a long-term transdisciplinary program of research and development.

The modus operandi for improvement at CEIC involves merging theory and research from fields as diverse as demography, ethnography, sociology, economics, political science, and education; and working across disciplines to seek broad, coherent patterns of linkage and collaborative working relationships with the practitioner's community, researchers, and policymakers. The child-family-school-community relationship is potentially rich with theoretical, methodological, and cross-disciplinary vitality. As a result, what is tried by the educator in the schools is tied to what the sociologist tells us about changing demographics in the city, connected to what the economist observes about shifts in employment opportunities, and linked with what the psychologist tells us about childhood and family stresses.

Indeed, the difficulties and rewards of bringing together researchers and practitioners from various disciplines and professional fields to collaborate in an intensive effort to find solutions to important educational problems of the inner cities is an evolving process of development and a distinct subject of study in its own right.

## Building on Existing Structures: One Center with Outreach

At CEIC's inception, concerns were raised over the fact that there was only one national research and development center, with limited resources, to address this nation's most pressing education problems. A guiding principle of our work, and indeed the challenge to CEIC researchers, is to ensure that the work of CEIC leads to practical applications in the near term, while pursuing long-term programs of research, innovative development, and dissemination/utilization. The work of CEIC has little value if the results and the wealth of information and experience are not effectively communicated in forms that are useful to policymakers and practitioners who can apply them directly in practical improvement initiatives.

CEIC researchers consider learning how to work across disciplinary and professional lines to find solutions to the multiple, co-occurring problems facing children and families in the urban communities an essential aspect of their own professional growth. Together, they work on finding ways to achieve (a) rapid dissemination of important findings; (b) mutual influence, collaboration, and exchange among practitioner colleagues and fellow researchers; and (c) the development of a research utilization process that involves a transdisciplinary team of researchers and practitioners to nurture healthy development and learning success in inner-city communities.

CEIC is expected to extend the impact of its findings and products by drawing attention to salient and emerging issues, as well as establishing formal and informal channels of communication and collaboration among diverse stakeholder groups. A first and most obvious approach to extending the work of CEIC is to carry out our work not only in Philadelphia, where CEIC is located, but also in other cities, building on the expertise and resources that already exist at major urban centers, including Chicago, Detroit, Houston, Los Angeles, Minneapolis, New York, Pittsburgh, and elsewhere.

A second outreach strategy has been the establishment of a dissemination and research utilization network through the dissemination and outreach mechanisms of standing structures such as urban universities, schools, human services-related agencies, and professional organizations. A major program of outreach at CEIC is the forging of ongoing collaborative relationships with professional organizations and other national, regional, and state education and related service provider agencies whose work is central to strengthening the capacity for healthy development and education of children and youths in urban communities.

## Next Steps and Prospects

What do we know after nearly 5 years of transdisciplinary research and development at CEIC to improve this nation's capacity for education in inner-city communities? Implications for next-step solutions are discussed in this section under four major topical areas: (a) fostering educational resilience, (b) implementing practices that are responsive to student diversity and resilience development, (c) forging school connections with family and community, and (d) building on existing structures for dissemination and scaling up efforts in using what is known to work from research and practical experience to significantly improve the capacity for education in this nation's inner cities.

### Fostering Educational Resilience

A major connecting theme of the work of all CEIC projects is the identification and description of conditions that contribute to resilience development and learning success among students, families, and schools in inner cities. Findings from our past 4 years of research provide a rich database for identifying processes that underlie adaptation and promote successful pathways that lead to educational resilience of students in inner-city schools, particularly those in circumstances placing them at risk of school failure.

CEIC's program of research on educational resilience was based on the concept of human capacity for successful adaptation despite developmental risk and adversity—a human phenomenon rooted in the work of developmental psychopathology (Garmezy, 1974). The center has advanced the resilience construct by applying it within an ecological framework that focuses on educational resilience (Wang & Gordon, 1994). This groundbreaking work has served a dual purpose: (a) It has broadened the meaning and understanding of resilience development, which, in the context of the work at CEIC, links the capacity for education in inner cities to the role of the schools, families, and the community; and (b) it has generated new approaches to studying and designing effective school and related services that serve children and families in inner-city communities with a high concentration of poverty and other risk factors.

Emerging from this aspect of CEIC's program of research is a construct of *educational resilience* that serves as a conceptual basis for studying and characterizing mechanisms by which educational resilience is developed by the individual students, the institutions such as schools that serve them, and the communities in which they live. This construct offers a provocative challenge to educational researchers and practitioners, suggesting several useful notions and priorities. For educators, it suggests the potential benefits of early experience, the need to mitigate adverse circumstances, and the importance of educationally facilitative and alterable protective factors in communities, homes, peer groups, schools, and classrooms. For educational researchers, it offers the intriguing hypothesis that alterable (possibly sustained) conditions fortify students to persist successfully through endemic difficulties. The next steps for researchers and educators, therefore, are (a) to identify individual and institutional attributes that contribute to resilience development, and (b) to determine how such development can be promoted through intervention.

Although fostering educational resilience is a relatively new area of investigation, the construct is gaining increasing attention in improvement-oriented research. Researchers at CEIC are beginning to be able to draw from the databases they have accumulated during the past 4 years to formulate better questions, and the concept of educational resilience is expanding. Perhaps of equal importance is the identification of situations and intervention strategies to develop students' ability for successful adaptation despite risk and adversity. These advances, a result of the work of every CEIC project during the past 4 years, are particularly salient in three specific areas: (a) understanding resilience in ecological contexts of inner-city development, (b) educationally resilient students in inner-city schools, and (c) characteristics of inner-city schools that are effective in promoting educational resilience.

## *Understanding Resilience: Ecological Contexts of Urban Redevelopment*

Taking into account the overlapping and multiple contexts surrounding an individual's development, CEIC has developed a rich research base that demonstrates how healthy development and learning success occur in the interactive contexts of a multitude of environmental, dispositional, and circumstantial influences, not as the result of a single precipitating event or innate personal characteristic. Three powerful and pervasive contexts influence children: the family, the school, and the community. The mix of environmental features, in combination with individual children's vulnerability to particular stressors, determines the impact of environmental adversities on children's educational accomplishments. This linking of resilience to ecological and contextual factors represents a critical conceptual advancement that has emerged from the center's work.

The ecological framework and the contextualized approach to the study of urban education have generated a national database that links economic and fiscal changes to education in urban communities. CEIC's database points to the conventional wisdom that a school is an institution embedded in a local community that exists within the structure of the larger society. The resources received by the school at the local, state, and federal level; the social and cultural capital brought to the classroom by its students; and the nature of the opportunities that await them after they leave provide the multilayered contexts within which schooling takes place. It is this multilayered and multidisciplinary perspective that CEIC's research on resilience development so effectively captures.

## *Educationally Resilient Students*

Resilient individuals are characterized in the literature for being proactively engaged in a variety of activities; having well-developed "self-systems," including a strong locus of control, high self-esteem, a clear sense of purpose, and healthy expectations; having the ability to successfully plan, change their environment, and alter their life circumstances; having strong interpersonal and problem-solving skills; and being capable of achieving learning success (Garmezy, 1974; Masten, 1994; Wang, Haertel, & Walberg, 1994). The CEIC database on educational resilience development provides information on the contexts of children and youths who live in circumstances that place them at risk of school failure. For example, a study by Peng, Wang, and Walberg (1992) found that resilient students (those from low socioeconomic status [SES] backgrounds in urban communities whose combined reading and mathematics test scores were in the highest quartile, based on national norms) showed higher self-concepts and educational aspirations and felt more internally controlled than did students with similar SES who were not considered educationally resilient. In a similar study, educationally resilient and nonresilient students in inner-city schools in Houston and Philadelphia exhibited consistent patterns of proactive participation and a high level of academic and social interaction with teachers and peers (Wang, Freiberg, & Waxman, 1994). Other CEIC studies on educational resilience have focused on race/gender differences in school performance and access to education (Rigsby, Stull, & Morse-Kelly, 1995), environmental factors that impede the transition of inner-city youths from school to work (Stull & Goetz, 1994), and the achievement performance levels of adolescents from minority backgrounds (Taylor, 1994; Taylor & Wang, in press).

## Characteristics of Urban Schools That Are Effective in Promoting Educational Resilience

Findings on what makes inner-city schools with a high concentration of students in high-risk circumstances effective in promoting educational resilience suggest quite consistent patterns of organizational and behavioral characteristics across studies (Anderson & Walberg, 1994; Freiberg, Stein, & Huang, 1995; Oxley, 1994; Wang, Freiberg, et al., 1994; Yancey & Saporito, 1995) and in congruence with the extant literature on effective schools (Edmonds, 1979; Levine & Lezotte, 1990; Purkey & Smith, 1983; Teddlie & Stringfield, 1993). For example, a series of studies of schools in Houston and Philadelphia (Freiberg et al., 1995; Wang, Freiberg, et al., 1994) showed significant differences between inner-city schools that were more effective in achieving student outcomes over time and those that were less effective. Teachers in effective schools spent more time interacting with students, whereas students spent more time working independently, expressed more positive perceptions about their school, and were more satisfied with their schoolwork and peer relationships. One common element across these inner-city schools that worked was a site-specific and ongoing in-service program for the school staff, based on implementation needs identified by teachers and administrators.

The CEIC database on resilience-promoting strategies provided the critical conceptual and practical basis for a field-based intervention program initiated in collaboration with school and community agencies in Houston and Philadelphia (Wang, Haertel, & Walberg, 1996). Initial implementation and program outcome data illustrate how the capacity for education can be expanded both inside and out of school contexts by building effective connecting mechanisms for productive communication, coordinated service delivery, and mobilization of latent energies and resources of communities. Its work not only draws from the expertise of local professionals to address site-specific improvement concerns but also provides demonstration of practical know-how in implementing the resilience-promoting concept.

## Practices and Policies That Are Responsive to Student Diversity

Schools in this country, particularly urban schools, are responsible for effectively serving an increasingly culturally diverse and economically heterogeneous student population. They are challenged to address issues of program specialization and social integration. It is estimated that within the next 10 years, more than 25% of the student population in this nation's elementary and secondary schools will be from racially, culturally, or linguistically diverse backgrounds (Garcia, 1995). Implementing effective school responses to student diversity requires major—in some cases, revolutionary—institutional changes and better approaches to school restructuring that can lead to significant improvements that are genuinely responsive to the diversity of needs of the students. This need is particularly pressing in inner cities, where schools enroll the highest number of children living in poverty and other adverse circumstances caused by a litany of modern morbidities, and where student diversity is the norm rather than the exception.

Harnessing all of the major resources and expertise to effectively meet the diverse needs of students has been a major focus of improvement initiatives at CEIC's collaborating schools in several cities. The central question of this work is, "How can we effectively implement and maintain innovative practices that have been shown from research and practical experience to work to achieve far-reaching changes in institutional policies and practices that are responsive to meeting the diverse learning needs of students?" These investigations show that although problems of schooling in inner cities may seem similar—low academic achievement, high rates of dropout and school violence—the causes and solutions vary from city to city and among school sites. For example, the flux of immigrant children from Mexico and the various Asian nations into Los Angeles contrasts sharply with the virtually all-Black schools of parts of Detroit, Philadelphia, and Chicago (Wang & Reynolds, 1995; Wong & Sunderman, 1995; Yancey & Saporito, 1994).

Findings from this aspect of the center's work point to three essential next steps in strengthening our capacity for urban education: (a) implementing effective practices that focus directly on classrooms and homes, where learning takes place; (b) eliminating educational segregation in schools and providing for student diversity within the context of one inclusive education system; and (c) forging greater school connections with families and the community. Each of these next steps is discussed below.

## *Implementing Effective Practices*

Much research has been conducted over the past half century on what yields better learning. This research deserves close attention from policymakers and educators as school programs are revised to better serve children. Center researchers consider the role of knowledge utilization a key to improving the capacity for rebuilding inner-city education, and a central consideration at every stage of their work. Thus CEIC researchers work with schools and other service-providing agencies to establish demonstrations of well-confirmed knowledge about what works to promote resilience development and schooling success of children in the inner cities. The challenge to CEIC is to work toward improved current practices as the center continues to seek advances that promote learning of all children.

CEIC's program on school implementation of effective practices has proceeded along several dimensions. They range from knowledge base synthesis to field-based experimentation and replication. Knowledge syntheses are an integral part of the work of all CEIC projects, partly to chart future work and partly to cull practical guidelines for innovative developments and program improvement from the extant research. CEIC researchers work to find ways to incorporate the confirmed knowledge base on effective practices to significantly improve schools' capacity for serving diverse student populations. Many schools that collaborate with CEIC have expressed special interest in scaling up efforts using results from CEIC studies. This aspect of the center's work has allowed researchers to adopt a unique role that directly impacts on the work of schools and their own research agendas and methodologies.

## *Effective School Responses to Student Diversity*

Each decade has seen an increased proportion of the population in school, a higher number of students from more diverse sociocultural and economic backgrounds, and a diversifying of the kinds of educational programs we offer. But these accomplishment have fallen far short of the educational vision of an inclusive system that provides all children with equal access to success in school. In principle, schools should provide for the diverse needs of all students, including those requiring special, remedial, or accelerated education in a coherent and coordinated system of service delivery. In practice, however, a disjointed and separatist second system has developed and continues to be the norm in responding to student diversity.

Studies of the categorical or second-system approach to providing for student diversity generally place these special programs on the "doesn't work" side of educational practice (Baker, Wang, & Walberg, 1994; Wang & Reynolds, 1995; Wong & Wang, 1994). Narrowly framed categorical programs designated to serve only specific categories of students contribute to the disjointedness and inefficiency that plague schools as they attempt to meet legislative mandates associated with implementation. This problem is particularly serious in urban schools with high concentrations of economically disadvantaged students. It is usual to find schools in which more than 50% of students are in pullout programs separated from regular and other categorical programs.

How do we restructure schools to implement and maintain an inclusive system of education that ensures that educational experiences in elementary and secondary schools are appropriate, meaningful, and encourage positive development and education for every student? A substantial knowledge base posits that public schools should be inclusive and integrated, and that separation by race, gender, language background, ability, or any other characteristics should require a compelling rationale. CEIC research indicates that schools that provide a common curriculum for all students, with a blending of resources and expertise directed at meeting the needs of the individual students and support needs of the teachers, show an overall improvement in student performance for all students—including those at the margins who require greater-than-usual support (Reynolds, 1994, 1995; Wong, 1994; Wong & Sunderman, 1995). Inclusive approaches allow schools to identify which students need extra support without resorting to costly and stigmatizing identification and classification, to empower the local school staff to redirect their resources toward implementing "what works" to improve instruction and learning for students, and to bring about collaborative efforts to provide for student diversity. Findings from these and other studies indicate that what all students need is powerful instruction delivered in the most adaptive and efficient manner possible.

## Forging School Connections with Family and Community

The ecological perspective, which views children and their families as a subsystem within a much larger ecosystem, has provided the conceptual framework for CEIC's program of research on school connections with families and communities. By linking families with multiple resilience-enhancing resources of their communities, CEIC research in this area has focused on (a) describing the design and implementation of school-family-community collaborating efforts to support student learning, (b) determining what is required to implement such collaboration, and (c) identifying approaches that effectively assess the outcomes of these collaborative ventures.

### *Schools and Communities Working Together*

New strategies for meeting the social, health, and educational needs of inner-city children and their families are essential—even the hallowed notion that schools are the sole providers of learning environments must be challenged. This aspect of the center's work focuses on two fundamental questions: (a) How can we provide social and health services to needy children and (b) how can we coordinate these services with those provided by school, and other educational institutions?

There is an emerging organizational, professional, and institutional movement to address the multiple and interconnected needs of children and their families by bringing together existing social and health agencies, schools and other educational institutions. Referred to variously as the "integrated collaborative," or "coordinated" service movement, its goal is to create learning environments that support learning success by focusing on meeting the physical and social wellness needs of students and their families by linking these families with the multiple resilience-enhancing resources of their communities (Rigsby, Reynolds, & Wang, 1995).

Although a variety of innovative programs have emerged across the country, all programs emphasize coherent and seamless child and family services that seek to improve education and life circumstances of children and youths placed at risk (Office of Educational Research and Improvement, 1995). Ranging from local, grassroots community efforts to state- and federal-level initiatives, these programs seek to transform fragmented, inefficient systems of service delivery into a network of coordinated partnerships that cross programmatic and agency lines.

Despite unprecedented national attention and a myriad of programmatic initiatives at all levels, solid information is glaringly lacking on ways to bring the knowledge base to bear on practice and implementation. Data from the series of core studies of collaborative school-linked coordination of services indicate that no single model for collaborative school-linked services exists; to the contrary, new programs emerge out of the needs of children and families in local communities. The crucial element is the way in which successful practices are combined in an integrated system of delivery that considers the needs of the students and the site-specific strengths and constraints of the staff, resource support, policy, and administrative levels.

Research suggests that coordinated services create as many dilemmas as they solve. Programs differ in organizational players; in critical issues addressed; in ultimate goals; in strategies of coordination, cooperation, and collaboration; and in the degree to which participants are conscious of the implied mandates for fundamental institutional change (Wang, Oates, & Weishew, 1995).

## *Schools and Families Working Together*

A general consensus presently exists that partnerships between schools and families, in sharing the responsibility for children's education, should be part of the solution to many social and economic problems facing our society today. Yet one of the most surprising findings from the center's knowledge base synthesis of parent programs shows the lack of data to substantiate or refute our present efforts to involve parents as partners in educational efforts (Iglesias, 1993). Present parent programs are a conglomerate of approaches that differ in their goals, formats, and durations with little or no regard to the interaction of parent characteristics and programs. The next step in this area should be rigorous evaluation of these programs—which, unfortunately, are often add-ons to other existing programs.

Parent programs should be designed to (a) address parents' individual needs rather than the generic needs assumed by program developers; (b) reevaluate the underlying philosophy of parent programs, shifting from a deficit framework to one that values and promotes cultural diversity; and (c) increase parent participation in ways that are consistent with individual families' beliefs of what is meant to be an active participant. We cannot afford to continue to base parent programs on unproven practices that are not sensitive to the specific needs of the families we are attempting to serve.

## *Field-Based Programs for Building Community Connections*

A key CEIC field-based project of particular relevance in building a database on how to establish and maintain family and community collaborative efforts is the Community for Learning project (CFL).

The Community for Learning project (formerly known as the Learning City Project) has a dual focus of harnessing and mobilizing family, school, and community resources and forging collaborative support for schools. This project seeks to achieve three major areas of student outcomes: (a) improved student achievement, particularly for those at the margins of the achievement distribution; (b) patterns of active learning and teaching processes that are consistent with the research base on effective classroom practices; and (c) positive attitudes by students and school staff toward the school learning environment.

Implementation of the CFL in inner-city schools in Philadelphia and Houston demonstrates that a school-linked coordinated educational and related service delivery program that addresses the multiple, interconnected, and co-occurring risks prevalent in the lives of many inner-city children has a significant impact on their academic progress. Overall findings from LCP implementation schools show a general pattern of positive progress in all three outcome areas, when compared with comparison schools. For example, students in CFL schools show more positive perceptions about classrooms and schools, better and more constructive feedback from teachers to students, an increased desire to learn, and an improvement in classroom behavior. The data show an increase in math and reading achievement among LCP students, who are also outperforming students in control schools (Wang et al., 1995).

CFL implementation provides a systematic database from which several policy and practical implications can be drawn. First, program implementation must be a shared responsibility of all stakeholder groups at the grassroots level. Local responsibility is essential to address the multiple, co-occurring risks prevalent in the lives and learning of many inner-city children and youths, who are placed further at risk by the inadequate and disjointed educational and related services they receive. Second, innovative programs evolve in stages of development, growth, and change. Procedures found useful in one city/community/school can be helpful in initiating similar programs elsewhere but adaptations and inventions are necessary to make the program work within the site-specific contexts. Third, few educational reforms have generated the same level of ground-swelling support as the comprehensive approach to coordinated educational and related services for children. Even though the research base and practical knowledge in implementing school-community connection programs require application of knowledge and expertise from many disciplines and professions, no system is in place to communicate and share the growing body of related research findings and innovative development experiences to practitioners and others who play major roles in education and health and human services delivery. This lack of access to information about program features and their implementation and effectiveness is a source of concern to CFL implementers as they engage in groundbreaking collaborative ventures.

## Building on Existing Structures For Dissemination and Research Utilization

Disconnectedness has been a major barrier to successful knowledge dissemination and utilization, both in terms of the substantive relevance of the information and how information is disseminated to reach the broad spectrum of education and related human services professionals and other stakeholders of inner-city education. Although there have been concerted efforts to improve the dissemination and utilization of knowledge at federal, state, and local levels, these efforts have had little impact on improving practice. This gap between the "state of the art" and the "state of practice" is particularly problematic in efforts to improve our capacity for education in schools with high concentrations of students from economically disadvantaged families in inner cities and rural areas.

CEIC's response to closing the chasm between the state of the art and the state of practice is to actively "reach out" and build on the resources and expertise of standing structures that extend to stakeholders—including researchers, practitioners, policymakers, and parents. The center, with its dual dissemination focus—to ensure that its work is known and used, and to receive feedback from the field for shaping and refining the work of the center—has established connections with field-based professionals to assure that the center addresses issues of genuine concern to individuals making everyday decisions about how to enhance the learning and quality of life of inner-city students.

### *A National Network of R&D Extension Stations*

CEIC has mounted collaborative R&D extension stations in major cities across the country through a network of collaborating universities in Chicago, Houston, Los Angeles, Minneapolis, and Philadelphia. The expertise and resources of the universities in those cities, as well as the respective collaborating links they have with local schools and related agencies, have greatly enhanced the center's ability to establish collaborating field-based projects. These collaborating R&D extension stations not only enhance the generalizability of CEIC's findings but also demonstrate how research can be brought to bear in local improvement efforts and extend the center's work through replication and scaling-up efforts. Thus a major outcome of the work of CEIC is the development of a national network of collaborating teams, researchers and practitioners who are in a strategic position to provide dissemination and utilization at their respective locales as well as nationally through their demonstration efforts.

## Partnership with Professional Organizations

The center has established working relationships with practitioner oriented professional organizations, permitting CEIC access to the dissemination and professional development mechanisms by which they reach their members at grassroots levels. By linking the work of CEIC with the ongoing work of the practitioner-oriented professional organizations, CEIC is able to serve the information and utilization needs of various organizations, thereby reaching individuals and local and state agencies with minimum effort and expense. More important, however, the center is given access to stakeholders who would not otherwise be exposed to the work of CEIC.

## Recruitment of Expertise

Although CEIC senior researchers collectively represent a major national R&D resource on inner-city education, the recruitment of expert input to further strengthen the institutional capability of CEIC has been instrumental in expanding the center's impact on research, policy, and practice. CEIC has enlisted a cadre of researchers, practitioners, and policymakers who are nationally known in their respective fields to provide ongoing input and serve as "dissemination agents" of the center's work. These experts have disseminated the center's work through the existing structures of their home institutions. This recruitment effort has been deliberately and strategically carried out through a variety of mechanisms, including the establishment of two ongoing advisory boards, the National Technical Advisory Board and the National Stakeholder Advisory Board. In addition to their overall advisory role in shaping the direction of the center's work, board members ensure that the work of CEIC is technically sound and useful for improving practice, and they advocate for the center to ensure that its work is known and used in efforts by the various stakeholder groups. Additional mechanisms established by the center to recruit expertise from leading researchers, practitioners, and policymakers include sponsorship of the practitioner fellows' program, the engagement of external reviewers for periodic review of CEIC products, the commissioning of special-topics papers, the hosting of discussion forums, and the Invitational Conference Series.

## Technical Assistance and Professional Development

CEIC's emphasis on research utilization has resulted in close collaboration and mutual support. Senior researchers at CEIC are integrally involved in the center's technical assistance and professional development activities. Field-based research projects are viewed as key elements for achieving research utilization and as ever-expanding opportunities for engaging practitioners in collaborative research. These projects use research tools to collect information on the implementation of interventions focused on particular improvement needs. The information they gather on the effectiveness and implementation of innovative practices contributes to an emerging procedural knowledge base that is essential to advancing our capacity for sustainable research-based reform.

# Conclusion

What can be expected from the next-step efforts to significantly improve our capacity for development and education in this nation's inner cities? We believe that research of the kind undertaken at CEIC—relating broadly to what families, schools, and communities can do to achieve a well-focused reform agenda to improve the educational resilience and learning success of all inner-city children and youths—must be at the center of our next-step efforts. To achieve this vision of improvement, we must target our efforts to accomplish the following:

- To seek useful connections across disciplines and professions so that cohesive, broad-based views and understandings emerge that provide coordinated and coherent resolutions to problems that hinder the improvement of the capacity for the development and education of children and youths in this nation's inner cities.
- To create a forum for the discussion of all stakeholders of inner-city education to establish practices and policies to overcome the barriers and procedural demands associated with multiple, narrowly framed solutions launched by federal, state, and local governments. However well-intentioned, an overwhelming number of implementation problems have resulted from some of these disjointed programmatic endeavors. Currently, ideas are being shaped for bringing about more coherent and accountable approaches, which should be extended in sustaining reform efforts.
- To create and test new modes of dissemination and outreach to ensure implementation by (a) providing assistance to educators and other service providers to conduct intensive microlevel research in their own settings to find site-specific, knowledge-based solutions; (b) creating a consortium of urban universities concerned with inner-city education and committed to providing a model of how researchers connect with community-based improvements; and (c) modeling and disseminating modifications in pre- and in-service programs that reflect new modes of education and social service delivery.
- To create new sources of information on inner-city education through innovative modes of communication (print, audio, and video). These information sources combine primary research, knowledge syntheses, theoretical treatises and practical reports from all disciplines central to inner-city education. Because the literature relevant to inner-city education is so scattered, it is impossible for researchers and field-based professionals to keep abreast of all developments. There is a need for new modes of information dissemination that focus on multidisciplinary and transdisciplinary treatments of inner-city education and related issues and progress. This will require much consultation with major professional groups, as well as reviews of all present publications and modification of their formats.
- To establish demonstrations of well-confirmed knowledge about ways to promote educational resilience and schooling success among inner-city children and youths in an effort to bring what is known to work in educational reform to scale.

Research and researchers can play key roles in forging significant improvements in inner-city education and development. Although the ubiquitous question, "What conditions are required to cause massive improvements in the learning and development of children and youth in this nation's inner cities?" must continue to drive and refine existing educational reforms, future reform must be directed toward practical applications in the short term, while pursuing long-term programs of research, innovative development, and dissemination and research utilization. At present, much research on educational reform results in what can be termed *declarative knowledge* relating to pedagogy, school climate, child development, family and community influences on child learning, and other related issues. Although important and potentially useful, declarative knowledge usually requires steps for transformation into procedural knowledge. That is, it must be further molded into forms that can be used in schools, homes, or other practical settings. This phase of the research-dissemination-knowledge utilization process is too often lacking, however, and must be recognized and "built into" the professional roles of staff members of schools, universities, professional societies, and others whose work directly and indirectly affects the healthy development and learning success of the young.

# References

Anderson, L., & Walberg, H. J. (1994). The applicability of data envelopment analysis to education research. In T. Husen & T. N. Postlethwaite (Eds.), *International encyclopedia of education* (Vol. 3, pp. 1381–1387). Oxford, UK: Pergamon.

Baker, E. T., Wang, M. C., & Walberg, H. J. (1994). The effects of inclusion on learning. *Educational Leadership*, 52(4), 33–35.

Edmonds, R. R. (1979). Effective schools for the urban poor. *Educational Leadership*, 37, 15–27.

Freiberg, H. J., Stein, T. A., & Huang, S. L. (1995). The effects of classroom management intervention on student achievement in inner-city elementary schools. *Educational Research and Evaluation*, 1(1), 33–66.

Garcia, E. E. (1995). The impact of linguistic and cultural diversity on America's schools: A need for new policy. In M. C. Wang & M. C. Reynolds (Eds.), *Making a difference for students at risk: Trends and alternatives* (pp. 156–180). Thousand Oaks, CA: Corwin.

Garmezy, N. (1974, August). *The study of children at risk: New perspectives for developmental psychopathology*. Paper presented at the 82nd annual meeting of the American Psychological Association, New Orleans, LA.

Iglesias, A. (1993). *Parent programs: Past, present, and future practices* (CEIC Publication Series No. 93-5f). Philadelphia: National Center on Education in the Inner Cities.

Levine, D. U., & Lezotte, L. W. (1990). *Unusually effective schools: A review and analysis of research and practice*. Madison, WI: The National Center for Effective Schools Research and Development.

Masten, A. S. (1994). Resilience in individual development: Successful adaptation despite risk and adversity. In M. C. Wang & E. W. Gordon (Eds.), *Educational resilience in inner-city America: Challenges and prospects* (pp. 3–25). Hillsdale, NJ: Lawrence Erlbaum.

Office of Educational Research and Improvement, U.S. Department of Education. (1995). *School-linked comprehensive services for children and families*. Washington, DC: Author.

Oxley, D. (1994). Organizing schools into small units: Alternatives to homogeneous grouping. *Phi Delta Kappan*, 75(7), 521–526.

Peng, S. S., Wang, M. C., & Walberg, H. J. (1992). Demographic disparities of inner-city eighth graders. *Urban Education*, 26(4), 441–459.

Purkey, S. C., & Smith, M. S. (1983). Effective schools: A review. *Elementary School Journal*, 83, 427–452.

Reynolds, M. C. (1994). A brief history of categorical school programs, 1945–1993. In K. K. Wong & M. C. Wang (Eds.), *Rethinking policy for at-risk students* (pp. 3–24). Berkeley, CA: McCutchan.

Reynolds, M. C. (1995). Funding for categorical programs: Where is the money taking us? In M. C. Wang, M. C. Reynolds, & H. J. Walberg (Eds.), *Handbook of special and remedial education: Research and practice* (2nd ed., pp. 345–369). Oxford, UK: Elsevier Science.

Rigsby, L. C., Reynolds, M. C., & Wang, M. C. (Eds.). (1995). *School-community connections: Exploring issues for research and practice*. San Francisco: Jossey-Bass.

Rigsby, L. C., Stull, J., & Morse-Kelly, N. (1995, April). *Resilience in schooling performances: Analysis by race/ethnicity*. Paper presented at the annual meeting of the American Educational Research Association, San Francisco.

Stull, W., & Goetz, M. (1994). *Secondary employment and education status of inner-city youth: Conventional wisdom reconsidered* (Atlantic Economics Society [AES] Best Papers Volume). Philadelphia: National Center on Education in the Inner Cities at Temple University Center for Research in Human Development and Education.

Taylor, R. D. (1994). Risk and resilience: Contextual influences on the development of African American adolescents. In M. C. Wang & E. W. Gordon (Eds.), *Educational resilience in inner cities: Challenges and prospects* (pp. 119–130). Hillsdale, NJ: Lawrence Erlbaum.

Taylor, R. D., & Wang, M. C. (in press). *Social and emotional adjustment and family in ethnic minority families*. Hillsdale, NJ: Lawrence Erlbaum.

Teddlie, C., & Stringfield, S. (1993). *Schools make a difference: Lessons* learned from a 10-year study of school effects. New York: Teachers College Press.

Wang, M. C., Freiberg, H. J., & Waxman, H. J. (1994, March). *Case studies of inner-city schools* (interim report). Philadelphia: National Center on Education in the Inner Cities.

Wang, M. C., & Gordon, E. W. (1994). *Educational resilience in inner-city America: Challenges and prospects*. Hillsdale, NJ: Lawrence Erlbaum.

Wang, M. C., Haertel, G. D., & Walberg, H. J. (1994). Educational resilience in inner cities. In M. C. Wang & E. W. Gordon (Eds.), *Educational resilience in inner cities: Challenges and prospects* (pp. 45–72). Hillsdale, NJ: Lawrence Erlbaum.

Wang, M. C., Haertel, G. D., & Walberg, H. J. (1996, October). *Revitalizing inner cities: Focusing on children's learning*. Paper presented at the National Invitational Conference on Development and Learning of Children and Youth in Urban America, Washington, DC.

Wang, M. C., Oates, J., & Weishew, N. (1995). Effective school responses to student diversity in inner-city schools: A coordinated approach. *Education and Urban Society*, 27(4), 463–503.

Wang, M. C., & Reynolds, M. C. (Eds.). (1995). *Making a difference for students at risk: Trends and alternatives*. Thousand Oaks, CA: Corwin.

Wong, K. K. (1994). The changing politics of federal educational policy and resource allocation. In K. K. Wong & M. C. Wang (Eds.), *Rethinking policy for at-risk students* (pp. 25–46) Berkeley, CA: McCutchan.

Wong, K. K. & Sunderman, G. L. (1995). *Linking federal Title I reform to inner-city classrooms: Impact on resource allocation and instructional practice*. Unpublished deliverable, Temple University, National Center on Education in the Inner Cities, Philadelphia.

Wong, K. K., & Wang, M. C. (Eds.). (1994). *Rethinking policy for at-risk students*. Berkeley, CA: McCutchan.

Yancey, W. L., & Saporito, S. J. (1994, October). *Racial and economic segregation and educational outcomes: One tale—two cities*. Paper presented at the invitational conference entitled "Social and Emotional Adjustment and Family Relations in Ethnic Minority Families," sponsored by the National Center on Education in the Inner Cities, Philadelphia.

Yancey, W. L., & Saporito, S. J. (1995). Ecological embeddedness of educational processes and outcomes. In L. C. Rigsby, M. C. Reynolds, & M. C. Wang (Eds.), *School-community connections: Exploring issues for research and practice* (pp. 193–227). San Francisco Jossey-Bass.

# Life in the Bricks

## CHARLES BRUCKERHOFF
*University of Connecticut*

> In Cleveland, the poorest children live in public housing projects called the bricks. The culture of the people there, along with some misguided public policy, has hindered efforts to reform the mathematics curriculum in Cleveland's intermediate-level schools. Students were poorly prepared for school, and their local culture was in conflict with teachers and school authorities. Teachers' efforts to change their curriculum from a textbook-dependent approach to an emphasis on problem solving were unsuccessful also because of their fears of being fired, their not wanting to appear ignorant to their students, and the high rate of student absenteeism. To make public schools integrated with the community and conducive to curriculum reform, they should cooperate with local religious and social organizations for the educational benefit of people in local cultures.

The primary purpose of the Cleveland intermediate-level teachers' problem-solving infusion project, funded by the National Science Foundation, was to enhance intermediate-level mathematics instruction. To reach this goal, teachers were to reorganize their mathematics curriculum through meetings, lectures, workshops, and demonstrations, giving special emphasis to improving their students' problem-solving ability. The grant proposal lists the project's two underlying motives as (a) to meet the new standards for school mathematics and (b) to empower teachers to make curriculum decisions.

The first year of the grant's operation was unsuccessful (see Bruckerhoff, 1990b). However, in the second year, the problem-solving infusion project succeeded in convincing some teachers to think differently about mathematics and teaching. Now the problem was at the classroom level. After 2 years of in-service training, these teachers' mathematical knowledge had improved, but problem solving was apparently a smaller part of their teaching. Why? The teachers' shift from their usual practice to problem solving encountered two obstacles: public policy and the school's social context.

The state's public policy and the school district's standard curriculum contributed substantially to the discrepancy between the project's intentions and its results. A district course of study, mandated by the state of Ohio, listed pupil performance objectives to micromanage how well mathematical topics or skills were taught and when. A textbook was acceptable to the district if

---

Charles Bruckerhoff, "Life in the Bricks," *Urban Education,* October 1995, Vol. 30, No. 3, pp. 317–336. Reprinted by permission from Sage Publications, Inc. (US), Corwin Press.
AUTHOR'S NOTE: The author expresses appreciation to Steven Gilbert and Shiva Tavana for commentary and editorial assistance. This research was sponsored by the Cleveland Education Fund and the National Science Foundation. An earlier version of this article was presented in a Division B paper session at the annual meeting of the American Educational Research Association held in Chicago, Illinois, April 3–7, 1991.

it fit the district's course of study and if it was used district wide. Adjustments would be made in the pupil performance objectives only after the purchase of a revised textbook. The other district policies that served to maintain a standard curriculum were crosstown busing, average daily membership (school enrollment data), and competency-oriented teaching, but these issues are not the subject of this study.

The school authorities' primary concerns were to meet the expectations of public policy for student achievement as well as to maintain safe, orderly schools. Following directives from the State Department of Education and the local school board, the Cleveland public schools' central administration established policies that were increasingly obtrusive and explicit. The district's principals organized their schools along traditional lines and standard procedures. The mathematics teachers shied away from process-oriented teaching and learning, relying instead on traditional methods and standard materials so as to be consistent with public school policy and meet expectations about classroom management. If they failed, they could lose their jobs.

Proficiency tests, mandated by the state of Ohio, are now administered annually in Grades 3, 6, and 9. It remains to be seen how the proficiency test results will influence public policy. The first proficiency tests were given in November 1990. The results showed that Cleveland had the lowest scores in the state in mathematics; in fact, 90% of the district's pupils failed (Rutti, 1991, sec. C, p. 4). When the results were published, an assistant superintendent with the Ohio Department of Education said that the proficiency tests would lead to more scrutiny of school administration at the district level and they would be the driving force for curriculum reform (Rutti, 1991). Some teachers have responded to the proficiency policy by teaching to the test for several weeks prior to the examination. As external pressures increase, this temporary adjustment could become, in effect, the mathematics curriculum.

Mathematics curriculum reform was also hampered by the social conditions in the school district. At the time of the study, the total Cleveland public school pupil population was approximately 72,000. A majority was Black (68%) and economically and educationally disadvantaged. Smaller percentages came from other minority groups, chiefly Hispanic, Puerto Rican, Asian, and Appalachian, and were also disadvantaged. Approximately 9% of Cleveland public school children lived in public housing projects, which had some of the worst living conditions in the city. The school district's standard curriculum is seen by Popkewitz (1988) as a colonization program for its ghetto children. Many Cleveland public school pupils had long family histories of welfare dependency. Many had suffered seriously from various forms of neglect and abuse. They entered school equipped with particular coping strategies and tendencies to behave confrontationally. Cleveland's pupils were not traditional and not accustomed to standard school-operating procedures. They were urban waifs, often pariahs. Despite the school's educational efforts, academic achievement was low, dropping out was common, and very few actually escaped from the conditions into which they had been born (cf. McDermott, 1974; Wilson, 1987).

Common elements of the urban children's lives, such as deception and violence, are seen by Ogbu (1988) as oppositional survival strategies inimical to school organization. In response to chronic student absenteeism and the abuse students endured at home, teachers had adopted a routine called the homework curriculum, which consisted of five phases: bell-work assignment, bell-work check, homework check, homework lesson, and homework assignment (see Bruckerhoff & Popkewitz, 1991). Despite the drawbacks of such standardization the homework curriculum at least gave teachers and students some assurance that some mathematics would be taught. So, naturally, the teachers had difficulty adopting a problem-solving approach. The problem-solving infusion project could play an important role in changing mathematics to meet the new standards and to empower teachers, but its successful implementation depends heavily on an environment (school and community) that is favorable to change: curriculum leadership, school reorganization, and public housing redevelopment. To the extent that plans for reform—including new school policy—neglect the social context of the urban poor, they will remain empty promises.

## Methodology

Fieldwork focused on the natural history of this curriculum reform project (see Bruckerhoff, 1991; Smith, 1986) during 10 months from April 1990 through March 1991. I used ethnographic methods, recording descriptive and historical data from observations and interviews with teachers, students, principals, public housing tenants, police officers, social workers, and health professionals. To help interpret the urban children's school behavior, I rode with Cleveland police officers who were assigned to the public housing area and accompanied social workers who were visiting public housing clients.

Seven teachers were chosen for this study on the basis of their active, long-term involvement in the problem-solving project, positive recommendation by their supervisors, and their willingness to allow themselves and their students to be research subjects. The teachers worked in one of four different intermediate-level public schools; each school has approximately 650 students. In addition to taking field notes, I recorded on videotape the housing projects where the children lived and mathematics classes in Grades 7 and 8.

## Life in the Bricks

Cleveland's public housing units are scattered around the city, mostly in poor areas. Usually referred to as the projects, they are densely populated, high-transfer, urban neighborhoods with names like the King-Kennedy Estates and Longwood Estates. Most are red or yellow brick buildings with apartments rented by single women—mostly Black and Hispanic—with several children. Some families living in the projects have been dependent on public assistance or welfare for five generations, sometimes even longer. The research concentrated on three housing projects on several city blocks on Cleveland's near east side.

According to the Cleveland Metropolitan Housing Authority, the total population living in the projects was approximately 16,000, of which 1,679 (10%) were under 5 years old; 2,855 (18%) were between 5 and 11 years old; and 1,658 (10%) were between 12 and 17 years old. Thus there were at least 6,192 (39%) children living in the projects. Unofficially, the housing authorities said that at least 1,000 additional, transient people lived there with families but were never included in the record. The tenant mobility rate was approximately 25% turnover per year.

In the chaotic flux that is life here, these bricks, along with the sort of inhabitants, remain one of the few constants. Year after year, tenants and vandals stripped and destroyed some apartments. Year after year, the Cleveland Metropolitan Housing Authority rebuilt them. Some project buildings were in bad condition, with broken doors and windows, leaking pipes, and so on. These were the worst apartments still occupied. Other apartments had collapsed internally, with doors and windows ripped out and floors and internal walls rotting. These were unoccupied, except temporarily by gangs and drug traffickers. Some project apartments were in good condition, having been recently rebuilt. A few were new, single-family, owner-occupied houses. Except for the new houses, most project yards were mud flats or dirt plots, depending on the season.

Dramatic contrasts among many apartments provided clues to the tenants' disparate lifestyles. One apartment would be clean and well furnished, with beds for adults and children, a dinner table and chairs, sofa, radio, television, and so on. The apartment next door would have little furniture, but dirt, grime, and debris everywhere, and cockroaches covering the walls. Mattresses, if there were any, lay on the floor and children were sometimes forced to sleep in the hallway. In the good apartment, a single mother would be using every available resource to provide for her family. In the hovel, another single mother would have a dysfunctional family with some combination of drug addiction, alcoholism, or physical abuse. Men, if around at all, were usually just temporary guests.

In the bricks, there was always some trouble at least brewing, and some tenants' lifestyles encouraged chaos and oppositional behavior. People were outside at all hours of the night, at any time of the year. Small groups hung out in doorways, alleys, and streets. Older men gathered regularly to talk, to smoke, and to drink. Children ran errands day and night, sometimes to sell drugs. Working men and women went to or from the bus stops. Men repaired cars in the parking lots as a front for their drug dealing. Female drug addicts, called strawberries, worked the streets as prostitutes. Suburban johns cruised the streets in new cars to pick up strawberries for sex or drug deals. Children were up most nights because the police arrived frequently at their apartment or next door to prevent domestic violence, to raid drug dens, to arrest fugitives, and so on.

In the bricks, people willfully did unhealthy and even dangerous things, perhaps because they believed that they had nothing to lose. For instance, at 3:00 a.m. during the summer months, mothers often took their children—even infants—to a nearby pool. Someone would cut through the protective fence, allowing anyone to enter and swim without a lifeguard. Other men and women joined them, many drinking and taking drugs. Broken bottles littered the pool. Fights broke out. People were seriously injured or perhaps someone drowned.

Many people who lived in the bricks took drugs and drank a lot, which explains not only the flourishing drug trade (which also served the suburban cities) but also various forms of physical, sexual, and emotional abuse—including self-inflicted injuries, child abuse, and neglect. For example, mothers—themselves children—occasionally left their infants with friends or relatives, then just disappeared. Authorities often found 2- and 3-year-old children wandering through the streets during the severest weather, unaccompanied, hungry, and poorly dressed. When these children were returned to their homes, their parents or guardians were usually so "zoned out" that they had been unaware of their children's absence.

Despite the daily efforts of some parents to shield their children from dangerous and illegal activities, they could lose out to the severely depressing local conditions and the endemic drug culture. As an example, one evening, police responded to a report from a woman, Abbey Hill, of a break-in at her apartment. Ms. Hill said that the thief had thrown a brick into the first-floor window, unlocked the door, and stolen $140 from her purse. She knew the thief; it was her 18-year-old daughter. Police and social workers reported that such burglary was common.

Abbey Hill was the mother of seven children, her oldest child was 18 and the youngest just 1. It was the first week of the month, and Ms. Hill had just received her welfare payment. Earlier that evening, she and her daughter had argued about money. Her daughter had demanded money to buy a dress so that she could attend a funeral. Ms. Hill had insisted that a new dress was not necessary, and—as always—there was scarcely enough money to provide for the family's basic needs. Ms. Hill told the police that her daughter was hanging out with a bad crowd, that she was most likely involved in the drug trade, possibly a user. Ms. Hill gave this further explanation:

> I hate to do this to my own daughter [initiate an arrest warrant], but look what she's done. Now we don't have any more money. She's out with her friends, and who knows what's happened to the money.
>
> I'm mad at her, but I am not going to let her ruin our chances. I've told her again and again: You're single, got no kids, got good looks, you're bright, and look at what you're doing to yourself.
>
> Last month I gave her more money than the other kids because she had to pay for graduation materials. Then I found out that she flunked out. She don't want to work; she don't want anything but money.
>
> I've pushed these kids to do well in school, to go in the right direction. Now my oldest one falls for this crap [drugs]. I'm not going to let it get me down. This daughter of mine will not ruin it for all of them. I'm getting out of these bricks by this time next year.

This woman was determined to improve her family's life chances in the midst of despair and hardship. Ms. Hill's apartment was well kept and had adequate furniture. On the tops of tables and shelves were many well-organized stacks of magazines, papers, and books. In her living room were a color television, typewriter, sleeper sofa, and study carrels. Bricks painted white trimmed the lawn outside her front door. In the opinion of police and social workers, Ms. Hill's apartment was unusual. Although Ms. Hill was doing her best for her family, her eldest daughter's conduct was threatening the whole family's chances to get out of the bricks and make something of themselves.

Ironically, the daughter's behavior was consistent with her mother's desire to get out of the bricks. Social workers called the daughter's entrance into drug or alcohol abuse a common but desperate and destructive escape mechanism. One social worker explained:

> Children see their moms getting beat up because they don't sell enough drugs or give enough as prostitutes. They find a dead body in the hall. They see a lot more than you or I do—or care to see, and it isn't pleasant.
>
> They have gone without soap because they only had enough money for bread. They have gone without supper. They see this and understand. They look for a way out—just like many of their parents—through drug abuse, alcoholism, and prostitution.
>
> So, when these students are sitting there in a classroom staring, it isn't that they are contemplating truancy; they may be wondering where they will sleep and whether they will get a meal that night. Their behavior is strictly from what is going on in the home.
>
> Other kids in school always know who's on welfare and who's living in the worst part of town or in the part that's worse than theirs. These kids see the difference. They are not stupid, just not educated. They see things in black and white and clear.
>
> Parents call here all of the time about children who leave home and don't come back. They want to escape. Most go for the "here and now": cocaine use. The situation is so bad, they want to get away, if even for a few minutes.

Children who lived in the bricks were aware that it was a different way of life from that which kids had outside the projects, that entailed homelessness, deprivation, and abuse. The common way to try to escape was through alcohol, drugs, and crime. The uncommon way was through education, jobs, and a stable family life. Local conditions made the dignified route appear an impossible ideal.

Anyone could suffer unprovoked and unmitigated violence at any time. Victims occasionally knew their attackers, but they might not reveal the attacker's full name or address to authorities. When victims did identify their attackers, they seldom appeared in court, thus undermining legal action. Almost no one carried identification, certainly not a photo ID. Deception and halfhearted efforts, often perceived by authorities as just so much more irresponsible behavior, were other ways in which these poor people attempted to protect themselves and further expressions of their hopelessness and despair. One evening a woman reported to the police that a 17-year-old girl had run up to her apartment door with a butcher knife and threatened to kill the woman's daughter. According to the mother, the would-be attacker was Laurice, her daughter's friend and was all messed up on drugs and liquor. When the police asked for the girl's last name, no one in the apartment could remember either it or her address. Before leaving, the police told the mother to keep her door locked at all times. Several days later, in what was most likely an unrelated event, the news media reported that a girl, who was waiting at a nearby bus stop, was stabbed repeatedly by another girl—a stranger—who said to her victim, "I don't like you." Almost everyone who lived in the bricks had suffered some kind of assault, sometimes deadly. On a nearby street, a boy wearing a Raiders football team jacket was assaulted and shot to death by a man who demanded the jacket as a part of a gang initiation rite.

According to social workers and emergency room staff, theft and violence occurred most likely during the first week of the month, when tenants received their welfare checks. Also, police officers and health professionals reported that a large number of project tenants were always high or drunk. Often desperate for money to buy drugs or alcohol, substance abusers commonly resorted to prostitution, theft, and violence. Cleveland's 1990 crime rate—already high—showed some increases since 1989: 144 murders (5%), 837 rapes (–1%), 2,939 aggravated assaults (15%), and 4,045 robberies (7%). Gang activity was increasing and certainly contributed its share.

In the bricks, drug dealing had replaced numbers running—the poor people's illicit lottery—as the most lucrative and dangerous economic activity. The drug trade discouraged honest business people from opening stores in the vicinity. In stores that were doing business, life-threatening conditions and greed on the part of some businessmen contributed to inflated prices for common household items, such as coffee tables and vacuum cleaners. Often gang-related drug dealing was the most attractive occupation to young males who lived in the projects, and many were involved. Despite great risk of injury or even death in drug deals gone bad and from rival gangs, some adults—including parents and grandparents—used children in drug trafficking. Life in the bricks taught children that to survive they had to deceive, to distrust, and to elude authorities—even family members.

For many Black males, life in the bricks included drug trafficking, car theft, assault and battery, time behind bars, and a criminal record. The main effect was Black-on-Black crime—a seemingly perpetual, self-destructive force within the urban Black community. Major players in the national drug trade operated in Cleveland's public housing projects, assuring a steady supply. Adolescents would rather sell drugs than hold ordinary jobs because a seller could make hundreds or even thousands of dollars in a single evening. It did not matter what had happened to people who used drugs and to sellers who went out and never returned. What did matter was making a bundle of money fast. To authorities—including those within the Black community—it was nightmarish.

The following incident, with all its consequences, is representative. One night, three boys, aged 15, 16, and 17, were walking across the yard outside an apartment building. Two police officers were waiting in a squad car across the street. One of the youths dropped a small container on the ground—a telltale sign. The squad car surged forward, jumped the curb, and raced onto the dirt lawn. The officers immediately grabbed the boys, frisked them, and found 79 rocks of crack cocaine with a street value over $3,000.00.

The boys were arrested, then jailed until their trials. At the time, each was on parole and had a criminal record, including grand theft auto and drug trafficking. They attended the same school and would not return there until they served time in a detention home. The Black police officer who made the arrest told the trio in emphatic street language that crack's effects on users were measured at the hospital in terms of Black people's overdoses, deaths, and addicted infants, and he held them personally responsible. In the bricks, illegal drug sales and substance abuse continue despite daily arrests and raids by police, counseling from social workers, and life support from hospital personnel.

The children in Cleveland's projects suffered from poverty and abuse. Existence depended on securing food, shelter, and clothing daily. Survival strategies, like theft and deception, helped these children to obtain basic necessities and to cope with ever-present life-threatening situations. Various forms of criminal activity—especially drug sales, burglary, and assault—were common in their poverty-ridden society. To "get out of these bricks" usually meant escape *from* the local culture—a self-destructive world of alcoholism drug abuse and crime. However, the local culture is based on street knowledge, spits on formal education, and destroys its members. A real escape, through education to a job, was like winning the lottery—rare and considered purely a matter of luck.

# The Social Context and the Urban Mathematics Classroom

The implications of the social context for teachers and for curriculum reform cannot be underestimated. The potential for violence in the school by students against students or teachers is very real. Under these circumstances, teachers and administrators are often more concerned with maintaining order and discipline than with such matters as process-oriented teaching and learning (see Bruckerhoff, 1990a). Thus the social context is a powerful conservative force, reinforcing the standard curriculum.

Children who lived in the projects came to school to get away from bad situations at home with their parents or the neighborhood. But they were poorly prepared for school, their ghetto culture colliding with that of the teachers and school authorities. Physical confrontations between students and teachers were not uncommon but much more frequent among students, most of them Black students attacking other Blacks. At all times, teachers had to be prepared for attacks.

For instance, one day when an intermediate-level mathematics teacher was giving his students a lesson on multiplication of fractions, a male student seated in the back row casually got up from his desk and walked to the front. The teacher assumed that the student wanted to sharpen his pencil or ask for a copy of the day's work sheet. Instead, the boy stopped beside a girl seated in the first desk and punched her in the face. The teacher went immediately to help the girl, pushed the boy away, and sent him to the office. The boy was suspended.

The next day, the boy's father came to the school and accused the teacher of mistreating his son. In the argument and discussion that followed, the teacher learned that the boy had already been scheduled to go to court the following day; he would avoid jail only if he could show good behavior. Therefore, as the father saw it, the teacher, by having the boy suspended, could have ruined his chances to get off. However, the father, besides being overprotective, misunderstood his son's intentions. The teacher's further inquiry revealed that the boy had struck the girl as part of a gang initiation rite that required gaining an official record for delinquency, school suspension, and jail time.

On another day, I was standing in the school's hallway during class time. Two boys, friends who were supposed to be in math class at the time, were racing at breakneck speed for the stairway. When they passed by me, their faces were distorted with crazed looks. The school's counselor reported that one of the boys was going out with a girl in a rival gang. He and his friend were running because the girl's angry brother and his gang had just threatened to kill him.

Despite pupils' desire for escape or at least relief at school, they frequently brought their problems with them there. These dangerous and life-threatening incidents disrupted the teachers' work. When the teachers imposed the homework curriculum, reinforced with traditional methods and row-by-row seating, they were attempting to ensure classroom order.

Teachers were haunted by the specter of classroom violence, as the following example shows: A mathematics teacher was helping her students solve a mathematical problem dealing with proportion and variability. Usually well prepared and self-confident, the teacher seemed to be having difficulty this period. She made some errors that were corrected by the students. The students made some errors that were not corrected by the teacher. The incorrect work produced answers that were different from those in the teacher's manual, necessitating rework.

Despite several repetitions, neither teacher nor students appeared to be bothered by the mistakes. Indeed, the students' behavior was typical—four girls had their heads down and appeared to be sound asleep, three boys in the back traded sports hero cards, three girls compared makeup, six students had no text and no paper, and two others stared at the floor. Only five students, who were sitting near the front and center, attended throughout the lesson and responded to the teacher's questions.

During class change time, I asked the teacher to comment on her lesson. She frankly declared that she needed to spend more time studying that material, but in this instance inadequate preparation was not what caused her confusion and mistakes. She has been disturbed all year about the next class period because most students in it were uncooperative and frightened her.

She said, "I'm not sure how much longer I can continue doing this. I've been teaching for over 20 years, and the problems with students are getting worse and worse."

The students who were enrolled for the next class filed in and slumped into their desks. Most of the male students were much taller and heavier than their teacher, who was 5 feet 2 inches tall, and thin.

The teacher abruptly stopped speaking to me and barked at a male student, "Who gave you permission to enter at this time? Where is your pass?" He had been suspended and had entered the classroom without giving his teacher a pass. Without speaking, he slowly groped into each pocket. Each thrust turned up the wrong piece of paper. The pass was in the last pocket. He handed it to the teacher

The teacher said to me, "There is a rule against coming in late without a pass. They [the administrators] never stick to the rules." At the sound of the bell, she started class by immediately telling the students to open their books to page 47. Her voice was louder than during the previous class, and she trembled slightly.

The teacher's reaction was understandable in light of prior instances when the principal had sent students to her class without a pass. Also, she was upset, worried about her safety and that of other students.

For this teacher, mathematics teaching required many preparation hours, including private study, active participation in the problem-solving infusion project, grading papers, exams, and so on. Partly because of the Cleveland Education Fund's mathematics collaborative, she was not naive about the work requirements for teaching urban students, mostly because she had been doing that for 20 years. Also, she expected to have some unruly behavior, and she used effective classroom management and discipline strategies. However, in recent years, the students' attitudes and behaviors made it appear too likely that a crazed, irrational student would do something to threaten her well-being and that of other, innocent students. The classroom's social atmosphere frightened her and unsettled her teaching.

It was apparent to this intermediate-level mathematics teacher (and others) that no amount of preparation could counteract the worsening social context. Her perception seemed to have some validity, for when I asked the trauma administrator at St. Vincent's Charity Hospital to give his perception of urban children, he said,

> The most pressing concern is teaching these young people about conflict resolution that is *short of violence*. This is a tremendous area of concern. They see violence. There will always be conflict. But they need to know how to resolve their differences without resorting to violent means.

The trauma administrator's staff routinely treated victims and attackers alike for gunshot and stab wounds, concussions, severe bruises, some fatal. Terrible incidents could occur in the projects at any time, and even family members were as likely as strangers to be victims. To the mathematics teachers, violence was equally likely in the classroom.

## Public Policy and the Urban Mathematics Classroom

To reform curriculum, public policy cannot afford to ignore social and cultural problems, and teachers' performance-based job evaluations. Almost regardless of the children's personal histories, school policy and practice emphasized standard academic achievement and classroom management and discipline. When a school recognized students' tragic histories, it treated these social and cultural problems as subject matter. For example, one school set aside one class period each week to improve the students' self-esteem with a lecture provided on video. Apparently, low self-esteem, like any other subject matter, was to be treated like a distant, cool subject.

The homework curriculum was the standard practice, but most teachers were unaware of its widespread use. Every teacher who participated in the problem-solving infusion project believed that the other teachers' work was different. In particular, each assumed that colleagues were incorporating the problem-solving materials much more. One teacher's remark was representative: "I'm sure that [a colleague] has been doing this every day and probably has no trouble fitting it in, but I can't figure out how I can do both problem solving and the course of study." My second-year observations of several teachers' classrooms revealed that mathematics teaching was similar across classrooms and schools and that it was typical to find light and inconsistent use of problem solving. The homework curriculum, tied to the district's course of study, was the standard instructional design. A problem-solving project, if there was one for a class period, was usually a substitute for bell work.

During the project's second year, the original teacher group to teach techniques of problem solving to other volunteers from the mathematics department, thereby increasing the program's scope. Observations and interviews suggested that this expansion was slow. One of the teachers explained:

> There's an abundance of problems—in addition to the social context—contributing to a loss of enthusiasm from the original group to those on the front lines. Don't mistake me; if it wasn't for our mathematics curriculum director and projects like this one . . . well, I would hate to think where we would be now.
>
> In this building, the principal is so puffed up with himself that he thinks everything good here happens because of him. When there's the least noise, he comes running to blame a teacher for starting a problem.
>
> Some principals make teachers worry that they will be transferred or nonrenewed for not meeting the competencies. If I'm fearful that my job depends on my performance, then I am not going to take a risk. If this is an extra task, then it will take away from my effectiveness. It's that simple.
>
> Veteran teachers are at fault because they resist change. They don't want to appear ignorant to their students. Some have a daily routine that they have been doing for years and years and require in-service that retrains them. Some teachers won't be able to do the problem solving without a recipe.
>
> Most teachers feel that they are unsuccessful just trying to teach the course of study because of interruptions, absences, and so on. It's almost impossible. Teachers feel that Cleveland's system—with all the requirements and responsibilities already in place—could make it impossible to implement another component beyond the course of study.

In this teacher's opinion (corroborated by others), there were three reasons why the problem-solving infusion project was not expanding on schedule: The teachers' performance-based job was often made even more difficult by principals with poor supervisory and public relations skills; teachers did not want to appear ignorant to their students, preferring to follow a fixed work routine; and a typical classroom period had many interruptions, making it difficult for teachers to complete a lesson.

Every teacher believed that problem solving was an important and long-neglected aspect of the intermediate-level mathematics curriculum. However, each also believed that problem solving interfered with the chances of completing course-of-study requirements. These teachers chose the standard curriculum over problem-solving activities because of intrusive public school policy and the urban school's social context.

## Discussion

A school's mathematics curriculum may respond to current calls for pedagogical reform but may fail because it neglects political and social problems of children in big cities. The children who lived in the projects had experienced hopelessness and violence. Most of these children came to school to escape from home rather than to gain from academic study. Curriculum reform cannot ignore this social context.

The problem-solving infusion project's purpose was to improve the students' mathematical skill through inquiry and discovery. Teachers were to bring the Cleveland mathematics curriculum into conformity with the new standards, and in so doing, empower themselves as decision makers. Mathematics teaching would include application and computation, using Cleveland-based problems. However, in regard to both administrative structure and curriculum, Cleveland's intermediate-level schools were organized along traditional lines. School district policy emphasized order and supported a standard curriculum. The mathematics course of study, including pupil performance objectives, was bureaucratic sclerosis. The state's proficiency test requirement reinforced the course of study and was expected by Department of Education officials to "drive the curriculum."

At the level of individual schools, principals followed central office directives. Some administrators threatened to fire teachers if they failed to demonstrate improved student achievement. Teachers responded by maintaining the homework curriculum, a daily routine that protected their jobs and ensured standard mathematics instruction. Teachers not among the first volunteers for the project were reluctant to adopt the problem-solving activities because they did not want to appear incompetent to supervisors or to lose face in front of students. Not surprisingly, proficiency tests showed that Cleveland's pupils scored the lowest in the state in mathematics.

The problem-solving infusion project, with its emphasis on teacher decision making and process-oriented learning, required a situation conceptually different from the standard curriculum. In particular, the project called for classrooms to become self-actualizing communities, where continuous student-teacher interactions regularly led to their using mathematics as a vital tool for solving real-world problems (West, 1991, p. 5). The realization of such classrooms necessitates a different kind of relationship between principals and teachers, among teachers, and between teachers and their students. Principals would have to build nonthreatening and supportive organizations encouraging the teachers' involvement in curriculum planning. Teachers would have to be sufficiently autonomous, self-confident, and effective in planning and implementing an educational program that balanced process-oriented with directed instruction (see Bruckerhoff, 1990a; Oliver & Gershman, 1989; Rosenshine, 1983). Students would have to be like their teachers—empowered as respectful, democratic people interested in mathematics (see Romberg & Carpenter, 1986).

If the process-oriented curriculum were implemented, schools would be converted into small, manageable, personal educational communities made up of principals, teachers, and students committed to the task of learning not in a distant, abstract context, but within the local culture itself. Local knowledge, embedded in local or folk culture, is the unique set of beliefs, activities, and values worked out over time and held in common by people. The characteristics of particular local knowledge originate in environmental factors, contributions of individual group members, and ties to sacred and secular traditions. Public education should be the formal, institutional means for all children to gain an understanding and appreciation of local knowledge in relation to the complex cultural heritage. Thus schools would not teach just subject matter, but values, which would provide children with the moral and ethical grounding needed for education, jobs, and a stable family life—the real avenues for escape from poverty—instead of drugs, alcoholism, and prostitution. With the cooperation and reinforcement of local religious and social organizations and committed parents, children would learn from school that money is not the answer to all their problems. The new school's staff would provide culturally appropriate and consistent spiritual, moral, and ethical guidance, including mentorship and regular visits by people who have escaped poverty and have thus become living examples of hope and to demonstrate that it can be accomplished. The new schools would develop relevant home and school services, such as parent and student training in conflict resolution, decision making, and communication skills. To the fullest extent possible for each child, the new schools would ensure a stable, safe, and caring home and school environment. They would develop as a result of the community's critical decision to give pride and history to its poorest children.

John Dewey (1938), encouraged by Jane Addams (1916), recognized the complex issues and problems of an experience-based educational program, especially for tenement-house children. Dewey's progressive organization of subject matter would serve well today as an intellectual guide for the education of ghetto children. Coupled with his conception of play and work in the curriculum (Dewey, 1916, pp. 228–242), this notion would provide a reasonable starting point for a new school, classroom, and curriculum.

For the ghetto children, educational opportunity—a real escape from poverty—requires the consolidated efforts of business, industry, government, community organizations, and educators to rebuild not just public housing or schools, but vital, symbiotic communities (see Comer, 1980; Levin, 1987; Oliver & Gershman, 1989; Wehlage, 1989). This idea of building the city for intimate, mutual cooperation is not a new idea (see Mumford, 1961, p. 89), but now our American cities need it more urgently than ever.

## References

Addams, J. (1916). *Democracy and social ethics*. New York: Macmillan.

Bruckerhoff, C. (1990a). Conversation as the centerpiece for reflective practice. *Curriculum and Teaching*, 5(1–2),14–24.

Bruckerhoff, C. (1990b, April). *Routines and mathematics curriculum reform*. Paper presented at the annual meeting of the American Educational Research Association, Boston.

Bruckerhoff, C. (1991). *Between classes: Faculty life at Truman High*. New York: Teacher, College Press.

Bruckerhoff, C., & Popkewitz, T. (1991). An urban collaborative in critical perspective. *Education and Urban Policy*, 23, 313–325.

Comer, J. (1980). *School power: Implications of an intervention project*. New York: Free Press.

Dewey, J. (1916). *Democracy and education*. New York: Macmillan.

Dewey, J. (1938). *Experience and education*. New York: Macmillan.

Levin, H. (1987). Accelerated schools for disadvantaged students. *Educational Leadership*, 44,19–21.

McDermott R. (1974). Achieving school failure: An anthropological approach to illiteracy and social stratification. In G. Spindler (Ed.), *Education and cultural process* (pp. 82–118). New York: Holt, Rinehart & Winston.

Mumford, L. (1961). *The city in history*. New York: Harcourt, Brace & World.

Ogbu, J. (1988). Diversity and equity in public education. In R. Haskins & D. MacRae (Eds.), *Policies for America's public schools* (pp. 27–39). Norwood, NJ: Ablex.

Oliver, D., & Gershman, K. (1989). *Education, modernity, and fractured meaning*. Albany: State University of New York Press.

Popkewitz, T. (1988). Culture, pedagogy, and power: Issues in the production of values and colonization. *Journal of Education*, 170(2), 77–90.

Romberg, T., & Carpenter, T. (1986). Research on teaching and learning mathematics: Two disciplines of scientific inquiry. In M. C. Wittrock (Ed.), *Handbook of research on teaching* (pp. 850–873). New York: Macmillan.

Rosenshine, B. (1983). Teaching functions in instructional programs. *Elementary School Journal*, 83(4), 335–351.

Rutti, R. (1991, January 31). Proficiency tests raise deficiency fears. *The Plain Dealer*, pp. 1, 10.

Smith, L. (1986). *Educational innovators: Then and now*. New York: Falmer.

Wehlage, G. (1989). *Reducing the risk: Schools as communities of support, and public policy*. New York: Falmer.

West P. (1991, March 13). Math groups urge changes in teacher preparation. *Education Week*, p.5.

Wilson, W.(1987). *The truly disadvantaged. The inner city, the under class, and public policy*. Chicago: University of Chicago Press.

# Child Rearing and Education in Urban Environments

## Black Fathers' Perspectives

JOSEPHINE A. BRIGHT
*Wheelock College*

CHRISTOPHER WILLIAMS
*California State University, Northridge*

In this article, the authors use in-depth interviews of African American fathers in an urban setting to explore their fathering attitudes and behaviors. From the qualitative data, four general themes emerge: paternal beliefs and modes of functioning, strength and resiliency, racial socialization of children, and fathers as teachers in the home. The findings lend support to the burgeoning reconceptualization of the role of the Black father in urban America. The self-descriptions of African American fathers provide evidence that urban African American fathers are important, vital, and active forces in their children's achievement development.

For many years, the role of the African American father has been overlooked in discussions of Black child development. Most traditional accounts of the African American family have overwhelmingly focused on the mother-only, female-centered, single-parent Black family (see, for example, Williams, 1992).

By contrast, little attention has been given to the many Black fathers who are an integral part of their families' life and functioning. The erroneous conclusion to be drawn is that the African American father has abandoned his position in the family (see, e.g., "A World," 1993).

More recently, researchers have begun to examine the roles of African American men in families. The stereotype of the Black father as an "invisible man," who is usually absent from the family, has little power in the family, and is not involved or interested in the socialization of his children is being challenged. Some studies have explored the meaning of the provider role

---

Josephine A. Bright and Christopher Williams, "Child Rearing and Education in Urban Environments: Black Fathers' Perspectives," *Urban Education,* September 1996, Vol. 31, No. 3, pp. 245–260.
AUTHORS' NOTE: This work was supported under the Educational Research and Development Center Program (R117QO0031) as administered by the Office of Educational Research and Improvement (OERI), US. Department of Education. The findings and opinions expressed in this report are the authors' and do not reflect the position or policies of OERI.

to African American fathers (see, e.g., Cazenave, 1979; Taylor, Leashore, & Toliver, 1988). Casenave's findings, for example, indicate that African American fathers see work as "a means to an end, i.e., providing, and providing itself is seen as a means to effectively carry out the husband and father roles" (p. 588). Whereas the role of economic provider was most salient to the self-perceptions of middle- and working-class African American men, there was also a trend toward more egalitarian, expressive, and involved relationships. Fathers were more involved with their families than their own fathers had been.

Other research confirms that a great deal of interaction occurs in the Black father-child relationship (see McAdoo, 1979, 1986, 1993; Mirande, 1991). McAdoo (1979, 1986) has argued convincingly that African American fathers are not absent or marginalized but actively participate in the socialization of their children. Fathers have warm, nurturing relationships with their children and share equally with their wives in child-rearing decisions. Mirande found that the overwhelming majority (99%) of African American fathers see child care as a responsibility of both parents, and 87% believe that their children have needs "only they, as fathers could meet" (p. 58).

A movement beyond the almost exclusive emphasis on father absence is both timely and encouraging. The present research effort seeks to join this promising line of inquiry in broadening our perspectives on fathering in urban African American families.

## The Study

This article focus on eight urban African American fathers of children who were identified by their kindergarten teachers as successful or having potential for success in school. The fathers, their spouses, and their children are taking part in a Center on Families, Communities, Schools, and Children's Learning (Center on Families) ethnographic research project that is exploring how 41 families of different racial, ethnic, and cultural backgrounds—11 African American and 10 each Chinese American, Irish American, and Puerto Rican—and a variety of family structures and income levels, prepare children for success in the early primary grades.

Of the 11 an American families in our Center on Families study, three are single-parent, mother-headed families, one is a single-parent, father-headed family, and seven are two-parent families. For this article, only the data obtained for one single-parent father and seven fathers in two-parent households are considered.

## Methods and Procedures

Two kindergarten teachers at a K–5 urban elementary school serving a predominantly African American student population (about 75% of the school's approximately 600 students) were interviewed on their definitions of success or having potential for success; the qualities that describe a successful student in their classroom; their expectations of children when they enter and leave kindergarten; and their thoughts concerning home/school partnership, among other topics. The teachers were asked to identify particular African American children in their classrooms who they thought to be poised for academic success. Home visits were arranged with the children's parents to explain the goals of the study and to invite their participation.

The interviews for the first- and second-grade teachers closely approximated the original interview schedule but were modified to include questions about the progress of the focus children in school.

Home interviews lasting from 1 to 2 hours were completed with parents and their children two or three times during each year of the 3-year data collection period. During the in-depth interviews, parents were asked to speak about their perceptions of support for their child-rearing roles; their child-rearing beliefs; socialization tasks and goals for their children; the things they did in the home to enhance their children's learning; and their attitudes about race/ethnicity, among other topics. All of the interviews were tape recorded and later transcribed.

The process of creating an organizational framework for the family data included reviewing the interview transcripts and supplementary field notes and highlighting patterns that emerged in the data, developing preliminary codes that seemed to suit the data, refining the codes and searching for new themes, and developing broader categories of interrelated pattern sets. Finally, the categories were compressed into 10 final categories that were judged sufficient to encompass the material:

1. The sociodemographic characteristics of the family (e.g., family size, structure, and composition; parent level of education; parental employment status);
2. The family's neighborhood/home environment (e.g., the physical condition of the neighborhood and the home; the resources available in the community; safety issues in the neighborhood);
3. Parental beliefs and child-rearing practices (e.g., family values for children; family beliefs about education; intrafamily functioning styles);
4. Social supports to families (e.g., extended family, friendship, and church support networks; parental coping strategies);
5. Racial/cultural identity (e.g., parental support for the child's racial/cultural identity development);
6. Parents as teachers in the home (e.g., parent(s) engage in teaching activities in the home; parent(s) provide verbal encouragement, motivation, and support for the child's academic learning);
7. Home/school relations (e.g., parent perceptions of and relationships with the teachers and the school);
8. Parental involvement in the school (e.g., the levels and types of parental involvement with school);
9. Information on the study child (e.g., the child's relationships with parents, siblings, teachers, and classmates; the child's feelings about school);
10. Other (e.g., information on the child's siblings and extended family members).

The qualitative analysis of the family interviews has yielded rich data on how the fathers of our sample view their family roles; their hopes, dreams, and aspirations for their children; how they promote and support their children's achievement; and the significance of racial/cultural identity in the fathers' own lives and in how they rear their children. Information on the sociodemographic characteristics of the sample is presented next. This is followed by discussions on four general themes—paternal beliefs and modes of functioning, strength and resiliency, racial socialization of children, and fathers as teachers in the home—as they emerged from analysis of the father interviews.

## Description of the Sample

The fathers whose views are reported here range in age from 34 to 52, with a mean age of 41 years. With the exception of one single-parent custodial father, all the fathers are currently married and living with their wives. The fathers' level of education ranges from high school graduate to 4 years or more of college. Two of the fathers hold high school diplomas, five have completed some college, and one holds a B.A. degree and is now completing a master's degree in psychology. All but one of the fathers reported current employment. (Contemplating a career change, this father is currently taking college-level courses in business administration.) The fathers' occupations are counselor, truck driver, detective police sergeant, printing press operator, school bus monitor, food services cash manager, and small business owner/operator. Of the eight families, two are middle class, four are working class, and two are low income.[1] All the families live in a lower income to middle-class Black community in a northeastern city.

## Paternal Beliefs and Modes of Functioning

Contrary to the prevailing stereotype of the African American father as distant, remote, and authoritarian in his interactions with his children, fathers in this study maintain relationships characterized by open communication, warmth, and nurturance toward their offspring. They do not see themselves as kings who rest on a throne in a distant corner of the house but rather believe that a good father spends quality time with his children. From interviews and (informal) observations conducted in their homes, one can see that there is much positive interaction between these men and their children.

In some of the households, fathers indicated that they enjoy closer relationships with their children than they remember with their own fathers. They want less formal relationships with their children. Recounting his childhood experiences, one father said

> My father would come home, he'd come home from work, he'd have his newspaper, he'd sit in his favorite chair in the living room, and that was it. There was no communication, you was a child and that was it. Not at all. And I want my son to be able to come to me. And if you don't break down that barrier a little bit, your kid will never come to you.

Other fathers appeared to be continuing traditions of close father-child relationships learned within their families of origin. In the following excerpt, a father shares with us his feelings of admiration for his own father and the hope that his children will admire him in the same way.

> My parents, they raised six of us. And my father, I got a lot of respect for him, the same way I feel like I want my children to respect me. He went to [law] school and worked and raised six kids all at the same time. And I got a lot of respect for him. All the time he was in school during the day . . . he worked at night. He had a heck of a schedule.

In discussing the open lines of communication he maintains with his (then) 6-year-old son, the same father said

> He always comes up with the right question. And if he does have a problem, we have a relationship like, he's not afraid to ask me anything.

In speaking about the many things that he and his son do together, the father quoted above commented, "I can't go nowhere without . . . my shadow. He loves going anywhere. If I'm just going to the store, he's got to go."

Another recurring theme in the family interviews concerns the father's role as protector and guide of his children. Acknowledging that things can happen anywhere, these fathers attempt to shelter their children, as much as it is possible, from the increased risk and vulnerability associated with living in an urban environment. When asked about his concerns with safety issues related to living in the inner city, and ensuring that his (then) kindergarten-age son grows up to manhood, one father responded as follows:

> He [study child] was going to a school for Kung Fu so he could take care of himself and mainly have more confidence, you know, in how to deal with people. . . . I teach him every chance I get. . . . I explain to him about the world and how people . . . you can't always go by smiling faces and stuff.

Suggesting that children need to develop self-discipline and learn how to handle themselves in all types of situations, another father said that it is important for his sons to play outdoors with other children on their street even though they might sometimes hear undesirable things. This father allows his children to interact with the neighborhood "street kids" occasionally because he believes that this provides a valuable learning experience. He tells his two young sons:

> You don't have to feed into the things you're hearing. You're gonna hear some things you might not like. . . . You're gonna have to understand what's OK for you, what's not. In other words, to be around some negative situations and to be able to remain positive. And if you can't, you need to remove yourself from that. Come on in. And if I notice you can't do it, then I'll help you do it.

Like many formerly thriving urban Black communities, the area in which these families live has been ravaged by high levels of unemployment, an influx of drugs, and an increase in crime in recent years. A father who has lived in urban Black communities all his life provides an analysis of the origins and historical background of the current drug crisis. He offers this assessment of the situation:

> This is my theory, right. I go back to 1967 . . . when we were first starting to get together, you know. Being proud of being Black and all that. And our young people, we were moving, you know. . . . And then I think, I think that the powers that be decided that they would, they would quiet down these young Black people . . . then they started making drugs available.

In discussing his concerns about his wife and children's comings and goings in the neighborhood, another father shared with us how he supports his wife in accomplishing her civic and social responsibilities.

> I don't let her go and send her out at night and knowing that she's going to get home late at night with the kids out there in the dark, not the way this world's going. . . . I'll go with them, you know. I go with them because I wouldn't feel comfortable laying here watching the basketball game, when I could have went to their cousin's party, or to the drum and bugle thing and participated with them, than laying here hoping that everything is going all right while they're traveling up and down. . . . I wouldn't enjoy the game, I'll tell you that. . . . I don't feel comfortable, so I'll just get up and give up the game for that night, and go with them. And, it's not just for their safety, but to make me feel comfortable, too.

Does this father regard his behavior as extraordinary? The answer is clearly no. He says that his own father set a similar example, adding, "Hey, I mean, my father did it for me."

The fathers we interviewed are particularly sensitive to the elements in society (e.g., television, movies, the mass media) to which their children are being exposed. They make earnest attempts to monitor the popular films and music videos their youngsters find entertaining and to protect them from the unwholesome images resulting from the ever-declining standards of family television programming. One father commented that

> When we're watching TV and he [study child] comes into the room, and there's a particularly gruesome story on there, I want to tell him to get out of the room. And he always comes in when there's something on TV I don't want him to see, murders or something.

In discussing his family's television policy, another father said that he does not permit his children to watch television by themselves.

> No, because it's too suggestive and if I don't go in there and watch what they watch, I don't know what kinds of suggestions they're getting. There's always something in between [e.g., commercials, previews] that you didn't know was going to come up at that time. It's easier on the weekend to monitor, because we spend most of the weekend together.

## Strength and Resiliency

Many fathers from diverse socioeconomic backgrounds are feeling the pangs of the economic recession in the United States. African American fathers, however, are experiencing greater strife from the problems of the economic situation as they have been historically on less firm economic and occupational standing than their Anglo-American cohorts (see, e.g., Staples, 1991).

In spite of their often difficult social and economic conditions, the fathers in our Center on Families study have not lost hope. As reflected in their behavior and by their statements, they are exhibiting strength, resiliency, and perseverance and have become more skilled at wielding limited resources. What gives these fathers such strength? One father, who shared that he is no longer an active church member, stated simply, "I have a daily connection with God. It's not a Sunday thing, it's an everyday thing." Emphasizing the importance of family church attendance, another father said

> We're church-going people, we go to church. . . . You got to believe in something. That's what kept me on the right track. Believing in the higher, you know, the other, having faith in something. And I have faith. . . . and I think that had a lot to do with the way I grew up. And the same with them [his children]. I say, well they got to have some kind of faith, they got to believe in something . . . whatever, as long as they believe in something, have faith in something.

When asked how his life was going, the same father, commenting that he works 9 months a year as a school bus monitor and is trying to "land a better job. Or to land a job period, cause I don't think I want to go back to the school bus this year. It's just not hitting it financially," responded thus:

> "I'm alive, healthy and happy. Things could be better, but I'm not complaining. . . . I get up every morning, brush my teeth, wash my face, and out the door. And every day's a new day. Like I tell them, as long as I can get up and go out the door, there's always a chance, opportunities, you know, to accomplish whatever I'm trying to. So far, I've [got a] beautiful little family, and they're happy, and I'm happy. All we're trying to do now is improve our financial situation.

Although the father quoted above did not explicitly state that he relies on his extended family as a source of strength, it is evident that it serves as such. When asked how he defined family, he included his parents, his wife's parents, aunts, uncles, and cousins. Other fathers included their extended family as a support system and to some extent various members within the community.

Another father, in speculating on the source of his drive and motivation, identified his feelings of never feeling satisfied or content. In fact, he believes that contentment provides a false sense of security.

> Well, one thing you can't do is be content because contentment breeds complacency. And it's a vicious cycle. You never get any place. You know, my philosophy is . . . to never be content. Always be trying to make it better. It can always be better.

## Racial Socialization of Children

We have learned that racial identity is an important factor in the fathers' own lives and in how they socialize their children. Despite differing degrees of explicit racial socialization in these families, all the fathers want their children to have positive self- and ethnic identity. One father said about being a Black parent that

> It's a little tougher because you've got more things to worry about for your kids. . . . It would be nice to be in the suburbs, you know. . . . You could raise your kids nice out there and not worry about nothing. . . . I think that for Black kids we're going to have to instill in them that . . . the same sun that woke you up this morning, will wake all the rich people up.

This same father tries to teach his son that, all too often, Blacks are consumers rather than producers of industry. He tells his son to start his own business one day. He believes that White parents do not need to teach their children this lesson as explicitly because business ownership is already evident in White children's communities and society at large. As far as racism is concerned, his philosophy is to allow his (then) kindergarten-age son to remain innocent for as long as possible.

> His world is pretty sheltered right now. It's good for it to be that way. We have to prepare him verbally for it [racism]. . . . Over the years, you just tell them little things. You always measure it out.

Other fathers stressed the importance of preparing their sons and daughters early on to negotiate potential barriers to Black achievement. (The findings on racial socialization practices within the study families are further explored in Bright, 1994.)

## Fathers as Teachers in the Home

The fathers of the African American cohort have many valuable lessons about life to impart to their children. One recurring theme in the fathers' statements is their desire to teach their offspring a strong sense of responsibility. These men want their children to learn very early that meeting one's responsibilities is inextricably linked to success in many domains of life. In the following excerpt, a father of three children (then) 2, 4, and 6 years of age articulated his belief that children are never too young to begin taking on a measure of responsibility.

> I believe that children at the earliest age, when they begin to talk, no matter how old you are and all that stuff, they should be given some responsibilities. If it's every night or every day that they have to take their pillow and put it in their bed, or any little thing.

Another father, in discussing how he prepares his sons to succeed as Black men in America, said that he tries to teach his sons that it is important to stay focused on their own goals and to not allow other people to stand in the way of their success.

> I would tell them to be successful, take responsibility for your actions. That you can do whatever you put your mind to. However, I know there will be times when people will try to stop you from doing the things that you want to in life. You just have to find a way around that.

Teaching their sons and daughters that they must anticipate the consequences of their actions also ranks high on the fathers' agenda of lessons to be imparted. Modeling appropriate behavior is one technique the fathers used to reinforce their teachings.

As the father of three youngsters commented:

> I tell them all the time, "You don't see me doing this, so don't you do it." And if they catch me in the wrong, don't think they won't speak up. Oh, all three of them will get me!

Another prominent theme concerns the importance of teaching children, by word and deed, the value of hard work, perseverance, and effort. Emphasizing that these qualities are essential to achieving success in the educational and occupational arenas, one father said

> I'm trying to show [my sons] that you have to work. . . . You have to work at things to win. . . . You have to do something again and again to get it like you want it. . . . I bring them in [my office] and show them that I am trying to plan something. It's not going to be an accident. What did you want? Did you plan it? [I tell them] "I'm not sure I'm doing it very well, but that is my intention."

Fathers continue a long tradition of parents as educators in the African American home. Given the amount of preparation African American children must have to negotiate the many elements they will encounter as adults, the fathers realize that their teaching time is crucial. As one father put it:

> I'm trying to get him to think of education, and my philosophy is you start them young with education and they get hooked on it.

These fathers see education as playing a crucial role in their children's future life chances, as evident in the following father's statement.

> I try to motivate him about school, to make it important so he knows that it will be an important part of his survival. . . . Every chance I get, I try to motivate him to get that education.

Speaking in earnest and with great intensity, another father says that the high premium placed on education is an ongoing tradition in his family. Reflecting the values and attitudes about education that his own father transmitted to him, this father conveys to his children the message that for Black Americans education has value in its own right and is also a means for liberation.

> School was automatic. You were supposed to go to school. What else are you going to do? How else will you get out of this situation, and how else will you be able to pull someone else out of this?

All the fathers were active in their efforts to promote and support their children's academic success. Paternal support for education takes a variety of forms and includes helping the child with schoolwork at home, showing an interest in the child is doing at school, participating in parent-teacher conferences, PTA, and Parent Council meetings, attending school performances and events, and contributing time and labor to improve the Physical plant of the school.

Finally, the data illuminate that fathers in these households work collaboratively with their wives to foster the academic achievement of their children. In discussing how he and his wife reinforce each other's efforts to support their child's school success, one father identified himself as the math teacher in the family.

> I teach him math and [my wife] teaches him English [and] reading. . . . When he finishes reading, then he comes to me and I give him some problems. So we work together on that.

Another father spoke at length about how the school enriches his family's life and, in turn, how he and his wife support the school.

> Oh, yes, and we're really a part of the school. . . . It's more than just a school. All the after-school activities and all that. . . . And we never knew until we started going to Parent Council meetings of all the activities they had, and they fit right in our schedule perfectly. . . . And my wife is there at least two or three times a week. . . . Then, when the school let out over the summer, we were there maybe on three different occasions to clean up the school and paint the walls. . . . it's a daily part of our lives.

## Summary and Discussion

A review of the literature on African American families reveals that research focusing on the family roles of Black fathers is exceedingly rare (see Taylor, Chatters, Tucker, & Lewis, 1990). Further, when the Black father has been studied, "there has been more concern with his absence than with his presence" (Cazenave, 1979, p. 583). This overwhelming emphasis on the absent Black father has deflected attention away from the many Black fathers who are present in the home and actively involved on all fronts of their families' lives.

The study reported here represents an attempt to explore urban African American fathers' perspectives on their family and children, issues of ethnicity and child rearing, and what fathers do in the home to support their children's healthy growth and learning during the early elementary years. The ethnographic interviews conducted with eight fathers over a 3-year period only begin to fill the knowledge gap in this area. Much more work needs to be done in the documentation of the roles urban African American fathers play in their children's development.

The findings from our study lend support to the view that the African American father is not marginal or inconsequential in the family but is an important, active, and vital force in his children's achievement development (McAdoo, 1979). Open and candid in expressing their thoughts about education and what African American children need to set them on a course toward achievement and success, the fathers of our sample believe that a good father is more than just a breadwinner for his child but is there to guide, help, love, protect, teach, and be a positive role model for his children.

As evidence of the importance these fathers place on spending quality time with their youngsters, the fathers described a list of activities they do with their children from reading together and helping them with schoolwork at home to having a good time in the park and going to Drum and Bugle Corps and Boy Scout meetings. The fathers are aware that they are doing a good job in raising their children, yet they do not regard themselves as out of the ordinary. Providing for and caring for their children, and nurturing their youngsters' unique talents and gifts, is seen as a paternal duty and as giving meaning and joy to life.

There are at least two ways in which these warm, nurturing, father-child relationships may influence our study children's achievement development. First, it is possible that positive parental affect enables the parent to present tasks to the child in a positive manner and to be more available as a resource to the child (Arbuckle & MacKinnon, 1988). Second, satisfactory parent-child relations could have a positive effect on the child's social competence, which has been linked to performance on cognitive tasks (see McAdoo, 1979). In either case, the positive father-child interactions that characterize our study families may be optimally important in producing instrumentally competent and motivated children.

The fathers we interviewed want their children to have high levels of self-confidence and self-esteem and positive attitudes about being Black. These qualities, fathers believe, are essential for laying the foundation for children to do well in school and in other aspects of life. Previous research highlights the need for Black children to develop a positive attitude toward their ethnicity (Hale, 1991). After all, a socially competent child would be expected to have a positive self- and ethnic identity to do well in school and to lay the groundwork to be successful in future occupational endeavors (McAdoo, 1979).

The data reported, taken from a larger Center on Families, Communities, Schools, and Children's Learning research study of how families of different cultural backgrounds prepare children for success in the early primary grades, were based solely on the perceptions of eight low- to middle-income African American fathers residing in a northeastern city. Although our sample is admittedly small, the ethnographic interviews giving voice to African American fathers can broaden our perceptions of fathering in urban Black communities and provide valuable insights into the factors that enable urban African American fathers to successfully fulfill their parenting responsibilities, areas that have been neglected in previous family research.

The positive findings for the fathers of the African American cohort of our Center on Families study not only challenge prevailing myths and stereotypes of all urban African American fathers but highlight the need for educational researchers to examine the factors that might enable urban African American fathers to promote and support their children's successful achievement. This information, in turn, could be used for the development of programs and policies that acknowledge the strengths and better support the needs of urban African American fathers and their families.

## Note

1. In this study, various subjective indexes of financial status were relied upon to assess the socioeconomic status (SES) of the families. First was the number of parents fully employed. Two parents working compared with one parent or no parents working would be assigned to a higher socioeconomic status. The second factor considered was home ownership. If the parents owned their place of residence, the family received a higher SES rating. A third factor was whether the parents mentioned experiencing financial hardship. In situations in which financial stress was mentioned, a lower SES rating was assigned. Finally, and most subjective, SES rating was determined by taking note of the conditions of the family dwelling (e.g., immediate neighborhood, condition of exterior and interior of home, and such durable items as stove, refrigerator, etc.).

# References

A world without fathers. The struggle to save the Black family. (1993, August 23). *Newsweek*, 16–29.

Arbuckle, B. S., & MacKinnon, C. E. (1988). A conceptual model of the determinants of children's academic achievement. *Child Study Journal*, 18, 121–147.

Bright, J. A. (1994, Winter). Beliefs in action: Family contributions to African-American student success. *Equity and Choice*, 10(2), 5–13.

Cazenave, N. A. (1979, October). Middle-income Black fathers: An analysis of the provider role. *Family Coordinator*, 583–592.

Hale, J. E. (1991, September). The transmission of cultural values to young African American children. *Young Children*, 7–15.

McAdoo, J. L. (1979, January). Father-child interaction patterns and self-esteem in Black preschool children. *Young Children*, 46–53.

McAdoo, J. L. (1986). A Black perspective on the father's role in child development. In R. A. Lewis & M. B. Sussman (Eds.), *Men's changing roles in the family* (pp. 117–133). New York: Haworth.

McAdoo, J. L. (1993). *Understanding fathers: Human services perspectives in theory and practice* [Brochure] (Family Resource Coalition Report, No. 1, pp. 18–20). Chicago: Family Resource Coalition.

Mirande, A. (1991). Ethnicity and fatherhood. In E. W. Bozett & S.M.H. Hanson (Eds.), *Fatherhood and families in cultural context* (pp. 53–82). New York: Springer.

Staples, R. (1991). Changes in Black family structure: The conflict between family ideology and structural conditions. In R. Staples (Ed.), *The Black family: Essays and studies* (pp. 28–36). Belmont, CA: Wadsworth.

Taylor, R. J., Chatters, L. M., Tucker, M. B., & Lewis, L. (1990, November). Developments in research on Black families: A decade review. *Journal of Marriage and the Family*, 52, 993–1014.

Taylor, R. J., Leashore, B. R., & Toliver, S. (1988). An assessment of the provider role as perceived by Black males. *Family Relations*, 37, 426–431.

Williams, B. (1992). Poverty among African Americans in the urban United States. *Human Organization*, 51, 164–174.

# Safe Schools: Policies and Practices That Impact Teachers and Teacher Training Programs

DAWN KUM-WALKS, PH.D.

*Assistant Professor, Old Dominion University*

> Violence is a safety condition that results from individual, social economic, political, and institutional disregard for basic human needs. Violence includes physical and non-physical harm that causes damage, pain, injury, or fear. School violence disrupts the school environment and results in the debilitation of personal development, which may lead to hopelessness and helplessness.
>
> (California Commission on Teacher Credentialing, 1994, p. 7)

## Introduction

In a 1993 national survey conducted by the Metropolitan Life Insurance Company (i.e., "The American Teacher: Violence in America's Public Schools"), it was found that 11% of the 1,000 teachers surveyed have been victims of violence that occurred in or around school. Ninety-five percent of those incidents involved students (p.7). One-third of the teachers surveyed reported feeling their schools provided only fair to poor education, and 16% of the teachers in schools with all or a majority of students of color reported being victims of violence in or around their schools.

During the Reagan/Bush administration, a bill was introduced in response to a document that highlighted the State of America's Schools. The bill was "American 2000, National Educational Goals." Since its introduction, American 2000 has been revised, renamed, and signed by the Clinton Administration in October 1994. It is now referred to as Improving America's School Act of 1994. This legislation reauthorizes the existing Elementary and Secondary Education Act of 1965 to provide schools greater flexibility to use federal dollars to teach students with special needs. It also promotes safe and drug free schools. Additionally, the bill allows resources for improved use of technology and federal seed money for local charter school initiatives. A total of $11 billion is allocated for school districts and individual schools to improve teaching and learning for economically disadvantaged children. But most importantly, this bill proposes a broad approach to the problem of violence, not just another layering of punitive interventions. For instance, goal six of the Act states, "By the year 2000 every school in America will be free

---

Dawn Kum-Walks, Ph.D., Assistant Professor, Old Dominion University.

of drugs and violence and will offer a disciplined environment conducive to learning." This goal has three areas of focus: (1) disciplined school environments; (2) drugs and schools; and (3) violence in schools. Safe schools are important for the achievement of all other national and local educational objectives; therefore, this is a very crucial goal (National Education Goals Panel, 1994).

## Factors Contributing to Violence in Schools

The occurrence of school violence is a multifaceted phenomenon. According to both teachers and students, the components of school violence include the presence and use of weapons (the most serious aspect); verbal and physical assaults; fights; and robberies. Intruders and self-identified gang members who are students also pose the risk of bodily harm or threats on school grounds.

### Weapons in Schools

Students may carry weapons to school for various reasons. The 1993 Met Life survey finds that students carry weapons for four main reasons: protection to and from school; to impress their friends; for self-esteem; and for protection in school. Although knives are found most often on campuses, the presence of powerful firearms are increasingly evident. Weapons in schools are a reflection of their easy access in the community, presence in homes, and the apparent widespread attitude of American society that violence is an effective way to solve problems (Butterfield & Turner, 1989).

### Intruders

Prohibiting unwanted and potentially threatening individuals from entering school campuses continues to concern stakeholders. Metal detectors, locking outside doors, and security guards have been used to deter intruders. Historically, the use of metal detectors and security guards has been controversial and their effectiveness unproven (Gottfredson, 1985). Exit doors can, however, be fitted with electromagnetic locks set to open when a fire alarm is set off. It is recommended that schools modify their structures to improve security for students and staff (Rich, 1981).

### Fights and Assaults

Fights and assaults involve bodily harm and threats of harm. Robberies, as a form of assault, also involve threats or harm to individuals in the process of taking things. No programs or studies presently exist to solely monitor or address robberies in schools. There is research, however, on curricula to prevent violence and implement conflict management training for both students and teachers to reduce the occurrence of fights and assaults (Cohen & Wilson-Brewer, 1991; Dryfoos, 1990; Mendel, 1995).

From a survey of 51 school- and community-based programs, it was found that none of the programs being studied were rigorously evaluated (Cohen & Wilson-Brewer, 1991). It was also found that there is no comprehensive tool to evaluate the effectiveness of programs for preventing violent behaviors among adolescents. Even though 20% of the programs had an outcome evaluation, they are mostly pre- and post tests of knowledge and attitudes, rather than behavioral changes.

## Gangs

Youths may become involved with gangs for various reasons, (e.g., to gain power and prestige, peer pressure, self-preservation, adventure, money and limited opportunities) (Prothrow-Stith, 1991). When asked whether they thought gang or group membership, or peer pressure was a major, minor, or non-contributor to violence in their schools, teachers and students responded:

- Teachers:
    4% = Major factor
    43% = Minor Factor
    22% = Not a Factor

- Minority Students:
    54% = Major Factor
    37% = Minor Factor
    9% = Not a Factor (Met Life, 1993)

Teachers, students, and law enforcement officials more frequently name incidents of lesser violence (such as pushing, shoving, grabbing, slapping, verbal insults, and stealing) as major problems in schools than the threat or use of weapons and other more severe violence. Additionally, in comparison to their counterparts in suburban and rural schools, teachers, students, and police department officials in urban areas are more apt to believe that any incidence of violence is a major problem.

Based on focus groups, students feel devalued and ignored by the school community, and therefore care very little about images and outcomes (CTC, 1995; Kum-Walks, 1994). It is difficult for someone to care about something that he or she does not value. Many students wish to change adult perceptions that they do not care about their personal well being. They want to be heard, in classrooms and in the decision making process. They desire and deserve a positive alternative to invest their energies.

## Developing Safer Schools

In their 1995 publication, *The Prevention of Youth Violence: A Framework for Community Action*, the Center for Disease Control (CDC) offered global strategies for educational approaches, legislative or regulatory change, and environmental modification to prevent youth violence.

### Common Educational Approaches

- Mentoring: the one-on-one pairing of adults with youths in order to give young people wholesome role models, supportive attention, and guidance.
- Conflict Resolution Training: programs for teaching young people to manage potentially violent situations through peaceful, nonviolent means.
- Training in Social Skills: programs for helping young people relate to others in positive and friendly ways. These skills include maintaining self-control, communication, forming friendships, resisting peer pressure, and establishing good relationships with adults.
- Firearms Education: training to prevent injuries from firearms that provides youths with the information necessary to avoid injuries if firearms are present.
- Parenting Education: training to improve parenting skills, which in turn improves parent-child interaction and supports family life and development.

- Peer Education: strategies focusing on the powerful influence of other youths to persuade and encourage adolescents to resolve conflicts peacefully.
- Public Information and Education: strategies and programs for raising the awareness of youth violence in the community; providing basic information on prevention; and modifying the social and behavioral norms of the community.

The process for developing safe schools and classrooms requires a collaborative effort. Stakeholders including policymakers and community members must become involved and be represented at all stages of developing, implementing, and evaluating safe schools. Suggestive approaches at the policy and community levels follow.

## Policy and Legislative Approaches for Change:

These approaches are for communities choosing to enact new laws or regulations regarding access to guns or alcohol or to strictly enforcing existing regulations.

- Regulation of the use of and access to weapons prohibits weapons in schools or set requirements concerning gun ownership or storage.
- Regulation of the use of and access to alcohol prevents the sale of alcohol to minors, may prohibit or control the sale of alcohol at special events at which youths are likely to be present, and may require that beverage servers and sellers be trained to identify minors and those who have had too much to drink.

## Community and Environmental Contributions:

- Changes in the social environment in which children or adolescents live can relieve or eliminate problems that foster violent behavior now or later in life. Some avenues for changing the social environment are providing lay home visitors who teach parenting skills and encourage a nurturing environment for children, preschool programs (like Head Start), therapeutic and group counseling activities for children and adolescents who may be at risk for behaving violently, recreational activities, and work or academic experiences.
- Changes in the physical environment can help make the occurrence of violence less likely. These changes include improving visibility in public areas, increasing the use of an area by community people, controlling access to certain areas, and creating a sense of ownership of an area by community members.

Comprehensive approaches by teachers, schools, teacher training institutions, and the community that incorporate multiple strategies and intervention are most promising. Such approaches should address multiple risk factors and provide protective conditions against these risks. Strategies will have to be adapted to the needs of the community. It will be necessary to commit to longitudinal studies, for implementation, assessment and modification of safety programs.

## Creating a Safe Classroom Environment

As indicated by the CTC (1995) research, teachers need and are requesting more training in effective communication skills. These skills are not only for the practice of teaching, but for enabling cooperative student behaviors so that they have the opportunities to teach. Unfortunately, many teacher-training programs are inadequately preparing teachers to diffuse violent incidents occurring with greater frequency in schools. It is extremely difficult to predict such acts since they are not restricted to students from a specific racial background, a particular socioeconomic status, or academic achievement level. These acts are neither gender nor age-based. Many classroom incidents result from a complicated set of events and relationships with multiple causes, with few originating in school (Short & Short, 1987).

Teachers have to negotiate and manage complex classrooms with the skill and charisma of an orchestra conductor. This role places teachers in the center of the classroom, causing them to occupy a significant place in the school as authority figures. This authoritative role is derived from five areas of teacher power (Roetter, 1987): (1) power of position; (2) reward power; (3) coercive power; (4) power of personality; and (5) power of expertise. When these powers are exercised daily, any actual or perceived abuse of them may result in retaliatory behavior by students. Frequently, it is this authoritative role that makes teachers convenient targets of two types of violent behaviors, direct aggression and displaced aggression. Direct aggression is demonstrated when the behavior is aimed directly at the teacher as a reaction to some exercise of his or her legitimate authoritative powers. Displaced aggression happens when a teacher is the recipient of violence due to the behavior of a third party (Houston & Grubaugh, 1989). In the classroom, teachers are exposed to a variety of violent outbursts. If their actions are perceived incorrectly, this increases the potential for such violent behaviors.

Howard Muscott (1987) created a conceptual outline for a school-wide proactive approach for teachers and administrators. They should:

1. examine their personal value systems—refine and articulate their philosophies on acceptable and unacceptable behaviors in school;
2. analyze the school's environment based on the historical occurrence of violent behaviors, whether problems currently exist, and the potential for future problems;
3. examine the common values of the school board and central office; and
4. develop policies, rules, and procedures based on the three previous steps.

These steps require a school-wide commitment, a concerted investment of time with supportive resources that ensure appropriate training, and allotted time for reflection and discussion.

## Teacher Management Techniques

Before the successful implementation of any technique, teachers need to have an ample understanding of the values, cultural norms and practices, and belief systems of their students. This understanding enables teachers to sensitively approach students as individuals—not as members of a group or gang—and provide the necessary reinforcements within the learning environment. This practice encourages higher academic performance and mutually acceptable behavior modification. The combination of appropriate psychological preparation, consistent support, and equitable practice of school policies may reduce the frequency of violent behaviors.

Teachers may effectively utilize a variety of basic techniques as they routinely interact with, guide, and direct students within the educational setting. These techniques include:

- Value Students: The most important thing a child can learn is to value his or her self. Even though multicultural education plays a role in teaching children self-worth, it is best accomplished through modeling. Teachers demonstrate their value of students by developing relationships with them that show they are cared for and respected. This demonstration reinforces to students that they are valuable. Students, in turn, internalize this value. In other words, adults teach children to care for themselves by caring for them.
- Student Voice: Teachers should create safe and honest environments for students to be heard and to share information. Allowing students to voice their options should not be used for punitive purposes. In special circumstances where the shared information may

have legal implications, (i.e. child abuse), students should be informed of the teacher's responsibilities and possible outcomes.

- **Positive Attention:** Due to the teacher's awareness of the effects of contagious behaviors, positive behaviors should be recognized and praised in a consistent and equitable manner.
- **Physical Cuing:** Teachers may use facial expressions and other recognizable body languages to convey personal messages to individual students. This strategy reduces the likelihood of contagious negative behaviors and the potential of publicly embarrassing students.
- **Physical Proximity:** Even though the physical presence of the teacher encourages some students to control their emotions (Bullock, et al., 1983), this situation may also increase the level of tension. Teachers must be aware of and respect students' physical space.
- **Alternative Instruction and Assessment:** Students who experience difficulty with directions or lessons should be given the opportunity to temporarily work on another task that better helps them focus and achieve success. Classroom centers and cooperative learning activities are alternate strategies. Portfolio assessment also encourages students to self-select assignments that may better enable them to feel a sense of accomplishment, ownership, and success.
- **Reduce Distracters and Stimuli:** Classrooms are viewed as veritable playgrounds of optical stimulation (Bullock, et al., 1983). Many students are able to function with bright colors and other busy objects in the room, but others experience difficulty differentiating between relevant and irrelevant stimuli. Teachers must minimize distracters and stimuli that cause excitement and frustration among students.
- **Reality-based Instruction:** Research continues to show that relevant curriculum driven by the students' realities tends to reduce the occurrence of disruptive behaviors (Darder, 1991 & 1993). The students' lives (e.g., culture, primary language, gender, etc.) must be reflected and incorporated within the context of the Standards of Learning (SOLs).
- **Positive Time-out:** Whenever the teacher is able to circumvent a student who is losing control, the practice of positive time-out is most effective. This proactive strategy enables teachers to act before students become a threat to themselves or others. The primary intervention is to remove the student to an isolated area until both student and teacher agree that it is appropriate to return to the group.

These basic techniques serve dual purposes. They assist teachers in maintaining a conducive learning environment and are preventive measures for the reduction of violent confrontation in the classroom. It is imperative to emphasize that these techniques should be as individualized as a teacher's instructional program. Creating safe classrooms and schools are not merely a product but an on-going process.

# Reference

Children Defense Fund (1997) *The state of America's children yearbook*. Washington, DC: CDF Publication.

Center for Disease Control (1993). *The prevention of youth violence: A framework for community action*. Atlanta: CDC.

Cohen, S., and Wilson-Brewer, R. (1991). *Violence-prevention for Young Adolescents: The state of the Art of Program Evaluation*. New York: Carnegie Council on Adolescent Development.

Darder, A. (1991). *Culture and power in the classroom: A critical foundation for bicultural education*. New York: Bergin and Garvey.

Darder, A. (1993). A Call for Educational Justice: Bringing Community Voice to the Center. In A. Darder (Ed.), *Bicultural studies in education: The struggle for educational justice.* Claremont, CA: The Institute for Education in Transformation.

Dryfoos, J. (1990). *Adolescents at risk: Prevalence and prevention.* New York: Oxford University Press.

Gottfredson, G., & Gottfredson, D. (1985). *Victimization in schools.* New York: Plenum Press.

Rich, J. (1981). School violence: Four series explain why it happens. NASSP Bulletin: 65–71.

Houston, R., & Grubaugh, S. (1989). Language for Preventing and Diffusing Violence in the Classroom. *Urban Education*, 24(1), 25–37.

Kum-Walks, D. (1994). *The voice from within: A look at violence from the students' perspectives—A monograph on incarcerated youth.* Los Angeles, CA: Los Angeles County Office of Education.

Metropolitan Life Insurance Company and Louis Harris and Associates, Inc. (1993). *The Metropolitan Life survey of American teachers 1993: Violence in America's public schools.* New York: Louis Harris and Associates, Inc.

National Education Goals Panel. (1994). *Data volume for the National Education Goals Report, Volume One: National Data.* Washington, DC: Author.

Prothrow-Stith, D. (1991). *Deadly consequences.* New York: HarperCollins Publisher.

Roetter, P. (1987). The positive approach in the classroom. *High School Journal*, 70, 196–202.

Short, P., & Short, R. (1987) Beyond technique: Personal and Organizational Influences on school discipline. *High School Journal*, 71, 31–35.

# Section II

## *Vital Signs of American Schools*

# International Comparisons

## How do U.S. schools stack up? And are world rankings really fair and accurate?

### KEVIN BUSHWELLER
*Associate Editor, The American School Board Journal*

"Balderdash." That's the blunt response of psychologist David Berliner and Bruce Biddle to the widely circulated notion that the United States is falling behind the other industrialized nations of the world in education.

In their 1996 book *The Manufactured Crisis: Myths, Fraud, and the Attack on America's Public Schools*, Berliner and Biddle say politicians and others distort reality when they draw conclusions about American education on the basis of international comparisons of student performance. Critics of American education, they say, fail to consider the vast differences between U.S. and foreign schools.

"To condemn public education based on international comparisons is silly," Berliner told *The American School Board Journal*. "It doesn't make any sense."

Even so, international comparisons are an integral part of the debate about the condition of American schools—a debate that attracts a great deal of attention in the local and national news media. Politicians, reporters, and parents, as a result, are likely to bombard school board members and school administrators with challenging questions about how U.S. students stack up against their peers around the world. (The charts on these pages show a few examples of these world rankings.) As a school leader, you need to understand the uses and limitations of the tests on which these comparisons are based, because the results are often used as political justifications for major changes in public education policy and funding.

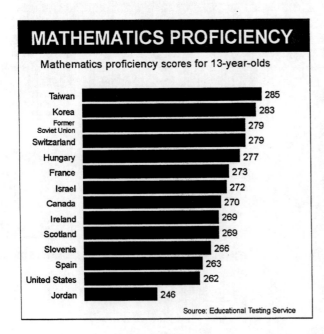

**MATHEMATICS PROFICIENCY**

Mathematics proficiency scores for 13-year-olds

|  |  |
|---|---|
| Taiwan | 285 |
| Korea | 283 |
| Former Soviet Union | 279 |
| Switzerland | 279 |
| Hungary | 277 |
| France | 273 |
| Israel | 272 |
| Canada | 270 |
| Ireland | 269 |
| Scotland | 269 |
| Slovenia | 266 |
| Spain | 263 |
| United States | 262 |
| Jordan | 246 |

Source: Educational Testing Service

---

Kevin Bushweller, "International Comparisons: How Do U.S. Schools Stack Up? And Are World Rankings Really Fair and Accurate?" *Educational Vital Signs*, December 1996, Vol. 183, No. 12, pp. A10–A31. Reprinted by permission from American School Board Journal.

## The Trouble with Averages

Berliner says the trouble with putting stock in U.S. averages on measures of student performance is that the nation is running two different school systems—one that is doing quite well educating middle-class and wealthy students, who are well represented in some states, and another that is failing miserably with poor students, many of whom are clustered in other states. A U.S. average, he says, overshadows the fact that students in some states are doing very well compared to their foreign peers.

Take mathematics. In *The Manufactured Crisis*, Berliner and Biddle point out that the average math achievement of eighth-graders in high-achieving states is about the same as that of students in high-achieving countries such as Taiwan and Korea. But the math achievement levels of eighth-graders in poor-achieving states are about the same as those of the low-achieving countries of the world, such as Jordan.

Yet those differences are not usually emphasized when the national debate about public education takes center stage in Washington, Berliner says. Rather, politicians and the public compare student achievement in foreign nations that have a national curriculum and national tests with overall achievement in the United States, with its 50 separate state education systems. A case in point was the widely cited 1983 report *A Nation at Risk*, which called for major reforms in education because American students never ranked higher than third, and came in last several times, on 19 international academic comparisons. And international comparisons continue to drive many education policy decisions at the national level.

There are signs of change, however. Researchers know it's more valid to compare like with like—for example, to compare areas that have similar poverty rates, per-capita incomes, and unemployment percentages. And because such figures vary considerably among states, some researchers are trying to provide state-by-state data in their international comparisons.

A report released in July by the National Center for Education Statistics (NCES) does just that. *Education in States and Nations* provides a plethora of data comparing each state with such countries as Canada, China, England, France, Hungary, Ireland, Israel, Korea, Spain, Switzerland, and Taiwan. The results are revealing: On a mathematics test for 13-year-olds, for example, U.S. students on the whole ranked 14th. Students in Taiwan and Korea ranked first and second, respectively, with average scores more than 20 points higher than those of U.S. students. But students in Iowa and North Dakota did just as well as the Korean students, and those in Minnesota scored only one point lower. And the scores of students in other states—including Maine, Nebraska, New Hampshire, and Wisconsin—were in the upper half of the international ranking. (The District of Columbia, on the other hand would have come in dead last on an international comparison, 12 points behind Jordan, and Mississippi would have tied for last.)

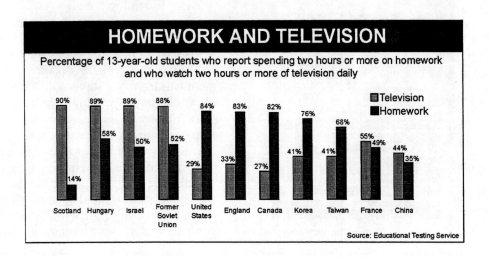

**HOMEWORK AND TELEVISION**

Percentage of 13-year-old students who report spending two hours or more on homework and who watch two hours or more of television daily

Source: Educational Testing Service

An added value of the NCES international comparison study, educators and researchers say, is that it provides more than just the results of achievement test scores. The report also includes information about students' lifestyles, which provides a better context for examining the overall performance of U.S. students in comparison to their international peers. You might assume that American teenagers are the biggest television watchers in the world, for example, but the report says that is not the case. In a survey, of the TV-watching habits of 13-year-olds, the United States ranked only fifth highest on the couch potato barometer, with 84 percent of those surveyed saying they watched two or more hours of television per night. Scotland came in first at 90 percent, followed by Hungary, Israel, and the former Soviet Union. At 35 percent, China had the lowest percentage of kids watching that much television.

But at 93 percent, the District of Columbia would have ranked first in the world for television watching, with Alabama and Mississippi following at 90 percent. Utah (72 percent) had the lowest percentage.

## Who's Included?

Another persistent problem is the question of which students are represented in the test data from each nation—all students, or only the top ones? On the NCES math comparison, for instance, the data for Brazil, China, England, and Portugal were not reported because those countries either excluded certain groups from test taking or had low participation rates.

Critics of such comparisons say it's good the data from those countries were not published. But they insist that even among countries with high participation rates and comprehensive samples, drawing comparisons can be misleading. In China and France, for instance, only an elite group of students makes it into the higher grades, while in countries such as the United States and Germany, most youngsters complete secondary school.

"To what extent are they comparing apples and apples, apples and oranges, or vegetables and animals?" asks Monty Neil, associate director of FairTest, a Boston-based group that analyzes the quality of standardized tests. "One has to take the results with a grain of salt."

To begin with, Neil says, politicians and reporters frequently emphasize the rankings of countries, rather than the point spreads among the average scores. As a consequence, even a small difference in scores looms large, with the frequent result that U.S. educators are reprimanded for not keeping up with the rest of the world. "International comparisons are used more as a political weapon than as a thoughtful tool for improving education," Neil says. "Will having this information really help anyone do anything better? That's a common problem with tests—you get numbers, but so what?"

Some researchers disagree. Among them is Harold W. Stevenson, a University of Michigan psychologist who has conducted international comparisons of math achievement of students in the United States, China, and Japan. Stevenson argues that international comparisons are very useful, especially in a world that is becoming much more globally competitive. Ignoring the importance of international comparisons, he says, creates an isolation-

### COMPUTER USE

Percent of 13-year-old students who report they sometimes use computers for schoolwork or homework

| Country | Percent |
|---|---|
| China | 6% |
| Former Soviet Union | 6% |
| Taiwan | 6% |
| Korea | 10% |
| Hungary | 31% |
| United States | 37% |
| Scotland | 38% |
| Canada | 42% |
| England | 44% |
| France | 57% |

Source: Educational Testing Service

ist atmosphere in American education that will not help students compete in the global marketplace.

Stevenson adds that some people in the education community don't like international comparisons because they are overly defensive about criticisms of their schools. But he believes analyzing American education in a global context will make our schools better.

What about the argument that average American children are being compared with "elite" students in other countries? "Absolutely incorrect," Stevenson says. "The other countries aren't cheats and liars. We have picked representative samples that include the best, worst, and average students in each country."

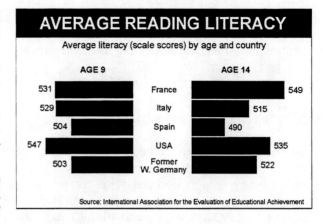

Stevenson and his colleagues say they picked representative samples for an international comparison they conducted that was published in 1993 by the American Association for the Advancement of Science. The longitudinal study, "Mathematics Achievement of Chinese, Japanese, and American Children: Ten Years Later," compared the mathematics achievement of students in Minneapolis, Taipei, and Sendai.

Among other things, it concluded that a heightened emphasis on math in this country between 1980 and 1991 did little to narrow the achievement gap between American kids and their Japanese counterparts. Japanese students continued to outperform American students at the same rate, and Chinese children improved more than American students over that decade, widening an existing achievement gap.

## The Attitude Factor

One of the more disturbing findings of that study, Stevenson says, was the suggestion of complacency on the part of American parents. About 5 percent of Chinese and Japanese mothers surveyed in 1980 reported they were "very satisfied" with their children's education, but more than 40 percent of American mothers said they were very satisfied. Those percentages remained relatively the same in the 1990 follow-up survey. (When fathers of 11th-graders were surveyed, the percentages in the three countries were relatively the same as those for mothers.)

Stevenson's study also showed differing cultural views about hard work. Is studying hard the most important factor in school success? Some 72 percent of Japanese students think so, and so do 59 percent of Chinese students. But only 27 percent of American students say the same. Among teachers, the pattern is the same: In Japan, 93 percent of teachers believe that studying hard is the most important factor for school success, but only 26 percent of American teachers agree. (Chinese teachers are between the two extremes.)

"Critics of the academic success of Chinese and Japanese students often suggest that their high levels of performance come at great psychological cost," the study says. But "American students re-

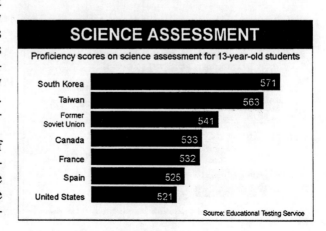

ported the most frequent feelings of stress, academic anxiety, and aggression."

And, the study says, "the most common source of stress was school. It was mentioned by 70 percent of the American students—more than three times as often as four other major sources (peers, family, sports, and jobs)."

## Motivation and Resources

Achievement on standardized academic tests might be influenced by more than just intellectual skill or hard work in the classroom, however. Several researchers argue that foreign students often do better on standardized achievement tests because higher stakes are attached to the outcome in other countries, such as determining which students will be accepted into college. In other words, they have more motivation to do well.

Harold F. O'Neil Jr., a University of Southern California education professor, says he always suspected that test motivation was a key factor explaining why kids in Asian countries fared better than U.S. students on many achievement tests. And that means the problem is not simply lack of knowledge among U.S. kids, but also lack of motivation.

Some of O'Neil's suspicions were confirmed when he investigated whether financial incentives lead to higher achievement. O'Neil's study found that students who were told they would be paid $1 for every correct answer on a national math exam averaged scores 13 percentage points higher than those of kids who were only encouraged to do well.

Now, he says, someone needs to do a similar motivational study of students in Asian countries. His guess is that the difference won't be as significant among Asian students.

"Testing is more important in their societies than it is in ours," O'Neil says. "Their scores are a good measure of what they know. They're not a good measure for us. They're only a good measure of minimum effort."

Even so, O'Neil says international comparisons should not be scrapped or ignored. They should serve as one of many tools for improving American education.

"You need benchmarks outside of our own environment," he says. "It's absolutely essential to have these external standards. I think they're flawed, but they're useful."

And, O'Neil says, the data used in international comparisons have noticeably improved over the years. The samples of students are more representative of average students in the countries surveyed, he says, and researchers have done a better job testing content that kids in all countries have covered. Test designers, he says, have also worked to eliminate the kind of cultural biases that plagued the use of international comparisons in the past.

One of the biggest persistent problems, he says, is the length of time required to collect the data for international comparisons. For instance, several of the international comparisons released this year are based on data collected in the late 1980s or early 1990s.

O'Neil and other researchers reluctantly admit it is likely the issue of how education should be assessed and compared across cultures will never be fully resolved. But a new report by the Center for Educational Research and Innovation in Paris sheds some light on the plethora of alternatives available to school officials.

The results of a seven-nation study, *Schools Under Scrutiny* examines the evaluation of individual schools. After looking at evaluation procedures used in England, France, Germany, New Zealand, Spain, Sweden, and the United States, re-

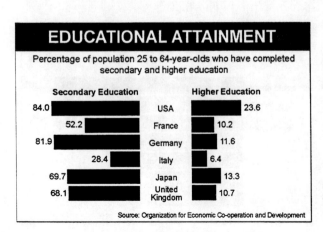

searchers recommend using a mix of school inspections and performance indicators such as achievement test scores to gauge a school's performance.

That would be a step in the right direction, say front-line educators like Danny Shaw, principal of Aiken Elementary school in Aiken, S.C. Shaw says he often hears local politicians call for an overhaul of American education because a recent international comparison showed that American students did not rank among the top achievers worldwide. But he says it's usually been a long time since those politicians have stepped inside a public school building during school hours.

International comparisons are valuable, Shaw says, if they "help us look at our schools from the perspective of instructional improvements. Are there gaps? How can we fill those gaps?" Instead, Shaw says, politicians too often take a "simplistic view" of achievement comparisons among nations, states, and local communities. And, he adds, "you can prove almost anything with statistics."

## Common Measures
### The theme for the year? Kids in distress

Several of the indicators *Education Vital Signs* tracks regularly show a common thread this year: kids in trouble. Drug use among students is up, and so is smoking. And students' behavior problems continue to frustrate educators, parents, and policymakers.

A federal report released this summer revealed disturbing increases in drug use

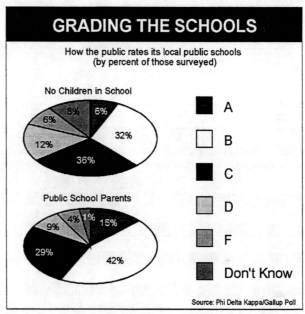

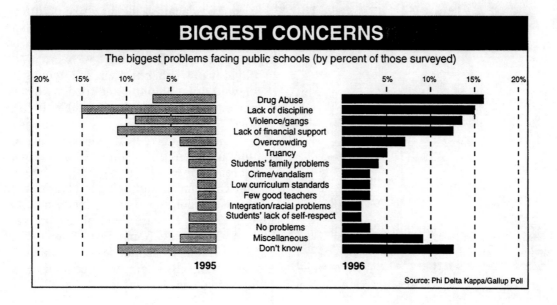

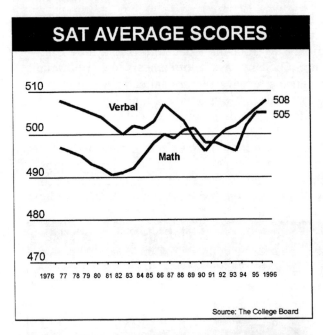

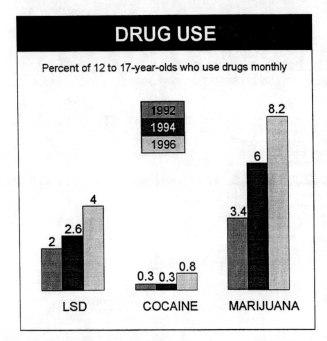

among teenagers. The *National Household Survey on Drug Abuse*, a U.S. Department of Health and Human Services report, found that marijuana and cocaine use among teenagers more than doubled between 1992 and 1995, and the percentage of 12th-graders using LSD increased from 2 percent to 4 percent in the same period.

Lax attitudes on the part of parents might contribute to the increase, substance abuse researchers say. The parents of today's teens grew up in the 1960s and 1970s, and many of them experimented with drugs themselves as young people. Given their own experience, these parents might consider it hypocritical to warn their children against drug use.

Lloyd D. Johnston, a substance abuse researcher at the University of Michigan, says most parents "do not want their children involved in drugs, marijuana, or otherwise, but many have fallen silent on the issues because they feel quite conflicted about it." Researchers and public health officials say educators can help arm parents with solid arguments they can use with their children—for example, that the drugs widely available today are much stronger and more addictive than the drugs available years ago.

Donna Shalala, secretary of the Department of Health and Human Services, says parents, teachers, coaches, clergy, and the government need to work together to warn kids against the hazards of drugs. Without such a broad-based effort, she says, "in a few years we're going to find ourselves right back where we were in the old days, when children and teenagers viewed drug, alcohol, and tobacco use as perfectly normal and acceptable behavior."

Unfortunately, many kids are already thinking along those lines. In addition to increased use of illegal drugs, tobacco smoking among high school kids is also on the rise, according to a study by the national Centers for Disease Control and Prevention. The study found that about one of every five high school seniors smokes tobacco daily, the largest percentage in almost two decades.

The use of alcohol remains a serious problem as well, even among younger students. In a study conducted by the Rhode Island health department, for example, one-third of the state's seventh-graders reported drinking beer, wine, or liquor in the month before the survey was taken. Equally troubling, 10 percent of seventh-graders reported coming to school under the influence of alcohol or drugs once or more in the three months preceding the survey.

## Public Opinion

Appropriately, the American public's perception of the problems facing local schools mirrors the results of these drug and alcohol surveys. Respondents to this year's Phi Delta Kappa/Gallup Poll of the Public's Attitudes Toward the Public Schools cited drug abuse as the Number 1 problem facing schools.

Presented with a list of 15 problem categories, 16 percent of respondents cited drug abuse as a problem, more than twice the percentage that chose drug abuse last year.

Lack of discipline came in a close second at 15 percent, and fighting/violence/gangs placed third with 14 percent.

The report points out, however, that student behavior might be perceived as a much greater problem than drug abuse because several separate categories relate to student behavior. Combine such categories as lack of discipline, fighting/violence/gangs, pupils' lack of interest and poor attitudes, crime/vandalism, and lack of respect for self and others, and the percentage citing these problems rises to 39 percent.

Only 3 percent of respondents say poor curriculum and low standards—a topic that often sparks firey debate among educators and politicians—is the biggest problem facing the schools. However, a significant percentage (13 percent) believe that lack of proper financial support is the schools' top challenge. (Board members agree: see "School Leaders," page 73.)

How should schools combat student behavior problems? Some 92 percent of the public supports efforts to remove troublemakers from classrooms; 88 percent would ban smoking on school property; 75 percent support the use of drug-sniffing dogs; and 63 percent favor using random drug tests to catch drug users.

Other findings:

- By a large margin—61 percent to 36 percent—the public rejects the idea of allowing parents to send

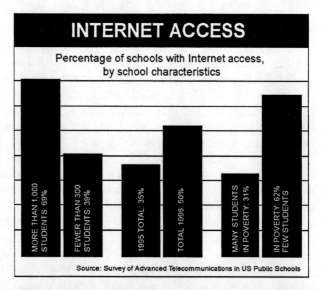

**INTERNET ACCESS**

Percentage of schools with Internet access, by school characteristics

- MORE THAN 1,000 STUDENTS: 69%
- FEWER THAN 300 STUDENTS: 39%
- 1995 TOTAL: 35%
- TOTAL 1996: 50%
- MANY STUDENTS IN POVERTY: 31%
- IN POVERTY: 62% FEW STUDENTS

Source: Survey of Advanced Telecommunications in US Public Schools

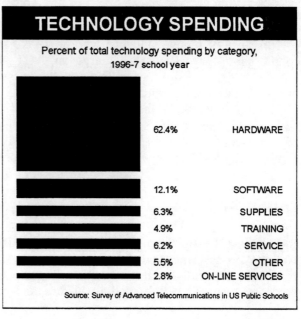

**TECHNOLOGY SPENDING**

Percent of total technology spending by category, 1996-7 school year

- 62.4% HARDWARE
- 12.1% SOFTWARE
- 6.3% SUPPLIES
- 4.9% TRAINING
- 6.2% SERVICE
- 5.5% OTHER
- 2.8% ON-LINE SERVICES

Source: Survey of Advanced Telecommunications in US Public Schools

their children to a private school at public expense.
- A small majority—54 percent—rejects vouchers that would let parents choose public, private, or church-related schools and have the government pay all or part of their kids' tuition.
- 43 percent grade their local schools *A* or *B* for performance, and 17 percent gave grades of *D* or *F*—relatively unchanged from last year.
- Public perceptions are not always accurate: 64 percent of respondents believe dropout rates are worse now than they were 25 years ago, but researchers say the opposite is true.

## Achievement

This year's Scholastic Assessment Test (SAT) scores are far higher than last year's, but the higher numbers have nothing to do with student performance. Rather, the 1996 scores are a product of new statistical procedures used to score the tests.

This year's average score on the math portion of the SAT is 508; the average score on the verbal portion is 505. Both are a far cry above last year's averages of 482 and 428, respectively. But when the 1995 scores are averaged using the new procedures, they become 506 for math and 504 for verbal—just slightly below the 1996 scores.

SAT officials are optimistic. "Compared to their predecessors," says Donald M. Stewart, president of the College Board, which administers the SAT, these students "have taken more honors courses and more precalculus, calculus, chemistry, physics, and other academic courses, and are more computer literate. They are also more ambitious, with over half planning to go beyond the bachelor's degree."

Stewart says average verbal and math scores are higher for students who take more classes in English, math, natural science, social sciences, history, foreign and classical languages, art, and music. Since 1987, he says, students who take the SAT have increased their study in all of those academic areas except English.

"Hard work pays off in school as anywhere else," he says.

Results from the American College Testing Program achievement test (ACT) were equally promising: The national average score increased for the third time in the past four years, and girls improved their scores and continued to narrow the gender gap. The national average on the test—which is taken by 60 percent of America's entering college freshmen—increased from 20.8 to 20.9. Girls averaged 20.8 and boys 21.0.

"The increase in the national average score this year can be attributed particularly to the performance of female students, who now take the ACT in greater numbers than their male counterparts," says Richard L. Ferguson, ACT president. "The average composite score for females has risen four times in six years, while the average composite score achieved by males during that period has remained stable."

Ferguson attributes the change to the fact that more girls are now taking larger loads of rigorous academic courses than they used to. "The percentages of males and females who report taking a core curriculum are now exactly equal," he says. "As recently as four years ago, significantly higher proportions of males took core courses, especially in math and science."

## Reading, History, and Geography

SAT and ACT officials might find reason for optimism, but the results of tests conducted by the National Assessment of Educational Progress (NAEP) are decidedly mixed. The federally funded NAEP—which calls itself the Nation's Report Card—this year released the results of tests of students' academic skill in reading, history, and geography.

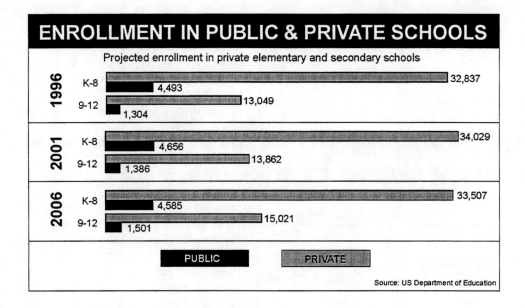

Results of the reading assessment, which included tests administered to students in grades 4, 8, and 12, were especially disturbing. The bad news: In just two years, from 1992 to 1994, the average reading proficiency of 12th-graders has declined significantly.

In all three grades, girls had higher reading achievement levels than boys. But the national decline in 12th-grade reading performance was evident among girls as well as boys, and for black and Hispanic students as well as whites.

Disturbingly, too, the NAEP history report card showed serious gaps in students' knowledge and understanding of history. Fewer than half of the 12th-graders tested (43 percent) scored at or above the basic level on the test. (The test has three achievement levels: basic, proficient, and advanced.) To take one typical example, only 29 percent of the 12th-graders could explain the effect of an economic or technological change on the nature of farming in America.

Students fared better on the NAEP geography assessment. About 70 percent of students at all three grade levels scored at or above the basic level on the test. Boys generally scored higher than girls at each grade level.

At grade 12, more than 90 percent of students successfully used a map to identify an area of earthquake activity, and 55 percent could offer at least two geographically correct reasons why a shopping mall should be located in a certain place. At grade eight, 90 percent of students could locate information in an atlas, and 48 percent could identify latitude on a polar map projection.

## Technology

Instruction is changing rapidly in many schools as new technologies are introduced into the classroom. What effect this influx of technology will have on student achievement remains to be seen, but researchers are tracking students' access to new technologies.

A report by the U.S. Department of Education—*Advanced Telecommunications in U.S. Public Elementary and Secondary Schools, 1995*—found a rapid increase in the number of schools with access to the Internet. A year before the survey was conducted, only 35 percent of the nation's public schools were on-line; now more than 50 percent are.

But the numbers might be a bit misleading: Although more than half of U.S. schools have access to the Internet somewhere in the school, only 9 percent of all classrooms, labs, and library media centers are connected to the Internet. Even so, this is triple the percentage of instructional rooms that were connected to the Internet in 1994.

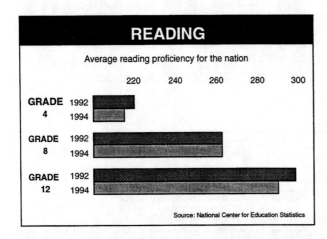

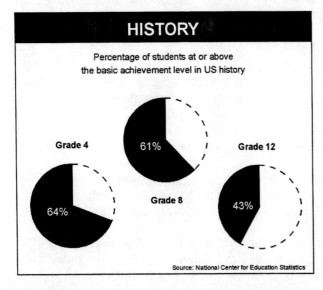

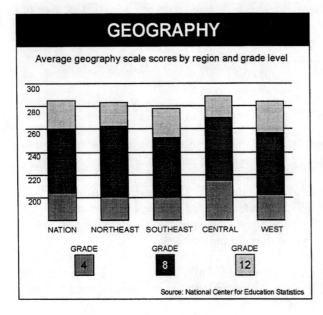

According to the report, today's public schools are just as likely to have a computer with a modem as they are to have cable television (76 percent have each). Some 71 percent of schools have access to broadcast television; 28 percent have closed-circuit television; 13 percent have one-way video with two-way audio or computer link; and 7 percent have two-way video and audio.

Not surprisingly, a school's socioeconomic characteristics largely determine what it has in the way of technology. Only 31 percent of poor schools have access to the Internet, compared to 62 percent of schools in affluent areas. (The report defines poor schools as those in which 71 percent or more of students receive free or reduced-price lunches; affluent schools are defined as those with less than 11 percent of students who are eligible for free or reduced-price lunches.)

The schools that have the money are spending more of it on technology than ever before, according to a report by Quality Education Data Inc. QED estimates that U.S. schools will spend about $4.1 billion on educational technology during this school year, an increase from the $3.9 billion spent during the 1995–96 school year.

The vast majority of that spending (62 percent) will be targeted for hardware. About 12 percent will be spent on software, 6 percent on supplies, 5 percent on training, 6 percent on service, 3 percent on on-line services, and the remaining percentage on miscellaneous needs.

QED researchers found that 55 percent of the planned computer purchases are for Macintosh computers, while 39 percent are for Windows-ready machines. QED says this is a decline for Macintosh from 61 percent during the last school year.

# School Leaders

Board members and superintendents alike feel the pressure of public scrutiny.

As the pressures of running a modern school system intensify, a school board's search for top-quality leadership becomes more urgent. In many places, however, a declining number of educators are willing to take the helm.

Joanne Rys, a spokeswoman for the Massachusetts Association of School Committees, says an average of about 35 candidates now apply for each vacant superintendency in the Bay State. Two or three years ago, she says, a vacancy would draw about 80 candidates.

"The position is one of the most satisfying in the world," says Gary Marx, a spokesman for the American Association of School Administrators (AASA).

"On the other hand, it's the hottest kitchen in America."

It gets pretty hot in the school board kitchen these days, too. Like superintendents, school boards are often the targets of parents, politicians, and special-interest groups who expect the nation's public schools to solve America's troubles. The result: an atmosphere in which it's becoming tougher to get re-elected and easier to be recalled.

## Worries on Board

School board members have a long list of worries, many of which have to do with money. For the second year in a row, in fact, lack of financial support tops the list of concerns for board members, according to a survey by *The American School Board Journal* and Virginia Tech. Five years ago, the No. 1 worry was state mandates.

### Increase in School Executives' Salaries

| Position | 1994–95 (average) | 1995–96 (average) | Percent Increase |
|---|---|---|---|
| Superintendents (contract & salary) | $90,198 | $94,229 | 4.5 |
| Assistant Superintendents | $75,236 | $77,007 | 2.4 |
| Subject Area Supervisors | $54,534 | $56,145 | 3.0 |
| Principals | | | |
|   (high school) | $66,596 | $69,277 | 4.0 |
|   (junior high/middle school) | $62,311 | $64,452 | 3.4 |
|   (elementary school) | $58,589 | $60,922 | 4.0 |
| Assistant Principals | | | |
|   (high school) | $55,556 | $57,555 | 3.6 |
|   (junior high/middle school) | $52,942 | $54,355 | 2.7 |
|   (elementary school) | $48,491 | $50,537 | 4.2 |

*SOURCE:* Educational Research Service

Rounding out the list of this year's top 10 worries are increasing enrollment, curriculum development, at-risk students, facilities, technology improvements, state mandates, parent involvement, management issues, and collective bargaining. It is worth noting that this is the first year technology has ranked among the top 10 concerns, a sign that the needs of school districts—and students—are changing.

The students are changing, too: More than a third (34.4 percent) of U.S. students are minorities, and that percentage is expected to increase steadily. But the number of minorities on boards is not keeping pace with that demographic change. Of the board members responding to this year's survey, only 3.2 percent are black, 1 percent are Hispanic, and 0.2 percent are Asian.

As a consequence, minorities sometimes feel shortchanged by school boards. In Dallas, for example, members of the New Black Panther party were arrested this fall after staging a protest at a school board meeting. At issue was the perceived lack of diversity among the district's top leaders. (Board president Bill Keever and vice president Kathleen Leos are both white.) The board later created a second vice presidency to be filled by an African American.

Women, on the other hand, have continued to narrow the school board gender gap. Five years ago, a little more than a third of board seats were filled by women, but the percentage has now risen to 41 percent.

For the most part, board members continue to be a conservative group. The survey found that 65.4 percent of board members consider themselves politically conservative. Only 28.5 percent label themselves as politically liberal, and about 6 percent say they fall somewhere between those two camps.

More than 40 percent of board members have three or fewer years of experience on school boards. About 25 percent have from four to six years of board experience, however, and more than a third have served seven or more years on their boards.

## White Male School Chiefs

Like board members, the nation's top school administrators are not an ethnically diverse group. Ninety-six percent of school superintendents are white, about the same as five years ago. Only 1.7 percent are black, and 1 percent are Hispanic, according to a survey by *The American School Board Journal* and Xavier University.

The superintendency is also largely a male domain: This year's survey of administrators found that 89 percent of superintendents are men. Women are making gradual progress, however. Five years ago, 93 percent of superintendent jobs were occupied by men.

"The routes to the superintendency and the methods by which superintendents are chosen have really suited men," says Margaret Grogan of the University of Virginia, who is the author of *Voices of Women Aspiring to the Superintendency*. The situation, she says, "stems from stereotypes that women are not able to handle political conflict and political games that have to be played."

Grogan says many school boards often wonder whether women are tough enough for the superintendency because of its demanding schedule.

But AASA officials say fewer men, too, are choosing to follow the ambitious path to the superintendency because of the hours the position demands and the lack of job security.

In this year's survey, researchers found that almost a third of superintendents work more than 60 hours a week, and about 50 percent work between 51 and 60 hours a week. But these extra hours do not give them greater job security: Some 45 percent of superintendents say they have little or no job security. About a third feel "somewhat secure" in their jobs, and 22 percent feel very secure.

In comparison to principals, superintendents are far more satisfied with their salaries, according to the survey. About 65 percent of superintendents consider their compensation

## Profile of School Board Members

|  | 1991 (percent) | 1996 (percent) |
|---|---|---|
| **Sex** | | |
| Male | 65.3 | 59.0 |
| Female | 34.7 | 41.0 |
| **Ethnic Background** | | |
| Black | 2.2 | 3.2 |
| White | 96.5 | 94.3 |
| Hispanic | 0.8 | 1.0 |
| American Indian | 0.3 | 0.4 |
| Asian | 0.1 | 0.2 |
| Other | — | 1.0 |
| **Age** | | |
| Under 25 | 0.4 | 0.3 |
| 26–35 | 5.6 | 2.5 |
| 36–40 | 15.8 | 10.9 |
| 41–50 | 45.6 | 41.6 |
| 51–60 | 18.8 | 19.8 |
| Over 60 | 13.3 | 13.5 |
| **Income** | | |
| Under $20,000 | 1.7 | 1.0 |
| $20,000–$39,999 | 18.0 | 11.8 |
| $40,000–$59,999 | 28.6 | 21.3 |
| $60,000–$79,999 | 20.3 | 20.9 |
| $80,000–$99,999 | 12.3 | 14.7 |
| More than $100,000 | 16.2 | 27.6 |
| **Where Board Members live** | | |
| Small town | 25.5 | 20.1 |
| Suburb | 33.3 | 34.0 |
| Rural area | 29.2 | 32.3 |
| Urbana area | 8.8 | 8.4 |
| **Political Classification** | | |
| Conservative | — | 65.4 |
| Liberal | — | 28.5 |
| Other | — | 6.2 |
| **Years of Service on the Board** | | |
| 0–3 | — | 40.5 |
| 4–6 | — | 25.0 |
| 7 or more | — | 34.5 |

**Board Member's Worries**

| 1991 | 1996 |
|---|---|
| 1. State mandates | 1. School finance/budget |
| 2. Facilities | 2. Increasing enrollment |
| 3. Curriculum development | 3. Curriculum development |
| 4. Management/leadership | 4. At-risk students |
| 5. Large schools/overcrowding | 5. Facilities |
| 6. Collective bargaining | 6. Technology |
| 7. Parents' lack of interest | 7. State mandates |
| 8. Declining enrollment | 8. Parent involvement |
| 9. Personnel relations | 9. Management issues |
| 10. Use of drugs | 10. Collective bargaining |

SOURCE: The American School Board Journal/Virginia Tech

## Profile of Administrators

|  | Superintendents (percents) | High School Principals (percents) | Junior High/ Middle School Principals (percents) | Elementary School Principals (percents) |
|---|---|---|---|---|
| **Sex** | | | | |
| Male | 89.0 | 90.1 | 75.2 | 57.0 |
| Female | 11.0 | 9.9 | 24.8 | 43.0 |
| **Ethnic Background** | | | | |
| Black | 1.7 | 4.6 | 6.0 | 10.2 |
| White | 96.3 | 91.4 | 88.0 | 83.8 |
| Hispanic | 1.0 | 1.3 | 4.0 | 4.2 |
| Asian | — | 1.3 | 1.0 | 0.6 |
| American Indian | 0.7 | 0.7 | 1.0 | 0.6 |
| Other | 0.3 | 0.7 | — | 0.6 |
| **Years in Current Job** | | | | |
| Less than 1 year | 5.7 | 4.6 | 6.9 | 3.6 |
| 1–3 years | 32.0 | 27.8 | 22.8 | 21.4 |
| 4–5 years | 18.7 | 17.2 | 16.8 | 14.9 |
| More than 5 years | 43.7 | 50.3 | 53.5 | 60.1 |
| **Percentage Who Consider Compensation Adequate** | 65.0 | 46.6 | 50.5 | 48.8 |
| **Assessment of Job Security** | | | | |
| No security | 12.4 | 9.9 | 12.9 | 12.7 |
| Little security | 32.1 | 19.9 | 12.9 | 20.0 |
| Somewhat secure | 33.1 | 40.4 | 33.7 | 40.0 |
| Very secure | 22.1 | 29.8 | 38.6 | 27.3 |
| **Political Affiliation** | | | | |
| Democrat | 37.6 | 38.9 | 43.3 | 41.5 |
| Republican | 36.6 | 32.6 | 28.9 | 29.9 |
| Independent | 19.1 | 20.1 | 19.6 | 18.3 |
| Other | 0.3 | 0.7 | 1.0 | 1.8 |
| None | 6.0 | 7.6 | 7.2 | 8.5 |
| **Average Hours Worked per Week** | | | | |
| 30–40 | 0.7 | — | — | — |
| 41–50 | 21.6 | 5.4 | 47.8 | 26.4 |
| 51–60 | 48.3 | 52.4 | 53.5 | 46.6 |
| More than 60 | 29.5 | 42.2 | 28.7 | 27.0 |
| **Highest Degree Earned** | | | | |
| Bachelor's | — | 0.7 | 1.0 | 0.6 |
| Master's | 31.3 | 72.2 | 60.0 | 66.7 |
| Specialist | 28.3 | 15.9 | 28.0 | 22.0 |
| Doctorate | 40.0 | 11.3 | 11.0 | 10.7 |

SOURCE: *The American School Board Journal*/Xavier University "adequate," compared to 47 percent for high school principals, 51 percent for junior high and middle school principals, and 49 percent for elementary school principals.

According to the Education Research Service, the average salary for superintendents is $94,229, and for high school principals, it is $69,277. The National Education Association estimates that the average teacher salary is now $37,846.

As a group, school administrators tend not to be as politically conservative as board members. About 38 percent of superintendents are registered Democrats, 37 percent are Republicans, and 19 percent identify themselves as independents.

## Buildings and Bonds
## Enrollment is up, and so are school bonds

School leaders across the country are bracing for an enrollment "echo boom" that will put greater pressures on them to build new schools and renovate old ones. The big question mark is whether taxpayers—many of whom will have no kids in the public schools—will agree to pay higher taxes to foot the bill.

The good news is that the American public was willing to contribute $2 billion more for K-12 school bonds in 1995 than the year before, approving $21.1 billion in school bonds, according to *The Bond Buyer*, a New York based newsletter that covers the bond industry. And so far, this year looks to be as good or better than last year. In the first six months of 1996, voters approved $13.3 billion in school bonds, an increase of more than $4 billion over the first six months of 1995.

California schools are slated to get a good chunk of that money. By a 20 percent margin, state voters passed a proposition this spring that provides $3 billion to upgrade aging school and university buildings.

And at the local level, some districts are actually breaking records for bond amounts despite the perceived anti-tax mood among voters. In Austin, Texas, for instance, voters passed a $369 million school bond, the largest in district history. And in Cobb Country, Ga., voters approved a $221 million bond, which was five times larger than the last bond the county approved five years earlier.

Other districts are hoping voters will sympathize with the infrastructure problems their schools are facing and give them the financial support they need to build new schools and renovate others. The Flint, Michigan, schools are hoping for such support in 1997, when the board plans to bring a bond issue to voters.

In Flint last winter, a section of a ceiling collapsed at an elementary school that was built in 1916. Nobody was injured, but workers had to install wood braces on all of the school's ceilings to prevent more collapses. A district report estimates it will cost $40.3 million to repair all the maintenance and structural problems in the school buildings.

Other schools are even worse off. The *New York Times* described the basement of Public School 54 in Bedford-Stuyvesant as

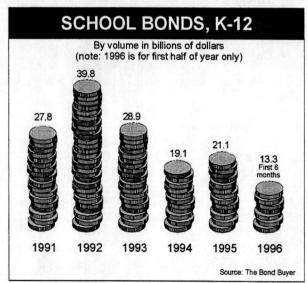

SCHOOL BONDS, K-12
By volume in billions of dollars
(note: 1996 is for first half of year only)

1991: 27.8
1992: 39.8
1993: 28.9
1994: 19.1
1995: 21.1
1996: 13.3 First 6 months

Source: The Bond Buyer

## Bond Issues by State, 1995–1996

| States | Principal Amount (in millions of dollars) | Rank | Market/ Share | Number of Issues |
|---|---|---|---|---|
| Alabama | 448.1 | 17 | 1.3 | 23 |
| Alaska | 176.1 | 31 | 0.5 | 4 |
| Arizona | 589.4 | 46 | 1.8 | 58 |
| Arkansas | 202.3 | 27 | 0.6 | 86 |
| California | 4,529.9 | 1 | 13.6 | 255 |
| Colorado | 1,016.5 | 10 | 3.1 | 60 |
| Connecticut | 195.7 | 29 | 0.6 | 27 |
| Florida | 2,652.2 | 4 | 8.0 | 41 |
| Georgia | 406.3 | 23 | 1.2 | 22 |
| Idaho | 123.6 | 37 | 0.4 | 14 |
| Illinois | 1,456.5 | 7 | 4.4 | 204 |
| Indiana | 1,183.3 | 8 | 3.6 | 140 |
| Iowa | 428.1 | 21 | 1.3 | 51 |
| Kansas | 194.5 | 30 | 0.6 | 24 |
| Kentucky | 380.4 | 24 | 1.1 | 90 |
| Louisiana | 441.2 | 18 | 1.3 | 30 |
| Maine | 13.8 | 44 | 0.0 | 2 |
| Maryland | 5.0 | 46 | 0.0 | 1 |
| Massachusetts | 224.7 | 26 | 0.7 | 43 |
| Michigan | 2,162.5 | 6 | 6.5 | 165 |
| Minnesota | 964.7 | 11 | 2.9 | 125 |
| Mississippi | 51.9 | 40 | 0.2 | 8 |
| Missouri | 607.6 | 15 | 1.8 | 134 |
| Montana | 27.6 | 42 | 0.1 | 15 |
| Nebraska | 47.4 | 41 | 0.1 | 29 |
| Nevada | 432.3 | 20 | 1.3 | 6 |
| New Hampshire | 73.0 | 38 | 0.2 | 13 |
| New Jersey | 860.2 | 13 | 2.6 | 102 |
| New Mexico | 150.8 | 34 | 0.5 | 37 |
| New York | 2,905.4 | 3 | 8.7 | 794 |
| North Carolina | 142.5 | 35 | 0.4 | 12 |
| North Dakota | 16.9 | 43 | 0.1 | 12 |
| Ohio | 1,158.2 | 9 | 3.5 | 189 |
| Oklahoma | 156.4 | 33 | 0.5 | 128 |
| Oregon | 436.1 | 19 | 1.3 | 55 |
| Pennsylvania | 2,598.0 | 5 | 7.8 | 231 |
| Rhode Island | 4.9 | 47 | 0.0 | 3 |
| South Carolina | 425.4 | 22 | 1.3 | 45 |
| South Dakota | 60.4 | 39 | 0.2 | 18 |
| Tennessee | 196.4 | 28 | 0.6 | 25 |
| Texas | 2,935.7 | 2 | 8.8 | 206 |
| Utah | 174.3 | 32 | 0.5 | 24 |
| Vermont | 6.3 | 45 | 0.0 | 2 |
| Virginia | 277.4 | 25 | 0.8 | 12 |
| Washington | 670.2 | 14 | 2.0 | 78 |
| West Virginia | 3.4 | 48 | 0.0 | 1 |
| Wisconsin | 920.5 | 12 | 2.8 | 129 |
| Wyoming | 127.5 | 36 | 0.4 | 15 |
| Top 48 Totals* | 33,261.5 | — | 100.0 | 3,788 |

*Hawaii and Delaware had no bond issues for K–12 education in this year.

"straight out of Dickens—dark, dank, and filled with hills of coal that reach for the ceiling."

That's right, coal. The newspaper reported that 300 of the city's 1,100 schools are heated with coal, a dirty-burning fuel that is especially unhealthy for children or adults with respiratory problems.

Farther south, in Washington, D.C., the library at Taft Junior High was closed in the spring after a good portion of the library ceiling collapsed and leaky water pipes gave the room and its dust-covered books a mildewy smell. House Speaker Newt Gingrich called the condition of the District's school buildings "terrible."

The problem, to a large degree, is money. Across the country, dollars spent on school construction dropped 3 percent—down from a record high $10.7 billion in 1994 to $10.4 billion last year, according to *American School & University* magazine's 22nd annual report on education construction.

The magazine also surveyed schools to find out just how much it costs to build a new school these days. An elementary school, according to the survey, averages about $6.3 million; a middle school, $9.8 million; and a high school, $15.4 million. In new construction, the survey found, the average square feet per student for elementary school students is 111; for middle school kids, the average is 129; and for high school kids, 149.

## State of the States
## Schools nationwide feel the impact of rising enrollment

A majority of states are now feeling, or preparing for, an enrollment boom that is expected this year to eclipse the record number of postwar baby boomers who filled the nation's schools 25 years ago. Dubbed the "echo boom"—many of these kids are the offspring of baby boomers themselves—this surge in enrollment might put more financial pressure on schools than ever before.

To meet the demand, says U.S. Secretary of Education Richard Riley, public schools will have to hire about 190,000 more teachers and build 6,000 more schools by the 2006–07 school year.

This year, the U.S. Department of Education estimates that about 51.7 million students will be attending public and private elementary and secondary schools—some 400,000 more than the 51.3 million baby boomers in school in 1971. And school enrollment is projected to reach 54.6 million by 2006. During the next decade, in fact, high school enrollment is expected to increase by more than 20 percent in nine states: California, Virginia, North Carolina, New Jersey, Maryland, Nevada, Washington, Alaska, and Massachusetts.

In many areas, immigration is a major reason for the surge in enrollment. Such is the case for the Los Angeles Unified Schools, which expect to have 657,495 students this year and about 10,000 more two years from now. The district's budget director, Marty Varon, says the enrollment surges are happening at the same time that state education officials are starting a program to reduce class sizes in elementary schools.

"We're already crowded to begin with, and all of a sudden you have this incentive to reduce class size," says Varon. "What we're doing is going out and getting about 500 double portables."

Stella Shelton, a spokeswoman for the Wake County Public Schools in North Carolina, is equally concerned. The enrollment in Wake County is expected to increase from 86,400 this year to more than 101,000 by the turn of the century.

"It's such a juggling act," says Shelton. "What it means is that we have classes in the media centers, classes in multipurpose rooms, and classes that are larger than the state believes is a good idea."

## State-by-State Highlights

Other changes also show up in the state-by-state statistical snapshots in this year's *Education Vital Signs* (see the charts on the following pages). Here are a few highlights:

- Special education. The number of special education students served by U.S. public schools reached 10.3 percent of the total enrollment, up from 10.2 percent last year. This slight but steady increase is putting a strain on school systems across the country—especially where the percentage of special education students is highest, as in the states of Massachusetts, New Jersey, Rhode Island, Connecticut, and Florida. A survey of elementary and middle school principals by the National Association of Elementary School Principals found that teaching special education students was one of the top concerns among its members.

    That concern is a major one in Florida, where a study by three education professors concluded that the state was facing a severe shortage of special education teachers because of the high turnover rate and lack of newly trained special education teachers.

    "School districts are caught in a Catch 22," says Mary Brownnell, a former special education teacher and one of three University of Florida education professors who conducted the study "They have to have a person to teach the children in their district with disabilities. When they don't have someone, however, they transfer a general education teacher into that classroom or hire an uncertified teacher. This person is not prepared for the demands, experiences a great deal of stress, and ends up leaving the classroom; and the district is back to square one."

- **Teacher salaries.** The National Education Association reports that the average teacher salary continues to increase—from $32,977 to $37,846 over the past five years. But teachers in some districts say they should be paid more to deal with the myriad problems facing public schools.

    In San Diego, for example, more than 5,000 teachers walked off their jobs this fall because they wanted a new contract with a 15 percent pay raise over three years instead of the 11 percent district officials offered. The teachers were on strike for a week but returned to work after an agreement was reached that gives the teachers the 15 percent raise they asked for. The average teacher salary in San Diego is about $43,000.

    At the state level, the highest average teacher salary was in Connecticut, which averaged $50,400. The lowest average—$26,346—was in South Dakota.

- **Per-capita income.** Which state has the most money to spend on education? That, of course, depends on a host of factors, but one very important factor is per-capita income. Connecticut, with the highest average teacher salary, has the second highest per-capita income ($30,303) among the states. Ironically, the District of Columbia, which has one of the most financially troubled school systems in the nation, has the highest per-capita income in the nation at $32,274.

    Mississippi has the lowest per-capita income in the nation, $16,531, and a correspondingly low average teacher salary, $27,689.

- **Pupil-teacher ratios.** For educators caught in the daily grind of classrooms filled with 30 or more students, the U.S. average of 17.3 students to each full-time teacher might seem like an ideal world—a touch more ideal, in fact, than last year's 17.4 average. But with record school enrollments projected for about 33 states over the next several years, that average could increase significantly unless more teachers are hired. For instance, California has the second highest pupil-teacher ratio in the nation at 24.0 (second to Utah's 24.3). The California average is up from 22.4 students per teacher just five years ago, and the state is expected to experience major enrollment growth over the next 10 years.

- **Per-pupil expenditures.** As schools work to improve student achievement by hiring more teachers and purchasing more educational technology—and as they respond to a growing list of social problems—it stands to reason that the average cost-per-student will rise. Last year, the average per-pupil expenditure in the United States was $6,098, an increase of nearly $900 from five years before. Alaska had the highest average ($10,156), with New Jersey and New York close behind. Utah, on the other hand, had the lowest average ($3,908).

  How is that money spent? A Wisconsin study showed that roughly half is spent on teacher salaries and benefits and on classroom materials such as textbooks and supplies.

  "That pretty much tracks with the national average," says Allan Odden, a researcher at the University of Wisconsin's Center for Education Research.

  The Wisconsin study found that about 9 percent went for administrators, 10 percent for special education programs, and 10 percent for operating and maintaining buildings. Smaller percentages went for categories such as transportation, extracurricular activities, and debt payments.

- **Technology.** The number of students per multimedia computer—a new entry on the state charts on the following pages—is one measure for evaluating how schools are doing in their efforts to bring sophisticated technology into classrooms.

  North Dakota averaged the fewest number of students sharing a multimedia computer (15.1), according to Quality Education Data Inc. Louisiana, on the other hand, had the highest average number (89) of students sharing a multimedia computer.

## Northeast

| Finances | Conn. | Del. | Maine | Mass. | N.H. | N.J. | N.Y. | Pa. | R.I. | Vt. |
|---|---|---|---|---|---|---|---|---|---|---|
| **Per-pupil expenditures** | | | | | | | | | | |
| 1990–91 | $7,902 | $5,994 | $5,683 | $6,351 | $5,470 | $9,159 | $8,463 | $6,544 | $6,415 | $5,740 |
| 1995–96 | $8,705 | $7,507 | $6,478 | $7,465 | $6,408 | $9,967 | $9,877 | $7,402 | $7,730 | $7,474 |
| **Average teacher salaries** | | | | | | | | | | |
| 1990–91 | $43,808 | $35,246 | $28,531 | $36,090 | $31,273 | $38,411 | $42,080 | $36,057 | $34,997 | $30,986 |
| 1995–96 | $50,400 | $40,533 | $32,869 | $43,756 | $35,792 | $47,910 | $48,115 | $46,916 | $42,160 | $36,295 |
| **Funding percentages** | | | | | | | | | | |
| 1990–91 | | | | | | | | | | |
| Federal | 4.0 | 8.4 | 6.4 | 5.2 | 2.7 | 3.5 | 5.0 | 4.7 | 4.0 | 5.0 |
| State | 41.6 | 68.0 | 51.1 | 36.9 | 7.5 | 38.3 | 42.0 | 46.7 | 41.0 | 37.0 |
| Local | 54.4 | 23.6 | 42.5 | 58.0 | 89.8 | 58.1 | 53.0 | 48.6 | 54.9 | 58.0 |
| 1995–96 | | | | | | | | | | |
| Federal | 4.6 | 8.2 | 6.9 | 5.6 | 3.2 | 3.6 | 6.0 | 5.6 | 4.0 | 5.1 |
| State | 40.1 | 65.2 | 47.5 | 36.1 | 7.8 | 40.3 | 37.4 | 41.8 | 41.0 | 29.7 |
| Local | 55.2 | 26.7 | 45.5 | 58.2 | 89.0 | 56.0 | 56.6 | 52.6 | 55.0 | 65.2 |
| **Capital Outlay** | | | | | | | | | | |
| 1995–1996 (in thousands) | $226,494 | $44,608 | $63,400 | $188,492 | $12,583 | $301,285 | $1,562,209 | $167,786 | $9,888 | $50,173 |
| **State budget allocations for education (percent of total state budget)** | 15.3 | 24.2 | 13.9 | 12.8 | 6.9 | 21.7 | 19.3 | 20.0 | 14.9 | 16.7 |
| Per-capita income 1995 | $30,303 | $24,124 | $20,527 | $26,994 | $25,151 | $28,858 | $26,782 | $23,279 | $23,310 | $20,927 |
| **Classroom characteristics** | | | | | | | | | | |
| Students per multimedia computer | 43.4 | 53.9 | 30.5 | 52.7 | 46.1 | 41.9 | 33.5 | 38.6 | 49.2 | 28.5 |
| **Pupil/teacher ratios** | | | | | | | | | | |
| 1989–90 | 13.3 | 16.4 | 14.1 | 14.0 | 16.2 | 13.5 | 14.7 | 15.7 | 14.5 | 13.8 |
| 1994–95 | 14.4 | 16.6 | 13.8 | 14.8 | 15.6 | 13.8 | 15.2 | 17.1 | 14.7 | 13.8 |
| **Enrollment** | | | | | | | | | | |
| Fall, 1990–91 | 342,785 | 57,669 | 154,526 | 601,493 | 115,480 | 765,165 | 1,458,289 | 929,273 | 83,092 | 58,411 |
| Elementary | 130,284 | 41,989 | 55,674 | 232,666 | 57,305 | 324,481 | 1,140,048 | 738,561 | 54,854 | 37,819 |
| Secondary | 473,069 | 99,658 | 210,200 | 834,159 | 172,785 | 1,089,646 | 2,598,337 | 1,667,834 | 137,946 | 96,230 |
| Total | | | | | | | | | | |

|  | | | | | | | | | | |
|---|---|---|---|---|---|---|---|---|---|---|
| **Fall, 1995–96** | | | | | | | | | | |
| Elementary | 380,743 | 59,579 | 155,594 | 671,001 | 128,578 | 841,014 | 1,602,800 | 991,230 | 89,283 | 59,846 |
| Secondary | 138,544 | 48,882 | 58,614 | 236,474 | 60,521 | 351,795 | 1,224,900 | 810,740 | 60,519 | 45,701 |
| Total | 519,287 | 108,461 | 214,208 | 907,475 | 189,099 | 1,192,809 | 2,827,700 | 1,801,970 | 149,802 | 105,547 |
| **% Minority enrollment Fall, 1994** | | | | | | | | | | |
| Black | 13.3 | 29.1 | 0.7 | 8.0 | 0.8 | 18.6 | 20.2 | 13.9 | 7.0 | 0.7 |
| Hispanic | 11.4 | 3.6 | 0.4 | 9.0 | 1.1 | 13.1 | 16.9 | 3.4 | 9.5 | 0.3 |
| Asian or Pacific Islander | 2.4 | 1.7 | 0.8 | 3.7 | 1.0 | 5.2 | 4.8 | 1.7 | 3.2 | 0.9 |
| American Indian/Alaskan Native | 0.2 | 0.2 | 0.5 | 0.2 | 0.2 | 0.2 | 0.4 | 0.1 | 0.5 | 0.5 |
| Total % Minority | 27.3 | 34.6 | 2.4 | 20.9 | 3.1 | 37.1 | 42.3 | 19.1 | 20.2 | 2.4 |
| **% Special education enrollment** | | | | | | | | | | |
| 1993–94 | 12.3 | 11.9 | 11.9 | 15.0 | 10.7 | 14.2 | 10.8 | 10.1 | 13.5 | 8.8 |
| **Mean SAT Scores** | | | | | | | | | | |
| 1987 Verbal | 515 | 517 | 510 | 511 | 527 | 502 | 501 | 505 | 509 | 518 |
| Math | 499 | 496 | 493 | 500 | 512 | 493 | 495 | 491 | 492 | 500 |
| 1996 Verbal | 507 | 508 | 504 | 507 | 520 | 498 | 497 | 498 | 501 | |
| Math | 504 | 495 | 498 | 504 | 514 | 505 | 499 | 492 | 491 | 500 |
| **% of graduates taking SAT—1996** | 79 | 66 | 68 | 80 | 70 | 69 | 73 | 71 | 69 | 70 |

**Per-pupil expenditures, average teacher salaries, funding percentages, and enrollment:** National Education Association, *Estimates of School Statistics, 1995–96*; revised 1990–91 figures from 1991–92 edition

**Per-capita income:** U.S. Department of Commerce, Bureau of Economic Analysis

**Capital outlay:** National Education Association, *Estimates of School Statistics, 1995–96*

**Pupil/teacher ratios, minority enrollment:** U.S. Department of Education, National Center for Education Statistics

**State Budget Allocations:** National Association of State Budget Officers

**Special Education Enrollment:** U.S. Department of Education

**SAT Scores and percentage of graduates taking test:** The College Board

## Southeast

| Finance | D.C. | Fla. | Ga. | Md. | N.C. | S.C. | Va. | W.Va. |
|---|---|---|---|---|---|---|---|---|
| **Per-pupil expenditures** | | | | | | | | |
| 1990–91 | $7,875 | $5,186 | $4,453 | $6,266 | $4,790 | $4,327 | $5,416 | $5,078 |
| 1995–96 | $7,519 | $5,984 | $5,546 | $6,930 | $5,147 | $5,018 | $5,888 | $6,902 |
| **Average teacher salaries** | | | | | | | | |
| 1990–91 | $39,497 | $30,555 | $29,172 | $38,382 | $29,276 | $28,301 | $32,239 | $25,967 |
| 1995–96 | $43,700 | $33,320 | $34,307 | $41,215 | $30,564 | $31,568 | $34,687 | $32,155 |
| **Funding percentages** | | | | | | | | |
| 1990–91 | | | | | | | | |
| Federal | 10.0 | 6.4 | 6.1 | 4.8 | 6.2 | 8.5 | 4.8 | 8.1 |
| State | — | 51.8 | 53.3 | 37.9 | 66.7 | 52.4 | 35.5 | 66.9 |
| Local | 90.0 | 41.8 | 40.6 | 57.3 | 27.2 | 39.1 | 59.7 | 25.1 |
| 1995–96 | | | | | | | | |
| Federal | 13.4 | 7.2 | 7.7 | 5.8 | 8.6 | 9.3 | 5.3 | 7.8 |
| State | — | 49.5 | 50.9 | 39.3 | 66.5 | 46.5 | 36.3 | 58.5 |
| Local | 86.6 | 43.3 | 41.4 | 54.9 | 24.9 | 44.2 | 58.4 | 33.7 |
| **Capital Outlay** | | | | | | | | |
| 1995–1996 (in thousands) | $16,500 | $2,311,140 | $704,451 | $466,354 | $400,000 | $262,000 | $471,611 | $95,781 |
| **State budget allocations for education (percent of total state budget)** | — | 17.2 | 24.6 | 17.0 | 24.3 | 18.2 | 17.0 | 20.7 |
| Per-capita income 1995 | $32,274 | $22,916 | $21,278 | $25,927 | $20,604 | $18,788 | $23,597 | $17,915 |
| **Classroom characteristics** | | | | | | | | |
| Students per multimedia computer | 19.1 | 38.5 | 16.0 | 34.0 | 24.0 | 37.9 | 35.1 | 81.6 |
| **Pupil/teacher ratios** | | | | | | | | |
| 1989–90 | 13.4 | 17.2 | 18.3 | 16.8 | 17.1 | 17.0 | 15.9 | 15.1 |
| 1994–95 | 13.2 | 19.1 | 16.3 | 17.0 | 16.2 | 16.4 | 14.1 | 14.8 |
| **Enrollment** | | | | | | | | |
| Fall, 1990–91 | | | | | | | | |
| Elementary | 49,762 | 1,100,056 | 849,548 | 424,748 | 779,591 | 452,968 | 645,368 | 191,555 |
| Secondary | 30,932 | 761,615 | 302,838 | 290,428 | 302,967 | 169,650 | 353,095 | 131,466 |
| Total | 80,694 | 1,861,671 | 1,152,386 | 715,176 | 1,082,558 | 622,618 | 998,463 | 323,021 |

## Vital Signs of American Schools

| | | | | | | | |
|---|---|---|---|---|---|---|---|
| **Fall, 1995–96** | | | | | | | |
| Elementary | 51,350 | 1,278,511 | 965,707 | 470,269 | 857,703 | 454,817 | 695,726 | 188,134 |
| Secondary | 28,452 | 897,592 | 345,419 | 335,275 | 315,191 | 182,279 | 384,128 | 118,317 |
| Total | 79,802 | 2,176,103 | 1,311,126 | 805,544 | 1,172,894 | 637,096 | 1,079,854 | 306,451 |
| **% Minority enrollment Fall, 1994** | | | | | | | | |
| Black | 88.0 | 25.0 | 37.5 | 34.7 | 30.5 | 41.7 | 26.2 | 3.9 |
| Hispanic | 6.6 | 14.4 | 1.8 | 3.1 | 1.5 | 0.6 | 3.0 | 0.2 |
| Asian or Pacific Islander | 1.3 | 1.7 | 1.5 | 3.8 | 1.2 | 0.7 | 3.4 | 0.4 |
| American Indian/Alaskan Native | 0.0 | 0.2 | 0.1 | 0.3 | 1.5 | 0.2 | 0.2 | 0.1 |
| Total % Minority | 95.9 | 41.3 | 40.9 | 41.9 | 34.7 | 43.2 | 32.8 | 4.6 |
| **% Special education enrollment** | | | | | | | | |
| 1993–94 | 7.3 | 12.2 | 8.7 | 10.6 | 10.4 | 10.7 | 10.7 | 11.8 |
| **Mean SAT Scores** | | | | | | | | |
| 1987 Verbal | 482 | 501 | 478 | 513 | 477 | 474 | 511 | 534 |
| Math | 462 | 497 | 470 | 502 | 468 | 466 | 499 | 519 |
| 1996 Verbal | 489 | 498 | 484 | 507 | 490 | 480 | 507 | 526 |
| Math | 473 | 496 | 477 | 504 | 486 | 474 | 496 | 506 |
| **% of graduates taking SAT—1996** | 50 | 48 | 63 | 64 | 59 | 57 | 68 | 17 |

**Per-pupil expenditures, average teacher salaries, funding percentages, and enrollment:** National Education Association, *Estimates of School Statistics, 1995–96*; revised 1990–91 figures from 1991–92 edition.
**Per-capita income:** U.S. Department of Commerce, Bureau of Economic Analysis
**Capital outlay:** National Education Association, *Estimates of School Statistics, 1995–96*
**Pupil/teacher ratios, minority enrollment:** U.S. Department of Education, National Center for Education Statistics
**State Budget Allocations:** National Association of State Budget Officers
**Special Education Enrollment:** U.S. Department of Education
**SAT Scores and percentage of graduates taking test:** The College Board

## Northcentral

| Finances | Ill. | Ind. | Iowa | Ky. | Mich. | Minn. | Mo. | Neb. | Ohio | Wis. |
|---|---|---|---|---|---|---|---|---|---|---|
| **Per-pupil expenditures** | | | | | | | | | | |
| 1990–91 | $4,893 | $4,662 | $4,867 | $4,387 | $5,268 | $5,357 | $4,391 | $4,341 | $5,325 | $5,794 |
| 1995–96 | 5,530 | $6,229 | $5,737 | $5,972 | $7,131 | $6,148 | $5,078 | $5,538 | $5,749 | $7,213 |
| **Average teacher salaries** | | | | | | | | | | |
| 1990–91 | $34,605 | $32,434 | $27,977 | $29,115 | $38,326 | $33,126 | $28,290 | $26,592 | $32,615 | $33,209 |
| 1995–96 | $41,008 | $37,805 | $32,376 | $33,108 | $49,168 | $36,937 | $33,341 | $31,496 | $37,835 | $38,571 |
| **Funding percentages** | | | | | | | | | | |
| 1990–91 | | | | | | | | | | |
| Federal | 7.5 | 5.1 | 5.5 | 9.6 | 4.5 | 4.1 | 5.6 | 6.5 | 5.6 | 3.8 |
| State | 36.7 | 56.4 | 51.4 | 68.8 | 35.4 | 55.1 | 38.9 | 25.8 | 42.8 | 40.9 |
| Local | 55.8 | 38.5 | 43.0 | 21.6 | 60.0 | 40.8 | 55.6 | 67.7 | 51.6 | 55.3 |
| 1995–96 | | | | | | | | | | |
| Federal | 8.8 | 5.2 | 5.1 | 8.9 | 3.8 | 4.6 | 6.8 | 4.7 | 6.3 | 4.3 |
| State | 29.9 | 52.3 | 49.6 | 67.2 | 58.6 | 50.5 | 37.3 | 38.8 | 41.7 | 43.2 |
| Local | 61.3 | 42.5 | 45.3 | 23.9 | 37.6 | 44.9 | 55.8 | 56.5 | 52.0 | 51.5 |
| **Capital Outlay** | | | | | | | | | | |
| 1995–1996 (in thousands) | $634,314 | $526,578 | $214,389 | $237,506 | $891,626 | $800,527 | $487,823 | $123,623 | $824,600 | $424,805 |
| **State budget allocations for education** | | | | | | | | | | |
| (percent of total state budget) | 17.1 | 27.7 | 19.1 | 26.0 | 33.0 | 21.8 | 24.5 | 16.5 | 16.5 | 20.2 |
| Per-capita income 1995 | $24,763 | $21,273 | $21,012 | $18,612 | $23,551 | $23,118 | $21,627 | $21,703 | $22,021 | $21,839 |
| **Classroom characteristics** | | | | | | | | | | |
| Students per multimedia computer | 38.4 | 22.2 | 21.2 | 31.8 | 37.3 | 27.6 | 42.2 | 18.6 | 61.8 | 25.3 |
| **Pupil/teacher ratios** | | | | | | | | | | |
| 1989–90 | 16.9 | 17.5 | 15.7 | 17.7 | 19.7 | 17.2 | 15.7 | 14.7 | 17.4 | 15.9 |
| 1994–95 | 17.3 | 17.5 | 15.7 | 17.0 | 20.1 | 17.5 | 15.5 | 14.5 | 16.6 | 16.8 |
| **Enrollment** | | | | | | | | | | |
| Fall, 1990–91 | | | | | | | | | | |
| Elementary | 1,304,892 | 676,368 | 272,542 | 439,698 | 1,139,395 | 434,703 | 589,832 | 166,643 | 1,146,449 | 528,622 |
| Secondary | 516,515 | 276,835 | 211,110 | 190,393 | 439,873 | 321,718 | 226,726 | 106,339 | 624,640 | 268,999 |
| Total | 1,821,407 | 953,203 | 483,652 | 630,091 | 1,579,268 | 756,421 | 816,558 | 272,982 | 1,771,089 | 797,621 |

Vital Signs of American Schools    89

|  | | | | | | | | | | |
|---|---|---|---|---|---|---|---|---|---|---|
| **Fall, 1995–96** | | | | | | | | | | |
| Elementary | 1,374,949 | 520,580 | 260,358 | 437,648 | 1,180,416 | 454,754 | 620,878 | 171,130 | 1,188,621 | 602,964 |
| Secondary | 552,570 | 451,508 | 241,985 | 201,603 | 451,991 | 379,404 | 252,760 | 117,553 | 648,510 | 267,211 |
| Total | 1,927,519 | 972,088 | 502,343 | 639,251 | 1,632,407 | 834,158 | 873,638 | 288,683 | 1,837,131 | 870,175 |
| **% Minority enrollment Fall, 1994** | | | | | | | | | | |
| Black | 21.0 | 11.2 | 3.2 | 9.7 | 17.5 | 4.5 | 15.8 | 5.8 | 15.1 | 9.3 |
| Hispanic | 11.6 | 2.2 | 1.8 | 0.3 | 2.6 | 1.8 | 0.9 | 3.8 | 1.4 | 3.1 |
| Asian or Pacific Islander | 3.0 | 0.8 | 1.5 | 0.6 | 1.5 | 3.7 | 1.0 | 1.2 | 1.0 | 2.6 |
| American Indian/Alaskan Native | 0.1 | 0.2 | 0.4 | 0.1 | 1.1 | 1.9 | 0.2 | 1.3 | 0.1 | 1.3 |
| Total % Minority | 35.7 | 14.4 | 6.9 | 10.7 | 22.7 | 11.9 | 17.9 | 12.1 | 17.6 | 16.3 |
| **% Special education enrollment 1993–94** | 11.5 | 11.3 | 10.8 | 10.0 | 9.4 | 9.3 | 11.5 | 11.2 | 10.6 | 9.8 |
| **Mean SAT Scores** | | | | | | | | | | |
| 1987 Verbal | 539 | 492 | 588 | 554 | 534 | 548 | 549 | 563 | 532 | 550 |
| Math | 540 | 487 | 586 | 538 | 533 | 549 | 538 | 562 | 521 | 551 |
| 1996 Verbal | 564 | 494 | 590 | 549 | 557 | 582 | 570 | 567 | 536 | 577 |
| Math | 575 | 494 | 600 | 544 | 565 | 593 | 569 | 568 | 535 | 586 |
| **% of graduates taking SAT—1996** | 14 | 57 | 5 | 12 | 11 | 9 | 9 | 9 | 24 | 8 |

**Per-pupil expenditures, average teacher salaries, funding percentages, and enrollment:** National Education Association, *Estimates of School Statistics, 1995–96*; revised 1990–91 figures from 1991–92 edition.

**Per-capita income:** U.S. Department of Commerce, Bureau of Economic Analysis

**Capital Outlay:** National Education Association, *Estimates of School Statistics, 1995–96*

**Pupil/teacher ratios, minority enrollment:** U.S. Department of Education, National Center for Education Statistics

**State Budget Allocations:** National Association of State Budget Officers

**Special Education Enrollment:** U.S. Department of Education

**SAT Scores and percentage of graduates taking test:** The College Board

## Southcentral

| Finances | Ala. | Ark. | Kan. | La. | Miss. | Okla. | Tenn. | Texas |
|---|---|---|---|---|---|---|---|---|
| **Per-pupil expenditures** | | | | | | | | |
| 1990–91 | $3,642 | $3,382 | $5,007 | $3,989 | $3,232 | $3,749 | $3,683 | $4,444 |
| 1995–96 | $4,479 | $4,370 | $5,900 | $4,844 | $4,190 | $4,458 | $4,717 | $5,593 |
| **Average teacher salaries** | | | | | | | | |
| 1990–91 | $26,862 | $23,611 | $29,767 | $26,240 | $24,366 | $24,457 | $28,248 | $27,658 |
| 1995–96 | $31,307 | $29,322 | $35,518 | $26,800 | $27,689 | $28,909 | $33,451 | $32,000 |
| **Funding percentages** | | | | | | | | |
| 1990–91 | | | | | | | | |
| Federal | 13.2 | 9.5 | 4.9 | 9.9 | 16.6 | 8.1 | 10.1 | 7.1 |
| State | 66.6 | 60.0 | 44.1 | 56.5 | 53.5 | 61.4 | 47.4 | 44.7 |
| Local | 20.2 | 30.6 | 51.0 | 33.7 | 29.8 | 30.5 | 42.4 | 48.3 |
| 1995–96 | | | | | | | | |
| Federal | 10.0 | 8.5 | 5.7 | 13.2 | 15.3 | 9.0 | 8.7 | 8.8 |
| State | 70.9 | 65.4 | 60.4 | 54.4 | 55.6 | 64.0 | 50.3 | 43.5 |
| Local | 19.1 | 26.1 | 33.8 | 32.5 | 29.1 | 27.0 | 40.9 | 47.7 |
| **Capital Outlay** | | | | | | | | |
| 1995–1996 (in thousands) | $450,032 | $227,658 | $183,441 | $166,836 | $212,616 | $366,000 | $50,213 | $2,090,689 |
| **State budget allocations for education (percent of total state budget)** | 22.8 | 21.6 | 24.4 | 18.8 | 24.3 | 26.5 | 19.1 | 28.1 |
| Per capita income 1995 | $18,781 | $17,429 | $21,825 | $18,827 | $16,531 | $18,152 | $20,376 | $20,654 |
| **Classroom Characteristics** | | | | | | | | |
| Students per multimedia computer | 44.0 | 46.9 | 16.6 | 88.8 | 62.7 | 36.6 | 38.0 | 31.0 |
| **Pupil/teacher ratios** | | | | | | | | |
| 1989–90 | 18.1 | 17.0 | 15.0 | 17.6 | 18.2 | 16.2 | 19.1 | 16.7 |
| 1994–95 | 17.2 | 17.1 | 15.1 | 16.6 | 17.5 | 15.5 | 18.6 | 15.7 |
| **Enrollment** | | | | | | | | |
| Fall, 1990–91 | | | | | | | | |
| Elementary | 409,571 | 243,490 | 291,803 | 593,162 | 316,032 | 338,000 | 600,185 | 2,003,343 |
| Secondary | 316,587 | 191,189 | 145,231 | 201,293 | 184,090 | 241,600 | 233,405 | 1,374,975 |
| Total | 726,158 | 434,679 | 437,034 | 794,455 | 500,122 | 579,600 | 833,590 | 3,378,318 |

| | | | | | | | |
|---|---|---|---|---|---|---|---|
| **Fall, 1995–96** | | | | | | | |
| Elementary | 408,952 | 243,821 | 314,741 | 560,762 | 324,397 | 347,500 | 637,286 | 2,465,393 |
| Secondary | 326,995 | 204,961 | 148,793 | 223,897 | 179,205 | 266,500 | 243,674 | 1,274,867 |
| Total | 735,947 | 448,782 | 463,534 | 784,659 | 503,602 | 614,000 | 880,960 | 3,740,260 |
| **% Minority enrollment** | | | | | | | | |
| **Fall, 1994** | | | | | | | | |
| Black | 35.8 | 23.9 | 8.4 | 45.7 | 50.9 | 10.4 | 23.0 | 14.3 |
| Hispanic | 0.4 | 1.1 | 5.7 | 1.1 | 0.3 | 3.7 | 0.6 | 36.1 |
| Asian or Pacific Islander | 0.6 | 0.7 | 1.9 | 1.3 | 0.5 | 1.2 | 0.9 | 2.3 |
| American Indian/Alaskan Native | 0.8 | 0.3 | 1.0 | 0.5 | 0.4 | 14.3 | 0.1 | 0.2 |
| Total % Minority | 37.6 | 26.0 | 17.0 | 48.6 | 52.1 | 29.6 | 24.6 | 52.9 |
| **% Special education enrollment** | | | | | | | | |
| 1993–94 | 11.7 | 9.8 | 9.2 | 9.0 | 11.0 | 10.6 | 11.9 | 9.8 |
| **Mean SAT Scores** | | | | | | | | |
| 1987 Verbal | 553 | 556 | 572 | 548 | 561 | 560 | 563 | 493 |
| Math | 535 | 540 | 562 | 530 | 540 | 539 | 543 | 486 |
| 1996 Verbal | 565 | 566 | 579 | 559 | 569 | 566 | 563 | 495 |
| Math | 558 | 550 | 571 | 550 | 557 | 557 | 552 | 500 |
| **% of graduates taking SAT—1996** | 8 | 6 | 9 | 9 | 4 | 8 | 14 | 48 |

**Per-pupil expenditures, average teacher salaries, funding percentages, and enrollment:** National Education Association, *Estimates of School Statistics, 1995–96*; revised 1990–91 figures from 1991–92 edition.
**Per-capita income:** U.S. Department of Commerce, Bureau of Economic Analysis
**Capital outlay:** National Education Association, *Estimates of School Statistics, 1995–96*
**Pupil/teacher ratios, minority enrollment:** U.S. Department of Education, National Center for Education Statistics
**State Budget Allocations:** National Association of State Budget Officers
**Special Education Enrollment:** U.S. Department of Education
**SAT Scores and percentage of graduates taking test:** The College Board

## Northwest

| Finances | Alaska | Idaho | Mont. | N.D. | Ore. | S.D. | Wash. | Wyo. |
|---|---|---|---|---|---|---|---|---|
| **Per-pupil expenditures** | | | | | | | | |
| 1990–91 | $7,887 | $3,263 | $5,048 | $3,678 | $5,281 | $4,078 | $5,001 | $5,574 |
| 1995–96 | $10,156 | $4,449 | $5,882 | $4,710 | $6,390 | $5,070 | $6,114 | $6,105 |
| **Average teacher salaries** | | | | | | | | |
| 1990–91 | $43,435 | $25,485 | $26,774 | $23,574 | $32,300 | $22,376 | $33,079 | $28,988 |
| 1995–96 | $49,620 | $30,891 | $29,364 | $26,969 | $39,650 | $26,346 | $38,025 | $31,571 |
| **Funding percentages** | | | | | | | | |
| 1990–91 | | | | | | | | |
| Federal | 12.6 | 6.5 | 9.2 | 7.4 | 6.0 | 11.2 | 5.4 | 5.6 |
| State | 63.6 | 61.7 | 46.2 | 48.6 | 26.7 | 26.8 | 73.8 | 50.8 |
| Local | 23.7 | 31.8 | 44.6 | 44.0 | 67.2 | 62.0 | 20.8 | 43.6 |
| 1995–96 | | | | | | | | |
| Federal | 12.6 | 7.7 | 10.0 | 12.5 | 7.1 | 10.1 | 6.3 | 6.5 |
| State | 63.6 | 61.2 | 48.3 | 43.3 | 56.5 | 25.8 | 69.4 | 49.0 |
| Local | 23.9 | 31.1 | 41.8 | 44.2 | 36.4 | 64.1 | 24.3 | 44.5 |
| **Capital Outlay** | | | | | | | | |
| 1995–1996 (in thousands) | $36,568 | $101,767 | $60,000 | $30,000 | $400,000 | $65,000 | $1,617,499 | $49,000 |
| **State budget allocations for education (percent of total state budget)** | 19.0 | 27.7 | 20.1 | 18.4 | 15.1 | 14.4 | 26.2 | 26.4 |
| Per-capita income 1995 | $24,182 | $19,264 | $18,482 | $18,663 | $21,736 | $19,506 | $23,639 | $21,321 |
| **Classroom characteristics** | | | | | | | | |
| Students per multimedia computer | 16.9 | 34.8 | 22.5 | 15.1 | 22.1 | 15.4 | 19.2 | 16.5 |
| **Pupil/teacher ratios** | | | | | | | | |
| 1989–90 | 16.8 | 20.1 | 15.7 | 15.1 | 18.4 | 15.5 | 20.1 | 14.5 |
| 1994–95 | 17.6 | 19.1 | 16.3 | 15.3 | 19.9 | 14.4 | 20.2 | 15.0 |
| **Enrollment** | | | | | | | | |
| Fall, 1990–91 | 83,584 | 125,037 | 111,126 | 84,252 | 321,500 | 94,287 | 488,346 | 56,227 |
| Elementary | 28,606 | 95,803 | 41,631 | 32,882 | 163,200 | 34,348 | 352,829 | 41,999 |
| Secondary | | | | | | | | |
| Total | 112,190 | 220,840 | 152,757 | 117,134 | 484,700 | 128,635 | 841,175 | 98,226 |

Vital Signs of American Schools   93

| | | | | | | | | |
|---|---|---|---|---|---|---|---|---|
| **Fall, 1995–96** | | | | | | | | |
| Elementary | 97,756 | 129,570 | 116,305 | 82,335 | 351,130 | 95,693 | 529,610 | 52,346 |
| Secondary | 34,623 | 113,527 | 49,232 | 36,755 | 175,930 | 40,711 | 428,359 | 47,513 |
| Total | 132,379 | 243,097 | 165,537 | 119,090 | 527,060 | 136,404 | 957,969 | 99,859 |
| **% Minority enrollment Fall, 1994** | | | | | | | | |
| Black | 4.8 | not | 0.5 | 0.8 | 2.5 | 0.8 | 4.6 | 1.0 |
| Hispanic | 2.6 | available | 1.4 | 0.8 | 6.3 | 0.7 | 7.4 | 6.1 |
| Asian or Pacific Islander | 4.1 | | 0.8 | 0.7 | 3.2 | 0.8 | 6.3 | 0.8 |
| American Indian/Alaskan Native | 23.8 | | 9.6 | 7.6 | 1.9 | 13.6 | 2.6 | 2.8 |
| Total % Minority | 35.3 | | 12.3 | 9.9 | 13.9 | 15.9 | 20.9 | 10.7 |
| **% Special education enrollment** | | | | | | | | |
| 1993–94 | 11.8 | 8.1 | 9.6 | 8.9 | 10.6 | 8.4 | 9.0 | 10.0 |
| **Mean SAT Scores** | | | | | | | | |
| 1987 Verbal | 521 | 548 | 555 | 583 | 521 | 587 | 532 | 557 |
| Math | 504 | 524 | 548 | 573 | 509 | 577 | 519 | 551 |
| 1996 Verbal | 521 | 543 | 546 | 596 | 523 | 574 | 519 | 544 |
| Math | 513 | 536 | 547 | 599 | 521 | 566 | 519 | 544 |
| % of graduates taking SAT—1996 | 47 | 15 | 21 | 5 | 50 | 5 | 47 | 11 |

**Per-pupil expenditures, average teacher salaries, funding percentages, and enrollment:** National Education Association, *Estimates of School Statistics, 1995–96*; revised 1990–91 figures from 1991–92 edition.

**Per-capita income:** U.S. Department of Commerce, Bureau of Economic Analysis

**Capital outlay:** National Education Association, *Estimates of U.S. School Statistics, 1995–96*

**Pupil/teacher ratios, minority enrollment:** U.S. Department of Education, National Center for Education Statistics

**State Budget Allocations:** National Association of State Budget Officers

**Special Education Enrollment:** U.S. Department of Education

**SAT Scores and percentage of graduates taking test:** The College Board

## Southwest & Total

| Finances | Ariz. | Calif. | Colo. | Hawaii | Nev. | N.M. | Utah | U.S. |
|---|---|---|---|---|---|---|---|---|
| Per-pupil expenditures | | | | | | | | |
| 1990–91 | $4,231 | $4,644 | $5,058 | $5,229 | $4,612 | $4,446 | $2,983 | $5,217 |
| 1995–96 | $4,332 | $4,934 | $5,484 | $6,282 | $5,259 | $5,654 | $3,908 | $6,098 |
| Average teacher salaries | | | | | | | | |
| 1990–91 | $30,773 | $39,598 | $31,819 | $32,541 | $32,209 | $25,754 | $25,578 | $32,977 |
| 1995–96 | $32,484 | $42,516 | $35,364 | $35,807 | $36,167 | $29,349 | $30,452 | $37,846 |
| Funding Percentages | | | | | | | | |
| 1990–91 | | | | | | | | |
| Federal | 5.0 | 7.0 | 4.6 | 8.6 | 3.8 | 11.9 | 6.3 | 6.2 |
| State | 43.2 | 66.9 | 39.4 | 91.3 | 41.6 | 76.2 | 57.0 | 48.9 |
| Local | 51.7 | 26.1 | 56.0 | 0.1 | 54.7 | 11.9 | 36.6 | 44.9 |
| 1995–96 | | | | | | | | |
| Federal | 8.7 | 8.7 | 5.8 | 8.4 | 4.6 | 10.7 | 6.4 | 7.0 |
| State | 42.0 | 57.1 | 45.5 | 89.5 | 33.2 | 74.3 | 58.4 | 47.7 |
| Local | 49.3 | 34.2 | 48.7 | 2.0 | 62.2 | 14.9 | 35.2 | 45.3 |
| Capital Outlay | | | | | | | | |
| 1995–1996 (in thousands) | $810,865 | $1,742,656 | $403,284 | $116,503 | $330,260 | $424,257 | $288,554 | $22,744,244 |
| State budget allocations for education | | | | | | | | |
| (Percent of total state budget) | 21.1 | 20.1 | 21.3 | 14.6 | 12.4 | 21.8 | 33.8 | 20.9 |
| Per-capita income 1995 | $20,421 | $23,699 | $23,449 | $24,738 | $25,013 | $18,055 | $18,223 | $22,788 |
| Classroom characteristics | | | | | | | | |
| Students per multimedia computer | 24.3 | 52.0 | 16.3 | 74.4 | 24.6 | 22.5 | 20.6 | 35.2 |
| Pupil/teacher ratios | | | | | | | | |
| 1989–90 | 18.9 | 22.4 | 17.6 | 19.1 | 20.4 | 18.3 | 24.9 | 17.2 |
| 1994–95 | 19.3 | 24.0 | 18.4 | 17.9 | 18.7 | 17.2 | 24.3 | 17.3 |
| Enrollment | | | | | | | | |
| Fall, 1990–91 | | | | | | | | |
| Elementary | 470,406 | 3,604,382 | 336,166 | 100,071 | 119,980 | 164,707 | 322,736 | 27,037,316 |
| Secondary | 162,846 | 1,346,092 | 238,047 | 70,985 | 81,330 | 118,396 | 121,996 | 14,159,180 |
| Total | 633,252 | 4,950,474 | 574,213 | 171,056 | 201,310 | 283,103 | 444,732 | 41,196,496 |

| | | | | | | | |
|---|---|---|---|---|---|---|---|
| **Fall, 1995–96** | | | | | | | |
| Elementary | 560,436 | 3,972,170 | 374,656 | 107,254 | 155,157 | 171,079 | 323,355 | 29,175,887 |
| Secondary | 201,660 | 1,486,357 | 281,623 | 79,327 | 109,884 | 132,747 | 150,311 | 15,445,889 |
| Total | 762,096 | 5,458,527 | 656,279 | 186,581 | 265,041 | 303,826 | 473,666 | 44,621,776 |
| **% Minority enrollment Fall, 1994** | | | | | | | | |
| Black | 4.3 | 8.7 | 5.4 | 2.7 | 9.3 | 2.4 | 0.7 | 16.7 |
| Hispanic | 28.7 | 37.9 | 17.6 | 4.9 | 15.5 | 46.4 | 4.8 | 13.0 |
| Asian or Pacific Islander | 1.7 | 11.2 | 2.5 | 68.8 | 4.2 | 1.0 | 2.1 | 3.6 |
| American Indian/Alaskan Native | 7.0 | 0.9 | 1.0 | 0.4 | 2.0 | 10.4 | 1.4 | 1.1 |
| Total % Minority | 41.7 | 58.7 | 26.5 | 76.8 | 31.0 | 60.2 | 9.0 | 34.4 |
| **% Special education enrollment** | | | | | | | | |
| 1993–94 | 8.8 | 8.7 | 9.1 | 7.2 | 9.0 | 11.9 | 9.6 | 10.3 |
| **Mean SAT Scores** | | | | | | | | |
| 1987 Verbal | 539 | 500 | 542 | 481 | 516 | 559 | 577 | 507 |
| Math | 526 | 507 | 535 | 502 | 508 | 544 | 557 | 501 |
| 1996 Verbal | 525 | 494 | 536 | 485 | 508 | 554 | 583 | 505 |
| Math | 521 | 511 | 538 | 510 | 507 | 548 | 575 | 508 |
| **% of graduates taking SAT—1996** | 28 | 45 | 30 | 54 | 31 | 12 | 4 | 41 |

**Per-pupil expenditures, average teacher salaries, funding percentages, and enrollment:** National Education Association, *Estimates of School Statistics, 1995–96*; revised 1990–91 figures from 1991–92 edition.

**Per-capita income:** U.S. Department of Commerce, Bureau of Economic Analysis

**Capital outlay:** National Education Association, *Estimates of School Statistics, 1995–96*

**Pupil/teacher ratios, minority enrollment:** U.S. Department of Education, National Center for Education Statistics

**State Budget Allocation:** Association of State Budget Officers

**Special Education Enrollment:** U.S. Department of Education

**SAT Scores and percentage of graduates taking test:** The College Board

# Section III

## *Addressing Developmental, Sexual, Cultural, Cognitive, and Ableness Diversity*

# Meeting the Developmental and Instructional Needs of Middle School Students: What Do Teachers Need to Know?

M. LEE MANNING

*Old Dominion University*

Middle schools provide 10–14 year olds, commonly known as young adolescents, with developmentally appropriate educational experiences that emphasize learners' education and overall well-being. Teachers, counselors, administrators, and parents collaboratively work to address young adolescents' developmental needs and to ensure some degree of success for all learners. Program content, curricular materials, instructional methods, and the overall school climate demonstrate an understanding of the early adolescence developmental period and a commitment to the education of young adolescents. Educators recognize and address young adolescents' developmental diversity as well as cultural and gender differences. Young adolescents know educators value academic achievement, especially when instructional and educational experiences reflect physical, psychosocial, and cognitive development.

## The Middle School

For many years, the school in the middle, regardless of whether called intermediate school, junior high school, or middle school, did not fully understand its purposes. The K–5 school perceived its mission as teaching basic skills, while the secondary school perceived its mission as providing general, academic, or vocational education; however, the school in the middle lacked a "mission"—it was a school without a clear sense of purpose and accompanying direction. This feeling of lack of purpose has changed. Table 1 shows several selected publications that provide effective middle schools with a clear sense of mission.

The middle school can be defined as a school organization containing grades 6–8 (and sometimes grade 5) that: (1) provides developmentally appropriate and responsive curricular, instructional, organizational, guidance, and overall educational experiences; and (2) places major emphasis on 10–14 year olds' developmental and instructional needs. Serving a far greater role than just being a "transition school" between the elementary school and the high school, purposes of the middle school include:

---

M. Lee Manning teaches in the Darden College of Education, Old Dominion University, Norfolk, VA.

- providing unique educational experiences that reflect the developmental and instructional needs of 10–14 year olds;
- meeting young adolescents' educational needs by implementing proven middle school concepts such as advisor-advisee programs, exploratory programs, interdisciplinary teaming and organization, and positive school climates;
- continuing to refine young adolescents' basic skills originally learned in the elementary school; and
- offering opportunities for young adolescents to explore curricular areas as well as discover unique abilities and talents.

The National Middle School Association's (1995) position paper for developmentally responsive middle schools calls for several broad goals:

*Goal 1:* Educators committed to young adolescents—educators who have professional preparation to work with and who genuinely want to work with young adolescents.

*Goal 2:* A shared vision of all stakeholders—young adolescents, teachers, administrators, families, board of education members, and community members.

*Goal 3:* High expectations for all young adolescents—educators' gestures, casual remarks, and overall attitudes convey high expectations, both for academic achievement and behavior.

*Goal 4:* An adult advocate for every student—each young adolescent has one adult that knows and cares for her or him and who supports her or his academic and personal development.

## Table 1

### Selected Publications on Effective Middle Schools
### (listed alphabetically)

*Achieving Excellence Through the Middle Level Curriculum* (National Association for Secondary School Principals, 1993)

*An Agenda for Excellence at the Middle Level* (National Association for Secondary School Principals, 1985)

*Caught in the Middle* (California State Department of Education, 1987)

*Developmentally Appropriate Middle Level Schools* (Association for Childhood Education International) (Manning, 1993)

*The Forgotten Years: PRIME* (Florida State Department of Education, 1984)

*Great Transitions: Prepaying Adolescents for a New Century* (Carnegie Council on Adolescent Development, 1996)

*Restructuring Education in the Middle School Grades* (Virginia State Department of Education, 1989)

*This We Believe: Developmentally Responsive Middle Level Schools* (National Middle School Association, 1995)

*Turning Points: Preparing American Youth for the 21st Century* (Carnegie Council on Adolescent Development, 1989)

*What Matters in the Middle Grades* (Maryland State Department of Education, 1989)

*Goal 5:* Family and community partnerships—middle schools recognize and support families and community members by encouraging their roles in supporting learning.
*Goal 6:* A positive school climate—middle schools provide a safe, inviting, and caring school climate, one that promotes a sense of community and encourages learning.

Another seminal document, *Turning Points* (Carnegie Council on Adolescent Development, 1989), calls for similar goals: (1) creating a community of learners; (2) teaching a core of common knowledge; (3) ensuring success for all learners; (4) empowering teachers and administrators; (5) preparing teachers for the middle grades; (6) improving academic performance through better health and fitness; (7) reengaging families in the education of young adolescents; and (8) connecting schools with communities.

## Middle School Learners' Developmental Characteristics

Table 2 provides an overview of young adolescents' physical, psychosocial, and cognitive developmental characteristics.

### Table 2

### *Selected Developmental Characteristics*

Young adolescents experience physical development such as:

1. experiencing a growth spurt which marks a rapid increase in body size and readily obvious skeletal and structural changes;
2. experiencing the onset of puberty, a period of physiological changes that include the development of the sexual reproductive system;
3. experiencing considerable diversity in the rate of development, although the sequence of development remains the same.

Young adolescents experience psychosocial development such as:

1. seeking of friendships and wider social dealings;
2. shifting their allegiance and affiliation from teachers and parents to peers who become the prime source for standards and behavior;
3. becoming preoccupied with themselves during this developmental period and compare themselves physically and socially with peers;
4. seeking freedom and independence from adult authority;
5. experiencing changing self-esteems which might vary from situation to situation.

Young adolescents experience cognitive development such as:

1. beginning to develop from Piaget's concrete operations stage to the formal operations stage;
2. developing the ability to think hypothetically, reflectively, and abstractly;
3. developing the ability to make reasoned moral and ethical choices;
4. developing personal attitudes and perspectives toward other people and institutions;
5. becoming capable of engaging in social analysis during which they make judgments about people and institutions;
6. developing cognitive skills which allow them to solve real-life problems.

(Compiled from: Manning, 1993; Manning, 1994/1995).

Addressing Developmental, Sexual, Cultural, Cognitive, and Ableness Diversity 101

The three selected physical developmental characteristics can have profound effects on young adolescents and the issues affecting their lives. Young adolescents' tremendous physical diversity (e.g., a physically small fourteen year old and large 11 or 12 year old) can affect self-esteem and can result in young adolescents worrying about when growth will begin or end. Young adolescents often experience restlessness and fatigue due to growing bones, joints, and muscles. They might find sitting for long periods of time more difficult and perhaps even painful when forced to sit in small desks. Exercise should be developmentally appropriate and physical competition between early and late maturers should be avoided. Young adolescent females and males develop differently, with females developing at a faster rate. Females might feel self-conscious due to their faster growth, i.e., their height and weight. Middle school educators need to understand gender differences in physical development and recognize how these differences might affect females' psychosocial development. Additionally, young adolescents' attainment of development of puberty sometimes results in a sense of sexual awareness which can have dangerous consequences if sexual experimentation occurs. Developmentally appropriate topics can be discussed in health and family life classes as well as advisor-advisee programs and exploratory programs.

The selected psychosocial developmental characteristics identified in Table 2 can affect young adolescents in a number of ways, especially as one considers the issues that relate to development (and vice-versa). Young adolescents' development into puberty and the accompanying physical changes can affect their emotional and social development. Through health classes and advisement (both large class and small group), middle school educators need to convey to young adolescents that development in one area does not imply comparable development in other areas.

Young adolescents need educational experiences that address their need to make friends and maintain friendships. Middle school educators can plan small group (e.g., cooperative learning) sessions that allow friends to study and work together. Advisory and exploratory sessions can include topics such as ways to make friends; how to select "good" friends; and how to make cross-gender and cross-cultural friendships.

Young adolescents need educators' and parents' support even though shifts in allegiance and affiliation suggest a desire to move away from adults. Still, both educators and parents need to show support and caring attitudes toward young adolescents. Understanding young adolescents' motives and perspectives during this shifting process can actually contribute to positive relationships between younger and older generations. Adults acting in a negative manner or punishing young people for their shifting affiliation can have harmful consequences that can worsen relationships. Additionally, young adolescents need to learn that their quest for independence and freedom requires responsibility for actions and also that such quests should not include participating in risky adventures and behaviors. Educators need to teach decision-making skills, so young adolescents will be equipped with the ability to make informed decisions. Having the knowledge, however, often does not suffice. Young adolescents tend to feel immortal and often make poor decisions.

Young adolescents need to understand that most people experience preoccupations with their appearances and behavior during or around the early adolescence developmental period. Conveying that developmental differences should be expected, educators can address learners' preoccupations in advisor-advisee sessions, exploratory programs, and health classes.

Young adolescents' changing self-esteems require middle school educators to recognize how self-esteem dips and to take appropriate action. Changing from the elementary school to the middle school, developing bodies, making new friendships, and tackling more difficult subject matter can have negative effects on self-esteems. Middle school educators face a two-fold challenge. First, they need to teach young adolescents to make accurate assessments of their self-esteems and, second, educators need to provide educational experiences that contribute to positive self-esteems.

Finally, young adolescents need educational experiences designed to address peer pressure, especially pressure to engage in illegal activities and risky behaviors. Two factors about peer pressure need to be remembered. First, one should not assume that all young adolescents succumb to negative peer pressure and, second, peer pressure can also be positive. In fact, sometimes peers exert pressure to eat the right foods, avoid abusive substances, and behave appropriately. Still, the number of young adolescents engaging in risky behaviors (Carnegie Council on Adolescent Development, 1989; 1996) suggests educators need to teach and counsel about negative peer pressure and its consequences.

Several implications can be gleaned from the published findings on cognitive development during early adolescence. Middle school educators can:

1. determine cognitive readiness levels by judging students' thought processes and complexity of thought (Brooks, Fusco, & Glennon, 1983) and by using the Arlin Test of Formal Reasoning to determine performance levels (Toepfer, 1985).
2. plan organizational strategies such as continuous progress educational experiences which allow students to progress according to their own levels and rates, learning styles, and cognitive developmental characteristics (National Middle School Association, 1995).
3. recognize art, music, health, and physical education as powerful sources of academic growth and as means to enhance conceptualizations and understandings of other academic areas (National Association of Secondary School Principals, 1989)
4. provide educational experiences which challenge young adolescents to think and excel academically while carefully avoiding frustrating learners and lowering their self-esteems.
5. address increasing cognitive capabilities by implementing integrated curricular designs and interdisciplinary approaches for teaching broad concepts and relationships between subject areas.
6. implement exploratory programs which: (a) provide young adolescents with considerable flexibility and freedom in selecting curricular content and instructional practices and (b) address intellectual curiosity, rapidly changing interests, and diverse cognitive levels.
7. address young adolescents' cognitive development in a variety of meaningful and accepted ways, including: multiple intelligences (Gardner, 1993); proper academic counseling (California State Department of Education, 1987); and appropriate left brain/right brain educational experiences (Sperry, 1974).
8. avoid concluding that all young adolescents around the ages of 11 or 12 have developed from the concrete operations stage to the formal operations stage.

In their efforts to provide developmentally responsive educational experiences, middle school educators recognize and address young adolescents' cultural and gender differences and place emphasis on helping students to develop positive and healthy cultural and gender identities. Cultural and gender differences in areas such as motivation, school success, and competition need objective consideration. Being particularly careful to avoid stereotyping and over-generalizing, perceptive educators provide instructional and other educational experiences that reflect cultural and gender characteristics.

## Middle School Curriculum

The curriculum in effective middle schools reflects the interests, concerns, and thinking levels of young adolescents. More than a time to review elementary content or preview secondary content, responsive middle schools base program content on young adolescents' physical, psychosocial, and cognitive levels (Manning, 1993; Manning, 1994/1995) as well as on their need to achieve, to experience success, and to have continuous learning experiences. Middle school educators need to consider content that students learned in the elementary school and what they will learn in the secondary school. Then, they need thoughtfully to determine developmentally responsive middle school content.

Four documents (i.e., Allen, Splittgerber, & Manning, 1993; Beane, 1993; Carnegie Council on Adolescent Development, 1989; National Middle School Association, 1995) provide essential directions for educators wanting to design and implement a developmentally responsive middle school curriculum. Interestingly, while all four documents offer similar and helpful insights, they differ in some ways. Middle school educators can look to all four and draw the best from each document, thus designing a curriculum that meets the needs of young adolescents.

An appropriate middle school curriculum balances academic, personal, and social contexts that match the needs of young adolescents. The "academic" curricular context consists of mathematics, science, English, social studies, foreign languages, health, physical education, unified arts, technology education, and home economics. These curricular areas involve and challenge young adolescents in specific subject areas while strengthening the basic skills; they also broaden knowledge, social skills, and personal experiences.

The "personal" context integrates content from subject matter areas, personal experiences and interests, career education, and exploratory activities to assist young adolescents in constructing and creating knowledge, values, beliefs, and attitudes. The emphasis is on self-awareness through personal experiences. The personal curriculum integrates content, thinking, feelings, and actions of young adolescents; recognizes the importance of self; and develops the whole person within the school and society.

The "social" curriculum focus is on developing socially competent individuals. This context emphasizes citizenship and intergroup relations. The middle level curriculum provides needed opportunities for socialization through interaction with peers, teachers, and adults outside the school in classes and extracurricular and community activities. The social curriculum also includes the recognition and appreciation of cultural diversity in young adolescents as an essential part of a democratic society (Allen, Splittgerber, & Manning, 1993).

James Beane (1993) has been an influential advocate for a middle school curriculum that addresses young adolescents' needs and interests. Beane (1993) proposes that the middle school curriculum be based on a general education that focuses on the personal and social concerns (e.g., common needs, problems, and interests) of young adolescents.

The "personal" concerns that Beane (1993) suggests are experienced by young adolescents include: (1) understanding and coping with physical, intellectual, and socio-emotional changes; (2) developing a personal identity and a wholesome self-concept; (3) exploring values, moral, and ethical questions in social contexts; (4) securing and finding a level of status in the peer group; (5) developing a personality balanced between independence from adult authority and dependence on adults for security; (6) coping with commercial interests related to fashion, music, and leisure activities; (7) negotiating the expectations in home, school, community, and peer group; and (8) developing commitments to people in order to obtain a feeling of self-worth, achievement, and efficacy (Beane, 1993).

The "social" issues that Beane recommends for the curriculum framework include: (1) interdependence among peoples in the global society of today; (2) students' diversity (e.g., cultures formed by race, ethnicity, gender, and geographical region); (3) environmental problems; (4) political processes and organizations; (5) economic problems; (6) the significance of technology, and (7) the increasing incidence of self-destructive behaviors (Beane, 1993).

A curriculum designed to meet these personal and social needs includes reflective thinking, critical ethics, problem solving, valuing, self-concepting and self-esteeming, social action skills, and searching for completeness and meaning. Three essential curricular contexts include the idea of democracy, the concept of human dignity, and opportunities to explore and appreciate the workings and values of diverse cultures (Beane, 1993).

*Turning Points* (Carnegie Council on Adolescent Development, 1989) recommends a common core of knowledge for middle school students. *Turning Points'* four recommendations for a core curriculum include teaching students to: (1) think critically through mastery of an appropriate body of knowledge; (2) lead a healthy life; (3) behave ethically and lawfully; and (4) assume the responsibilities of citizenship in a pluralistic society. Young adolescents can be taught to think critically when educators use learning techniques that allow students to participate actively in discovering and creating solutions to problems (e.g., service learning activities). Teaching curricular areas through the integration of themes can help students to see relationships rather than disconnected facts. Teaching young adolescents to lead a healthy life style can be addressed in the health or science curriculum. Just telling students about self-destructive or dangerous behaviors probably will not suffice; educators will need to include training in coping skills such as collaboration, problem solving, and conflict resolution. Behaving ethically and lawfully, and assuming the responsibilities of citizenship in a pluralistic society can include youth service that teaches values for citizenship, compassion, regard for human worth and dignity, and tolerance and appreciation of diversity.

*This We Believe* (National Middle School Association, 1995) calls for a challenging, integrative, and exploratory middle school curriculum. The "challenging" facets include curricular experiences that engage young adolescents, emphasize important ideas and skills, provide relevant experiences, and emphasize developmental responsiveness. The "integrative" dimensions help young adolescents make sense of life experiences, focus on the integration of issues relevant to learners and are taught by individuals and teams through integrated courses and units. Through the "exploratory" components of the middle school curriculum, teachers provide students exploratory curricular experiences, opportunities to discover their interests and skills, and the latitude for young adolescents to become acquainted with healthy leisure pursuits.

Regardless of the curricular model(s) selected, a middle school curriculum that centers on the child includes interdisciplinary or integrated curriculum. This curricular approach combines subject matter that is usually taught separately (e.g., literature, history, or science) under a single integrated curricular theme. Rather than learning only isolated facts, young adolescents see relationships and interconnectedness among the different curricular areas.

Other selected curricular essentials include efforts to: increase young adolescents' self-concept; provide appropriate responses to cultural and gender diversity; demonstrate an understanding of young adolescents' physical, psychosocial, and cognitive development; and provide a balance between skills, academic content, and experiences.

Effective middle schools also provide an exploratory curriculum. Commonly called exploratory programs or mini-courses, this essential middle school concept provides a developmentally appropriate response to the young adolescent's shorter attention span, fluctuating motivational levels, and declining self-esteem. Exploratory programs also build interest, i.e., young adolescents learn what they might like and want to pursue. While exploratory programs vary with schools and young adolescents' needs, the length of time can be for six or eight weeks or a semester for 40–50 minutes a day. Typical examples of exploratory courses include business, keyboarding, choirs, homemaking, print making, drama, foreign languages, arts and crafts, independent study opportunities, dance, music, and nearly any other area that a young adolescent might want to explore.

## Middle School Instruction, Organization, and Climate

Middle school instruction, organization, and climate work both individually and in unison to address young adolescent development and learning needs. Educators placing emphasis on one area (e.g., instruction) might have little effect unless complementary efforts (e.g., school climate) also receive attention. Instructional decisions, organization patterns, and school climates should all reflect young adolescent development as well as educational practices suggested by research to be effective.

## Middle School Instruction

The instruction, organization, and learning climates in effective middle schools:

- recognize and accept differences in young adolescents' physical, psychosocial, and cognitive developmental patterns and rates by setting developmentally appropriate curriculum goals;
- place emphasis on thinking and learning how to learn rather than focusing only on isolated skills and content;
- view guidance, both counselors and teacher-advisors, as essential components of middle school education;
- place value on gender and cultural differences and provide classroom organization and instructional approaches that recognize these differences;
- provide curricular materials that enhance young adolescents' acceptance of self and others and enable them to accept differences and similarities among people;
- promote integrated curricular approaches, so young adolescents will perceive relationships among and between curricular areas;
- allow young adolescents to make significant choices and decisions about grouping, organization, curricular, and management practices;
- ensure some degree of success for all young adolescents in all aspects of the school program;
- recognize the importance of self-esteem and its influence on academic achievement, socialization, and overall personal development; and
- promote heterogeneous grouping and seek other alternatives to homogeneous ability grouping and tracking.

## Middle School Interdisciplinary Team Organization

"Interdisciplinary team organization" may be defined as an organization pattern of two or more teachers representing different subject areas and sharing the same students, schedule, and adjoining areas of the school. Team organization is a more fundamental structural change than the team teaching that was popular in the 1960s and early 1970s (Erb, 1997). McEwin (1997) contends that interdisciplinary team organization is now widely recognized as an essential component of developmentally responsive schools for young adolescents.

The terms "interdisciplinary team organization" and "interdisciplinary team teaching" have been used interchangeably in the literature. Sometimes, educators confuse interdisciplinary team organization with team teaching. The term "interdisciplinary team teaching" involves a team of two or more subject teachers who share students and planning time and who work to draw connections between their subjects. While these teachers might sometimes teach together, it is not a requirement for interdisciplinary team teaching. The real distinction between team teaching and interdisciplinary team teaching is a curricular one; that is, a team of teachers become an interdisciplinary team when its members engage in purposeful efforts to integrate learning from normally disparate disciplines (Wraga, 1997). In summary, on interdisciplinary teams, teachers plan together and share the same students and schedule; they have an opportunity to learn from one another about teaching and young adolescents; and they might or might not teach in the same classroom. In fact, in most middle schools today, teachers do not teach in the same classroom.

Interdisciplinary teaming provides a more effective means of meeting developmentally responsive needs and individual interests of young adolescents. It minimizes the number of young adolescents who feel unknown, that teachers do not know their progress in other classes, or that other students do not know them well enough to accept them as friends. Interdisciplinary teaming helps students build team spirit and improves attitudes and work habits because of the closer, more coherent supervision and caring that occurs on a team.

There are a few essentials to effective interdisciplinary organization and teams. First, Erb (1987) identifies four necessary features of effective team organizations at the middle school level: (1) common planning or meeting time; (2) shared students; (3) common block-time schedule; and (4) spatial proximity of team members' classes. Erb maintains that teachers sharing common planning times and students are two absolute necessities for teams to function.

Research and scholarly opinion (Erb, 1997; Dickinson & Erb; 1997; Merenbloom, 1991) have developed to a point where characteristics of effective teams can be identified with considerable accuracy. Selected characteristics include: (1) a balance in teachers' expertise, age, sex, and race; (2) team leaders having specific responsibilities; (3) an established team decisionmaking process (e.g., goals, grouping, scheduling, homework, and discipline); (4) agreed-upon procedures to assess students' strengths and weaknesses; (5) development of a team identity; (6) flexibility in student and master schedules; (7) support of school and district administration for the teaming concept and team efforts; (8) sufficient time for team planning; (9) adequate staff development; and (10) team members being proficient in human relations skills.

## Middle School Climates

Positive middle school climates provide opportunities for students to interact, to find meaning in schoolwork and relationships, and to feel a sense of recognition. To create such a climate, schools can provide ongoing opportunities for healthy interaction between school professionals and young adolescents. Smaller classes, communities of learners, and advisory programs are just a few ways educators can provide increased positive interactions and opportunities for young adolescents to feel recognized and accepted. Schools can hold young adolescents accountable for their behavior and learning. Holding students accountable for their behavior shows recognition. To ignore a misbehavior or fail to make a student complete an assignment or participate in class activities suggests teachers hold low expectations. Additionally, positive middle school climates recognize each young adolescent as an individual and provide opportunities for students to be noticed and connected to others in trusting relationships.

Healthy school climates also include positive verbal climates which have been reported as having particular significance in the overall development of learners. Negative verbal climates cause learners to feel unworthy, incompetent, or insignificant. Common negative behaviors may include showing little or no interest in learners, speaking discourteously, or dominating verbal exchanges that occur daily. Verbal comments in positive climates are aimed at satisfying learners' psychological needs and by making them feel valued. In speaking situations with young adolescents, adults should focus not only on the content but also on the affective impact of their words. To maximize healthy climates, teachers should use appropriate words to show care and affection. They should speak courteously, listen attentively, plan spontaneous opportunities to talk with each learner, and avoid making judgmental comments about learners (Kostelnik, Stein, Whiren, & Soderman, 1988).

## Middle School Guidance Programs

Effective middle schools provide effective guidance programs that are specifically designed to meet young adolescents' developmental needs. These guidance efforts at the middle level cannot be a slightly revised elementary or secondary program. Rather than adopting a "one guidance program fits all" philosophy, middle school guidance programs plan and implement programs to address ten to fourteen year olds' needs.

Classroom teachers play major roles in middle school guidance efforts. Rather than guidance being only one hour a week or students requesting appointments, teachers serve integral guidance and advisory functions all during the school day (Cole, 1992; MacLaury, 1995). Such an assertion does not downplay the role of guidance counselors; however, middle school teachers (such as in the advisor-advisee program discussed later) play major guidance roles in both planned advisory programs as well as in their daily interaction with young adolescents.

The advisor-advisee program (also called advisories, teacher advisories, or homebased guidance) can be defined as planned efforts in which each student has the opportunity to participate in a small interactive group with peers and staff to discuss school, personal, and societal concerns. The advisory program helps each student develop a meaningful relationship with at least one significant adult in the middle school (Allen, Splittgerber, & Manning, 1993). Advisories seek to promote young adolescents' social, emotional, and moral growth while providing personal and academic guidance. To reduce the student-teacher ratio in advisories, all faculty serve as advisors. The most successful advisories occur at the beginning of the day, lasting at least 25 minutes (Arnold, 1991). Advisors serve advisees as friends, advocates, guides, group leaders, community builders, liaisons with parents, and evaluation coordinators. They also provide a warm, caring climate; plan and implement advisory programs; assist advisees in monitoring academic progress; provide times for students to share concerns; refer advisees to appropriate resources; communicate with parents and families; maintain appropriate records; and encourage advisee's cognitive and psychosocial growth (James, 1986).

The teacher plans advisory sessions, preferably with the help of other team members and, if needed, the guidance counselor. Topics for advisor-advisee sessions include: peer pressure; substance abuse; friendships; health-related issues; career exploration; development (e.g., early and late maturers); school rules; understanding parents; contemporary issues; and leisure time activities. Other activities during advisor-advisee sessions include: meeting with individual students about problems; offering career information and guidance; discussing academic, personal, and family problems; addressing moral or ethical issues; discussing multicultural and intergroup relations; and helping students develop self-confidence and leadership skills (Epstein & Mac Iver, 1990).

## Summary

Events during the past decade or so have placed major emphasis on understanding the early adolescence developmental period and on providing effective middle school practices. The contemporary impetus to implement middle school practices that reflect developmental and instructional needs can be seen in many forms—the increasing number of books and reports, the position papers of professional associations, and the efforts of state departments of education. Three phenomena—the increasing acceptance of middle schools, the increasing knowledge of young adolescents and their developmental needs, and the increasing recognition that middle school teaching methods need to be developmentally responsive—suggest middle school educators should take deliberate initiatives to implement curricular, organizational, teaching, environmental, and guidance practices that reflect the developmental and instructional needs of young adolescents.

# References

Allen, H. A., Splittgerber, F. L., & Manning, M. L. (1993). *Teaching and learning in the middle level school*. Columbus, OH: Merrill.

Arnold, J. (1991). The revolution in middle school organization. *Momentum*, 22(2), 20–25.

Beane, J. A., (1993). *A middle school curriculum: From rhetoric to reality* (2nd ed.). Columbus, OH: National Middle School Association.

Brooks, M., Fusco, E., & Glennon, J. (1983). Cognitive levels matching. *Educational Leadership*, 40(8), 4–8.

California State Department of Education. (1987). *Caught in the middle*. Sacramento, CA: Author.

Carnegie Council on Adolescent Development. (1989). *Turning points: Preparing American youth for the 21st century*. New York: Author.

Carnegie Council on Adolescent Development. (1996). *Great transitions: Preparing adolescents for a new century*. New York: Author.

Cole, C. (1992). *Nurturing a teacher advisory program*. Columbus, OH: National Middle School Association.

Dickinson, T. S., & Erb, T. O. (1997). *We gain more than we give: Teaming in middle schools*. Columbus, OH: National Middle School Association.

Epstein, J., & MacIver, D. (1990). *Education in the middle grades: An overview of national practices and trends*. Columbus, OH: National Middle School Association.

Erb, T. O. (1997). Meeting the needs of young adolescents on interdisciplinary teams. *Childhood Education*, 73(5), 309–311.

Erb, T. O. (1987). What team organization can do for teachers. *Middle School Journal*, 18, 3–6.

Florida State Department of Education. (1984). *The forgotten years: PRIME*. Tallahassee, FL: Author.

Gardner, H. (1993). *Multiple intelligences: The theory in practice*. New York: Basic Books.

James, M. (1986). *Advisor-advisee programs: Why, what, and how*. Columbus, OH: National Middle School Association.

Kostelnik, M. J., Stein, L. C., Whiren, A. P., & Soderman, A. K. (1988). *Guiding children's social development*. Cincinnati, OH: Brooks/Cole.

MacLaury, S. (1995). Establishing an urban advisory program throughout a community school district. *Middle School Journal*, 27(1), 42–29.

Manning, M. L. (1993). *Developmentally appropriate middle level schools*. Olney, MD: Association for Childhood Education International.

Manning, M. L. (1994/1995). Addressing young adolescents' cognitive development. *The High School Journal*, 78, 98–104.

Maryland State Department of Education. (1989). *What matters in the middle grades*. Annapolis, MD: Author.

McEwin, C. K. (1997). Trends in the utilization of interdisciplinary team organization in middle schools. In T. S. Dickinson, & T. O. Erb (Eds.), *We gain more than we give: Teaming in the middle school* (pp. 313–324). Columbus, OH: National Middle School Association.

Merenbloom, E. (1991). *The team process: A handbook for teachers*. Columbus, OH: National Middle School Association.

National Association of Secondary School Principals. (1985). *An agenda for excellence at the middle level*. Reston, VA: Author.

National Association of Secondary School Principals. (1989). *Middle level education's responsibility for intellectual development*. Reston, VA: Author.

National Association of Secondary School Principals. (1993). *Achieving excellence through the middle level curriculum*. Reston, VA: Author.

National Middle School Association. (1995). *This we believe: Developmentally responsive middle level schools*. Columbus, OH: Author.

Sperry, R. W. (1974). Lateral specialization in the surgically specialized hemispheres. In F. O. Schmitt & F. G. Warden (Eds.), *The neuro-sciences third study program* (pp. 5–19). Cambridge, MA: MIT Press.

Toepfer, C. F. (1985). Suggestions for neurological data for middle level education: A review of the research and its implications. *Transescence: The Journal of Emerging Adolescence*, 13(2), 12–38.

Virginia State Department of Education. (1989). *Restructuring education in the middle school grades*. Richmond, VA: Author.

Wraga, W. G. (1997). Interdisciplinary team teaching: Sampling the literature. In T. S. Dickinson, & T. O. Erb (Eds.), *We gain more than we give: Teaming in the middle school* (pp. 325–344). Columbus, OH: National Middle School Association.

# Creating Safe and Open Schools for Gay, Lesbian, and Questioning Youth: A Rationale and Practical Strategies

HOWARD E. TAYLOR

*Old Dominion University*

> Young men and women struggling with their sexual orientation during a time of intense physical, social, and developmental change are failed by physicians, educators, mental health professionals, and clergy who breach their ethical and professional obligations by being uninformed and unresponsive to the special problems and needs of these youth.
>
> Karen M. Harbeck, Ph.D., J. D.

## Introduction

"What do we want? Gay Rights! When do we want them? Yesterday!" "All right already!" you might exclaim. "How many more special interest groups are teachers going to have to accommodate?" I don't have that answer. But, I'd like to introduce you to a couple of students who may have some insights. First, meet Jamie Nabozny. From the time he entered seventh grade in Ashland, Wisconsin until he dropped out in the eleventh grade, Jamie suffered repeated physical and verbal abuse. He has been verbally harassed, urinated on, and once he was beaten so badly by a gang of schoolmates that he had to be hospitalized and undergo surgery (*The Lambda Update*, 1996). Why? Because he was perceived to be gay. Jamie and his parents met frequently with school officials to make them aware of the assaults and indignities Jamie was enduring but "no meaningful disciplinary action ever was taken against the abusers by the school" (*The Lambda Update*, p. 17). Subsequently, Jamie developed post-traumatic stress disorder, ran away from home, and tried to kill himself several times (*The Lambda Update*, p. 17).

Now, meet Mark Iverson. Like Jamie, Mark faced anti-gay attacks from his peers beginning in seventh grade. Initially, Mark assumed (hoped) the "name calling," which he perceived as "a normal junior high thing," would soon end (Dinh, 1997). However, the name calling led to further harassment, threats, and eventually, physical violence. During ninth grade, Mark was pushed into a locker with a broomstick while being called "fag." According to Mark, two teachers witnessed the attack but did nothing (Dinh, 1997). In October, 1996, while a senior at Kentwood High School in Washington state, Mark was attacked by at least eight of his classmates. According to Mark, he was working on the school yearbook during lunch when his assailants came up to the room and started calling him "queer" and "faggot" (Dinh, 1997). Though Mark tried to ignore them, the verbal assaults escalated into a physical attack with eight students, maybe more, kicking and beating Mark. There were approximately 30 other students in the room, besides the attackers, and the school security office was right across the hall, "but no one did anything" (Dinh, 1997, 14).

Like Jamie's parents, Mark's mother went to school officials "quite a few times" but found the "principal would throw everything back onto Mark. . . . He'd ask, 'What did you do to provoke them?' . . . Students weren't disciplined. Nothing meaningful was done." (Dinh, 1997, p. 14).

Eventually, something meaningful was done for Jamie. Earlier this year, Jamie was awarded $900,000 in an out-of-court settlement with the Ashland school district, which was accused of violating Jamie's rights to equal protection and due process rights in that "the school denied him protection from abuse because he is gay [and] created a climate in which anti-gay abuse is tolerated" (*The Lambda Update,* 1996). Presently, Mark is also asking that something "meaningful" be done by the courts. On July 23, 1997, Mark and the ACLU filed suit in the U.S. District Court in Seattle, Washington against the Kent School District, "charging that it refused to enforce its anti-harassment policy" (Dinh, p. 14).

## Purpose

Many students like Jamie and Mark are turning to the courts seeking financial remuneration for the emotional and physical harm they have endured and the educational opportunities they have lost. Many other gay students, however, cannot turn to the courts because of the lack of the parent, peer, district, and societal support. And, why should students have to turn to the courts instead of their teachers and principals for the safety they are due?

Students, gay, lesbian, questioning, and straight, need for schools and classrooms to be open and safe. To assist teachers in attaining this goal, the author: (1) explains a rationale for creating classrooms and schools with climates that are tolerant and understanding of gay, lesbian, and bisexual issues and (2) presents practical strategies for creating open and safe schools.

## Rationale

For a number of practical reasons it makes sense to create a climate of tolerance and understanding of gay, lesbian, and bisexual issues in the classroom.

### Social Issues

In a society obsessed with increasing test scores, developing and implementing state-of-the-art methods of instruction, meeting the diverse needs of the physically, emotionally, and intellectually "challenged," identifying and fostering the development of "multiple intelligences," and expanding our children's "emotional I.Q.s," it is ironic that we blatantly ignore the needs of the students who may be the most at risk of failure, at school and in life.

Study after study indicates that, in comparison to their straight peers, gay, lesbian, and questioning youth are at a significantly high risk of falling prey to a number of social ills, including prostitution, AIDS, sexual abuse, drug and alcohol abuse, homelessness, and suicide. Sixty percent of gay males and 30 percent of lesbians report being verbally or physically attacked at least once between middle school and college because they are gay or are perceived to be gay (Gross & Aurand, 1992). As many as 28 percent of gay teenagers drop out of school because of harassment and the threat of harm (Remafedi, 1987). According to the Department of Health and Human Services Report on the Secretary's Task Force on Youth Suicide conducted in 1989, gays and lesbians between 15 to 24 years of age are two to three times more likely than straight adolescents to attempt suicide and may account for as many as 30 percent of all completed youth suicides. These data were confirmed by a study conducted by Uribe and Harbeck (1992) to clarify the needs of gay, lesbian, and questioning teenagers in relationship to their school experiences and, more specifically, their experiences with PROJECT 10, the first school-based counseling and educational program for gay, lesbian, and questioning students. From their study, Uribe and Harbeck (1992) found that half of the self-identified gays and lesbian participating in their study had attempted suicide.

Homelessness is also a problem for gay, lesbian, and questioning youth. Kipke, O'Connor, Palmer, & MacKenzie (1995) reported in *Arch Pediatric Adolescent Medicine* that as many as 30 to 40 percent of all runaways and homeless adolescents are gay, lesbian, or questioning. Likewise, Remafedi (1987) found from his study of gay and questioning adolescent males, that "an unusually high relationship [exists] between homosexuality and sexual abuse, drug abuse, homelessness, prostitution, feelings of isolation, family problems, and school difficulties" (cited in Harbeck, 1992). These findings confirmed data from an earlier study conducted by Remafedi (1987) indicating that nearly sixty percent of gay and bisexual male teenagers abuse drugs or alcohol.

Other data resulting from Uribe and Harbeck's study that bears on the social needs of gay male teenagers in particular include: (1) the average age for having their first sexual experience is 14 years old; (2) none of the males in their sample reported having their first sexual encounter in a "safe" manner; and most alarming, (3) "gay males are frequently initiated into homosexual activity in a 'date rape' situation involving substance abuse and a lack of safe sex practices" (Uribe and Harbeck, 1992, p. 26).

## Developmental Issues

Early adolescence, the developmental stage occurring between the ages of 10 and 14, can be one of the most turbulent times in a person's life, and for good reason. During early adolescence, youths experience more physical change than at any other time in their lives except at infancy (Manning, 1993). Youths experience rapid growth spurts that often result in having mismatched body parts and some youths mature sooner than others (Manning, 1993). Furthermore, and possibly most important to homosexual identity, sometime between 10 and 14, youths develop functioning reproductive systems, complete with sexual attractions and urges.

Specific to psychosocial development, during early adolescence, the peer group becomes all-important as a source for standards and models of behavior (Manning, 1993). Being accepted by peers is crucial to a youth's overall development of "self" as an individual as well as a vital member of the group and society (Manning, 1993).

Faced with the dramatic changes of early adolescence, youths spend a lot of time examining every aspect of "self" (Manning, 1993). They find themselves asking questions such as: "Will these zits ever go away?" "When will I get breasts?" "Why doesn't Mike look at ME!!!" "Will Mary invite me to her party this weekend?" "Where is all that hair I'm supposed to get? And, by the way, why is mine so much smaller than his?" The wrong answer to any one of these questions can send a student reeling.

So, what if a boy finds himself asking why he is sexually aroused by other boys, or one of our girls finds herself hoping that Sally will choose to dance with her even though the boys are all lining up to dance with Sally? The answer? These boys and girls will be and are in danger of drug abuse, sexual abuse, verbal and physical assault, parental rejection, dropping out of school, homelessness, prostitution, and suicide. Not because there is anything wrong with them, but because they are living in "a society that discriminates against and stigmatizes homosexuals while failing to recognize that a substantial number of its youth has a gay or lesbian orientation" (Gibson, 1989, p. 110).

Above any other goal, schools are responsible for meeting the developmental needs of our children. During adolescence, a time in which students are trying to integrate the autonomous self and the social self, teachers need to be sensitive to the fact that a "substantial number" of their students are gay, lesbian, or questioning or may develop a sense of self (being homosexual) that is rejected, feared, and hated by the social group (a homophobic society). If teachers and schools do not provide curricula and instruction to counter this hatred and fear toward homosexuals, a "substantial number" of our youth will be at risk, not only of falling prey to any number of the social plights discussed earlier but, of not achieving the developmental tasks necessary for healthy psychosocial maturity.

Because early adolescents need to develop close friendships, share inner thoughts and feelings, and disclose personal information with their peers without the fear of being rejected (Crockett, Losoff, and Peterson, 1984), healthy gay and lesbian early adolescent development is further thwarted by learning, in no time flat, that sharing "inner thoughts and feelings" and disclosing "personal information" about same-sex attractions may easily result in peer and parental rejection, social isolation, verbal assault (from teachers, parents, and peers), and physical harm (Uribe and Harbeck, 1992). Hence, for gay, lesbian, and questioning youth, the primary developmental task becomes adjusting to "a socially stigmatized role in isolation without adequate, honest information about themselves and others like them" (Uribe and Harbeck, 1992, p. 13). As the incidence of suicide, drug abuse, prostitution, violence, and homelessness indicates, this "adjusted" developmental task is understandably too much for our youth to bear.

Finally, because to be gay, lesbian, or questioning is to be a member of the "invisible" minority and because they are often forced to choose between befriending other homosexuals for the emotional and social support they need or distancing themselves from "visible" homosexuals to avoid enduring the rejection, harassment, and violence associated with being "out," gay, lesbian, and questioning youth need for teachers and schools to take actions for ensuring that students can be "out" and safe at school. Actions teachers and schools need to take include: making "visible" gay, lesbian, and bisexual role models; helping all students access positive information about being gay, lesbian, or bisexual; and providing information on the availability of support and social groups for gays, lesbians, and questioning students, all which are necessary for healthy psychosocial development.

## Practical Strategies for Creating Safe Classrooms and Schools

Because gay, lesbian, and questioning youth are at a disproportionately high level of risk at failing school, being victims of drug, alcohol, and sexual abuse and violence, attempting and completing suicide, and not achieving the developmental tasks necessary for healthy psychosocial development, teachers need to work toward creating classrooms and schools that encourage tolerance and understanding as youths seek to develop a strong sense of their autonomous and social selves. The following strategies are provided to help teachers in this endeavor.

Arthur Lipkin (1997) offers a variety of strategies for teachers that range from "low risk actions" to "greater risk actions." Some "low risk actions" include: (1) learning more about homophobia and gay, lesbian, and bisexual people, culture, history, and issues by reading books, journals, and periodicals; (2) creating a safe and equitable classroom by not assuming that, or talking as though, everyone is heterosexual and by using language that is inclusive of gay and lesbian possibilities (e.g., using spouse rather than husband or wife, parent rather than mother or father, and date rather than boy or girl friend); and (3) creating a safe and equitable school by establishing yourself as a role model of acceptance and by challenging name calling and harassment. Lipkin (1997) also presents "some risk" and "greater risk" actions, including: (1) attending gay and lesbian film series and lectures or organization meetings; (2) putting up gay- and lesbian-friendly posters, pictures, or signs; (3) inviting gay/lesbian speakers into the classroom; (4) working to form a gay/lesbian/straight alliance for students; (5) working toward establishing a "gay/lesbian/bisexual awareness day"; and (6) if you are gay, lesbian, or bisexual, "coming out" to your faculty, administration, and students.

There are a number of other practical approaches teachers can take to create safe classrooms and schools. They can assess the level of homophobia and heterosexism in their schools. An assessment survey that teachers and schools may wish to use to get an understanding of their attitudes and the attitudes of other teachers, their department, grade level, or school, is available on the World Wide Web through GLSEN (the Gay, Lesbian, and Straight Education Network) at *http://www.glstn.org/*.

Teachers, gay, lesbian, and straight, need to examine their own assumptions and attitudes about homosexuality and bisexuality and seek to counter prejudices they have internalized from being brought up in a homophobic society (Bass and Kaufman, 1996). Likewise, teachers can work toward sensitizing other faculty members to the needs of gay, lesbian, and questioning youth.

A practical means for addressing personal prejudice, developing sensitivity to gay, lesbian, and questioning students' needs, and receiving basic information on sexual orientation, homophobia, and gay, lesbian, and questioning people, culture, history, and issues is through staff development workshops. The outline for the "Anti-homophobia Training for School Staff and Students" used by the Gay, Lesbian, and Straight Education Network can be found on-line at *http://www.glstn.org/pages/sections/library/schooltools/031.article*. Additionally, an article explaining "Why Schools Need to Act, and What They Can Do" can be located at *http://www.glstn.org/pages/sections/library/schooltools/033.article*, and an outline for an advanced anti-homophobia training workshop can be down-loaded from *http://www.glstn.org/pages/sections/library/schooltools/032.article*.

There are many other "school resources" available to teachers through GLSEN's "Black-Board On-Line" (*http://www.glstn.org/pages/sections/library/schooltools*), including information on: strategies for making colleges and universities safe for gay, lesbian, questioning, and transgender students; civil rights for gay and lesbian teachers; and organizations, publications, and videos for gay, lesbian, and questioning students.

As previously suggested, providing support groups for gay, lesbian, and questioning youth is also a valuable strategy teachers can take. At *http://www.glstn.org/pages/sections/library/schooltools/014.article#2*, teachers can down-load GLSEN's "ten easy steps to starting a gay/straight alliance" in their school. Support group activities suggested by GLSEN, include: going on a hike; going to or showing movies/videos with gay/lesbian/questioning/transgender themes; participating in a local Gay Pride Day March; conducting an "And the Award Goes To . . ." Show; hosting a diversity panel at your school; and sponsoring a "Bring-A-Friend Night," a "Gay/Lesbian/Bi Alumni Go-Back-to-School Party," a "Parents' Night," an "Outreach to the Community Project," a "Teach the Teachers Day," or a play, such as "Angels in America."

Whenever appropriate, teachers can introduce information about gay, lesbian, and questioning people, including Leonardo da Vinci, Alan Turing, Willa Cather, Socrates, Marcel Proust, Virginia Woolf, Herman Melville, Peter Ilich Tchaikovsky, Frida Kahlo (Bass and Kaufman, 1996).

Some general curriculum ideas for integrating gay, lesbian, and questioning people, history, culture, and issues into preexisting curricula include: *at the elementary school level*—challenging sex role stereotypes in children's play, including references to non-traditional families while discussing families, including gay, lesbian, and questioning people in discussions on discrimination; *at the middle school level*—discussing gay, lesbian, and bisexual issues while covering current events and integrating into health and family life curricula information on lesbian and gay issues; and *at the high school level*—incorporating readings with gay, lesbian, and bisexual themes into the curricula and teaching about the struggle of gays and lesbians to attain equality in the U.S. and the persecution of homosexuals during the Holocaust (Bass and Kaufman, 1996).

Teachers can take a stand against harassment (Bass and Kaufman, 1996). By simply establishing guidelines for respectful behavior and including gay, lesbian, and questioning issues in discussions on tolerance, teachers help make their classrooms safer places for gay, lesbian, and questioning youth (Bass and Kaufman, 1996). Similarly, by working toward adding "sexual orientation" to their schools' non-discrimination and harassment policies, teachers help send a message to students, parents, and faculty that their schools value equality.

By coming out, teachers can serve as role models of healthy, caring, well-adjusted, and "normal" human beings, helping to counter the bleak and destructive future for gays, lesbians, and bisexuals that is often presented by the media and religious organizations.

Teachers can request that books with gay, lesbian, and questioning themes be ordered for the library and classrooms. Some of the most often recommended books include, *The Family Heart: A Memoir of When Our Son Came Out* by Robb Forman Dew; *Gay American History: Lesbians and Gay Men in the U.S.A.* by Jonathan Katz; *The Pink Triangle: The Nazi War Against Homosexuals* by Richard Plant; *Positively Gay: New Approaches to Gay and Lesbian Life* by Betty Berzon; *Free Your Mind: The Book for Gay, Lesbian, and Questioning Youth and Their Allies* by Ellen Bass and Kate Kaufman; *Two Teenagers in Twenty: Writings by Gay and Lesbian Youth* by Ann Heron; *Homophobia: How We All Pay the Price* by Warren Blumenfeld; *The Gay, Lesbian and Questioning Students' Guide to Colleges, Universities, and Graduate Schools* by J. M. Sherrill and C. Hardesty; *Families: A Celebration of Diversity, Commitment and Love* by Aylette Jenness; *Asha's Mums* by Rosamund Elwin and Michele Paulse; *Oliver Button is a Sissy* by Tomi DePaola; *Losing Uncle Tim* by Mary Kate Jordan; *Daddy's Roommate* by Michael Willhoite and *Be a Friend: Children Who Live with HIV Speak* by Lori S. Wiener.

Teachers can work toward appropriate health care and education being offered to students, including information on sexuality, sexually transmitted diseases, and gay, lesbian, and bisexual lifestyles and planning.

Other practical strategies teachers can take to provide safe classrooms and schools for their gay, lesbian, and questioning students include: providing counseling services, sponsoring social events and conferences, establishing a resource center, and creating a visible network with community agencies, task forces, parent groups, and educational organizations (Bass and Kaufman, 1996).

## Conclusion

We are losing many of our youth to prostitution, AIDS, sexual, drug, and alcohol abuse, homelessness, and suicide, and many of our youth lack the support of friends, family, and society that is necessary for attaining the developmental tasks needed for healthy psychosocial maturity. It is the hope of this author that the rationale and practical strategies presented in this article will help teachers create classrooms and schools that are safe for all students, so that sometime soon our students will not have to turn to the courts to right our wrongs and the 1,500 of our youth who choose to end their lives each year will no longer need the knives, guns, pills, and ropes they are now using to find a place of safety and peace they are hard pressed to find in our homes, schools, and society today.

# References

Bass, E., & Kaufman, K. (1996). *Free your mind: The book for gay, lesbian, and questioning youth and their allies.* New York: Harper Perennial.

Crockett, Losoff, and Peterson. (1984). Perceptions of the peer group and friendship in early adolescence. *Journal of Early Adolescence*, 4, 155–181.

Dinh, Viet. (August 29, 1997). I had to try to hang in there: Student sues his school district for failing to stop harassment. *Washington Blade*, p. 14.

Gibson, P. (1989). *Gay male and lesbian youth suicide* (Publication No. ADM 89-1623, vol. 3, p. 110). Rockville, MD: U.S. Department Health and Human Services. Secretary's Task Force on Youth Suicide Report.

Gross, L., & Aurand, S. (1992). Discrimination and violence against lesbian women and gay men in Philadelphia and the Commonwealth of Pennsylvania. Philadelphia, PA: Philadelphia Lesbian and Gay Task Force.

Harbeck, K. M. (1992). *Coming out of the classroom closet: Gay and lesbian students, teachers and curricula.* New York: Harrington Park Press.

Kipke, M., O'Conner, S., Palmer, R., & MacKenzie, R. (1995). Street youth in Los Angeles. *Arch Pediatric Adolescent Medicine.* 149, 513–519.

Lipkin, A. (1997). What you can do: Assessing the risks. Gay, Lesbian, and Straight Education Network [on-line]. Available: *http://www.glstn.org/pages/sections/library/school-tools/013.article*

Manning, L. (1993). Developmentally appropriate middle level schools. Wheaton, MD: Association for Childhood Education International.

Nabozny v. Podlesny (Federal Court, Wisconsin). (1996, Winter). *The Lambda Update* 13(1), 17.

Remafedi, G. (1987). Adolescent homosexuality: Psychosocial and medical implications. *Pediatrics,* 79, 331–337.

Remafedi, G., Farrow, J., & Deisher, R. (1991). Risk factors for attempted suicide in gay and questioning youth. *Pediatrics,* 87, 869–876.

Uribe, V., & Harbeck, K. (1992). Addressing the needs of lesbian, gay, and questioning youth: The origins of Project 10 and school-based intervention. In Karen M. Harbeck (Ed.), *Coming out of the classroom closet: Gay and lesbian students, teachers and curricula* (pp. 9–28). New York: Harrington Park Press.

# Supporting the Invisible Minority

## JOHN D. ANDERSON

A hidden minority group of gay and lesbian students attends our schools. Pioneering educators in Stratford, Connecticut, have provided a leadership model to meet the needs of these students.

In every school, there is a group of forgotten children—a hidden minority of boys and girls whose needs have been ignored, whose existence has been whispered about, whose pain is just beginning to surface. These are our gay, lesbian, and bisexual students.

A Harris Poll, released in June 1992, said that 86 percent of high school students would be very upset if classmates called them gay or lesbian. A 1989 report from the federal government suggested that gay and lesbian youth were three times more likely than their peers to commit suicide (Remafedi, 1994). Historically, these teens have been given derogatory names by society and ignored by their schools, except as the butt of jokes. As teens come out, or go public with their sexual orientations, this oversight is becoming increasingly unacceptable. Studies show that the mean age of coming out for sexual-minority youth is declining, at least in urban areas. For 1993 the mean age for males was 13.1 years; for females, 15.2 (Baily and Phariss, 1996).

In the two places where a child should feel safe and supported, gay and lesbian youth are routinely reviled: family and school. Gay or lesbian children who are taunted usually have nowhere to turn.

Educators have a clear professional mandate to address the needs of sexual minorities. Organizations such as the National Education Association, the American Federation of Teachers, and the National Association of State Boards of Education have passed resolutions that protect the rights of sexual-minority students and staff. Even most businesses acknowledge the necessity of respect for diversity of all kinds (Carson, 1993; Mickens, 1994).

Educational leaders often deny there is a problem when it comes to sexual minorities in the schools. In a statewide survey of Connecticut teachers and administrators in 1991, respondents indicated that they recognized the plight of gay and lesbian students, admitted that next to nothing was being done for them, and expressed hope for a solution. The results showed a dichotomy between the perceptions of teachers as a group and those of administrators: Teachers called for action; administrators claimed there was no need or that programs were already in place to address these issues (Woog, 1995). The good news is that even in such an environment, great strides can be made, as they have in Stratford.

---

John D. Anderson, "Supporting the Invisible Minority," *Educational Leadership,* April 1997, Vol. 54, No. 7, pp. 65–68. Reprinted by permission from Association for Supervision and Curriculum Development.
John D. Anderson is a teacher at Bunnell High School in Stratford, Connecticut. He may be reached at 3 Barberry Lane, Woodbridge, CT 06525.

## The Coming Out of Leadership in Stratford

What makes Stratford Public Schools unique in the state and perhaps in the United States? The school system has an openly gay high school teacher, an openly gay elementary school principal, an openly gay middle school teacher, and a middle school assistant principal who openly accepts that one of her sons is gay. In our district, the actions of a few have had unimagined consequences, even in the face of administrative reluctance or objection (Anderson, 1996).

I began teaching in Stratford in 1985. In 1991 three events converged to change my life as a teacher. First, in October, Connecticut extended civil rights protection to gay and lesbian citizens, one of nine states to do so to date. Second, during the summer of that year, I began writing a bimonthly column on gay and lesbian concerns for the *New Haven Register*. Third, as my official evaluation goal, I chose to examine the needs of gay and lesbian students and to evaluate how the school system was meeting those needs.

I found that little is being done in U.S. schools to ameliorate the educational opportunity for gay and lesbian students (Harbeck, 1992; see also Rienzo et al., 1996). Nevertheless, in Stratford, we have raised sensitivity levels and provided increased support for gay and lesbian students and staff. We have found that most educators are caring people, but they are fearful and look in vain for direction from our educational leaders (Anderson and Edwards, 1996).

In lieu of leadership and policies, many educators in Stratford and in other cities and states have taken steps on their own to help reduce discrimination in our schools against those who are gay, lesbian, or bisexual. The amazing thing is the progress made with this scattered approach. It is discouraging, however, to note that few schools are replicating successful programs.

## Fragmented State and National Leadership

Kevin Jennings (1994) and David Woog (1995) have chronicled the stories of U.S. educators who are gay, lesbian, or bisexual. Yet the stories illustrate that schools have no unified procedure or policy concerning sexual orientation of educators or students.

Another researcher, Harbeck (1992), describes a program called Project 10, which began at a dropout-prevention center founded in 1984 by Virginia Uribe at Fairfax High School in Los Angeles. Project members produced a video documentary on the plight of gay and lesbian students. It was entitled *Who's Afraid of Project 10?*

In another isolated example, the New York Public Schools in 1985 opened the Harvey Milk School, an alternative high school, to provide psychological, social, and academic support to gay, lesbian, and bisexual students. This school is part of the youth advocacy work of the Hetrick-Martin Institute, which also publishes posters and resource materials.

These are partial solutions to local manifestations of a national problem. Until recently, Project 10 and the Harvey Milk School were the extent of organized efforts to address the needs of sexual-minority students in concrete terms.

Two states have begun using data to make decisions about gay and lesbian students and education. In Massachusetts, the Governor's Commission on Gay and Lesbian Youth (1993) produced an excellent study of the issues affecting gay and lesbian students, from poor school performance to suicide. The study makes recommendations for schools, for families of gay and lesbian youth, and for state agencies and the Massachusetts legislature. An important result of this report is that the State Board of Education now requires all teacher certification programs in the state to include sensitivity training on gay and lesbian issues.

The Minnesota Department of Education (1994) published a resource booklet that addresses how to include issues related to homosexuality in school policy, instruction, and student services. This text challenges schools to examine their environment and then to develop a more sensitive, inclusive place of learning.

A third state, Connecticut, began addressing homophobia in the schools in 1991, through the *Sex Equity Newsletter* of the State Department of Education. This newsletter features resources, information, and guidelines similar to those of Massachusetts and Minnesota.

All three states are addressing five important issues (Anderson, 1994). Stratford has made progress in all five.

## Effective Approaches to Sex Equity

*Professional development.* In the past few years, Stratford has begun to provide workshops and forums that address sexuality issues. In the first, during the 1992–93 school year, about 30 teachers attended voluntarily; only one administrator participated. At a 1994 workshop on "The Invisible Minority," no administrators attended. One of the presenters was Garrett Stack, an openly gay elementary school principal—and my life partner. Another presenter was Ann Edwards. Other forums have included limited presentations—mostly on sexual harassment and cultural minorities—at the annual "World of Differences" workshops.

*Support staff and services.* Stratford guidance personnel have increasingly been supportive of gay and lesbian issues. Several guidance counselors displayed posters that I provided. One poster declared "Homophobia is a Social Disease." One counselor has a rainbow sticker, pink triangle, and other gay paraphernalia on her bulletin board. The school psychologists, school nurses, and social workers also provide support to students.

*Sexuality in the health curriculum.* Like most high school health textbooks (Baily and Phariss, 1996), Stratford's text contains only one paragraph on homosexuality. The high school health teachers, however, are supportive and articulate. One teacher, Lea Dickson (1995), has written a resource directory for ASCD's Gay and Lesbian Network. After two years of lobbying the central school administration for permission, two teachers recently invited me to address their senior health classes on homosexuality. At one class session, a panel of "adults in relationships" included both heterosexual and gay couples. I proudly participated, representing Garrett Stack and myself. Student response to these classes has been overwhelmingly positive.

*Library resources.* The Stratford High School library held a diversity exhibit in 1993. It included noted gay and lesbian Americans—James Baldwin, Martina Navratilova, Col. Margarethe Cammermeyer, Audre Lorde, and Walt Whitman—as well as members of other minority groups. The exhibit occurred without incident. One librarian posted a list of 20 possible topics for research papers, including "civil rights for gays and lesbians." One high school recently held a "Stop the Violence Week." The library exhibited Elaine Landau's (1986) *Different Drummer: Homosexuality in America*, as well as works from two series for adolescents by Chelsea House Publishers: *Lives of Notable Gay Men and Lesbians* and *Issues in Gay and Lesbian Life* (see Gough and Greenblatt, 1992).

*Curriculum support.* We have found that any teacher can transform the curriculum into an inclusive experience for students. For example, a social studies teacher included sex equity issues in a unit on civil rights; in an English class, a student wrote a research paper on gay parenting. In another English class, an Advanced Placement student developed ways to include gay and lesbian material in the curriculum. I used the TV drama *Serving in Silence* about Col. Cammermeyer, as well as some of my *New Haven Register* columns, as extra-credit activities in my English classes (see Resources).

## Conversations and Civility

Most of our efforts in Stratford have focused on *educating the educators*. The administration has been less comfortable with services and support for students (Anderson, 1996; see also Riddle, 1996; Rensenbrink, 1996). Despite the lack of policy, we are slowly creating a supportive environment for our gay, lesbian, and bisexual students and staff.

A highlight of our equity movement was a 1995 conversation with the superintendent. The superintendent listened to parents, teachers, and students express concerns about equal educational access. He encouraged us to continue in our support for equal education for all and cautioned moderation in our choice of actions. We have found that careful planning, mutual respect, and civility go a long way toward achieving real progress.

## Resources

American Federation of Teachers, Gay-Lesbian Caucus, 1816 Chestnut St., Philadelphia, PA 19103.
American Library Association, Gay and Lesbian Task Force, 50 E. Huron St., Chicago, IL 60611.
ASCD Network, Lesbian, Gay and Bisexual Issues in Education, P.O. Box 27527, Oakland, CA 94602.
Chelsea House Publishers, 300 Park Ave. S., New York, NY 10010.
Gay and Lesbian High School Curriculum and Staff Development Project, Arthur Lipkin, Harvard School of Education, 210 Longfellow Hall, Cambridge, MA 02138.
The Gay, Lesbian and Straight Teachers Network (GLSTN, 122 W. 26th St., Suite 1100, New York, NY 10001.
Hetrick-Martin Institute, Inc., 2 Astor Place, New York, NY 10003.
National Education Association, Gay-Lesbian Caucus, P.O. Box 314, Roosevelt, NJ 08555.
Parents and Friends of Lesbians and Gays (PFLAG), P.O. Box 27605, Washington, DC 20038.
Sexuality Information and Education Council of the United States (SIECUS), 130 W. 42nd St., Suite 2500, New York, NY 10036.

## World Wide Web:

The Gay, Lesbian and Straight Teachers Network (GLSTN): http://www.glstn.org
"Programs in Gender and Lesbian, Gay, and Bisexual Studies at Universities in the USA and Canada": http: //www.duke.edu/web/jyounger/ lgbprogs.html
University of Maryland "Sexual Orientation Specific Resources": http://www.inform.umd.edu:8080/EdRes/Topic/Diversity/Specific/Sexual_Orientation/

# References

Anderson, J.D. (1994). "School Climate for Gay and Lesbian Students and Staff Members." *Phi Delta Kapa* 76, 2: 151–154.

Anderson, J.D. (1996). "Out as a Professional Educator." In *Open Lives Safe Schools*, edited by D. Walling. Bloomington, Ind.: Phi Delta Kappa Educational Foundation.

Anderson, J.D., and A. Edwards. (1996). *Out for Life*. Las Colinas, Tex.: Ide House.

Baily, N., and T. Phariss. (1996). "Breaking Through the Wall of Silence: Gay, Lesbian, and Bisexual Issues for Middle Level Educators." *Middle School Journal* 27, 3: 38–46.

Carson, C. (1993). "Perspectives on Education in America: An Annotated Briefing." *Journal of Educational Research* 86, 5: 259–310.

Dickson, L. (1995). *Lesbian, Gay, and Bisexual Issues in Education, An ASCD Resource Directory, 1994–1995*. Fairfield, Conn.: Garden Gates Communication.

Gough, C., and E. Greenblatt. (1992). "Services to Gay and Lesbian Patrons: Examining the Myths." *Library Journal* 117, 1: 59–63.

Harbeck, K., ed. (1992). *Coming Out of the Classroom Closet: Gay, and Lesbian Students, Teachers, and Curricula*. Binghamton, N.Y.: The Haworth Press,

Jennings, K. (1994). *One Teacher in Ten*. Boston: Alyson Publications.

Landau, E. (1986). *Different Drummer: Homosexuality in America*. Englewood Cliffs, NJ.: J. Messner.

Massachusetts Governor's Commission on Gay and Lesbian Youth. (1993). *Making Schools Safe for Gay and Lesbian Youth: Breaking the Silence in Schools and in Families*. Boston: Author.

Mickens, E. (1994). *The 100 Best Companies for Gay Men and Lesbians*. New York: Pocket Books.

Minnesota Department of Education. (1994). *Alone No More: Developing a School Support System for Gay, Lesbian, and Bisexual Youth*. St. Paul, Minn,: Author.

Remafedi, G. (1994). *Death by Denial*. Boston: Alyson Publications.

Rensenbrink, C. (1996). "What Difference Does It Make? The Story of a Lesbian Teacher." *Harvard Education Review*, 66, 2: 257–270.

Riddle, B. (1996). "Breaking the Silence Addressing Gay and Lesbian Issues in Independent Schools." *Independent School* 55, 2: 38–47.

Rienzo, B., J. Button, and K. Wald. (1996). "The Politics of School-Based Programs Which Address Sexual Orientation." *Journal of School Health* 66, 1: 33–40.

Woog, D. (1995). *School's Out: The Impact of Gay and Lesbian Issues on America's Schools*. Boston: Alyson Publications.

# Let's Stop Ignoring Our Gay and Lesbian Youth

## ANN T. EDWARDS

Whether we realize it or not we as educators are dealing with a hidden minority of gay and lesbian students, as well as gay and lesbian parents. People have to self-identify (come out) as homosexual. Gay and lesbian infants are not born with tiny inverted pink triangles on their foreheads. Most of our gay and lesbian youth will come out in high school after spending the middle school years questioning their sexual orientation in silence.

Here are some effects of that silence: Twenty-six percent of gay youth are forced to leave home because of their sexual identities, and 68 percent of young gay men and 83 percent of lesbians report using alcohol and other drugs on a regular basis. Forty-one percent of lesbian and gay youth suffer violence from their families, peers, and strangers. It is estimated that 30 percent of youth suicides are committed by lesbian and gay people. In 1992, 20 percent of all persons with AIDS/HIV were most likely infected as teenagers (see Hershberger, 1995; Hunter, 1990; Savin-Williams, 1994; see also Anderson, this issue, pp. 65–68).

In Stratford, Connecticut, we are working to break the silence and to provide a safe, supportive climate that includes *all* students. I have played a personal role in this awakening. When my son told me of his homosexuality, I decided to be an *open parent*, to help gay and lesbian students and their parents. I freely tell teachers and staff that one of my three sons is gay.

## Parents Seeking Understanding

My candor opens the door to many parents, including staff members. A year after I began this dialogue, an educator at my school started talking to me about her son. In her own time and way—despite her initial discomfort—she told me that she thought he might be gay. I lent her some books and offered to take her to a meeting of Parents and Friends of Lesbians and Gays (PFLAG).

One day I met the parent of a former student. During our friendly conversation, we discussed our children, including whom they were dating. I talked about the women in the lives of two of my sons; then I said that one was gay and not yet serious with any young man. She then told me she thought her son was gay but had not yet come out. I thought of some of the conversations I had with her son when he was a student. I thought of the journey that probably lay ahead for this family, but I had reason to hope that it might be a little easier as a result of our connections.

---

Ann T. Edwards, "Let's Stop Ignoring Our Gay and Lesbian Youth," *Educational Leadership,* April 1997, Vol. 54, No. 7, pp. 68–70. Reprinted by permission from Association for Supervision and Curriculum Development. Ann T. Edwards is Assistant Principal at Flood Middle School in Stratford, Connecticut. She may be reached at P.O. Box 322, Greenwich, CT 06836 (e-mail: edwardsl@idt.net).

Same-gender parents also feel more comfortable in talking with people who are openly accepting. A gay father with whom I had become acquainted finally came out to his 8th grade son, who had long suspected his father's orientation and had suffered teasing from classmates. Once he had that honest conversation with his father, the boy proudly told his friends that his father was gay and that now he has two fathers. The teasing eased. Both parents and student benefited from an open acceptance of diverse sexual orientations.

## Students Finding Allies

A few years ago, a girl at my school began questioning her sexual orientation. I first met her when she was a frightened 7th grader brought to my office before school. She had been intimidated by some kids on the bus, and her mother wanted her bus changed. The girl later returned alone to my office. She told me why kids were teasing her.

As an 8th grader, the student often sought me out and once spotted a book on my desk titled *Young, Gay, and Proud* (Alyson, 1980). It opened the conversation about lesbians and gays, and I told her of my son. She told me she had a lesbian friend and that she thought she might be a lesbian. She asked me not to tell her mother, but told me that she had let her mother know that she was questioning her sexual orientation. Later, I let her know about a gay and lesbian youth group, and I mentioned PFLAG. She asked me how old you had to be to join.

In the spring, the 8th grader's mother called me. She thanked me for the interest I had shown in her daughter and brought up her concerns about her daughter's sexual orientation. It was great to hear the mother say that if her daughter's orientation turned out to be same gender, it would be fine with her because she loved her daughter dearly. She then begged me not to relate our conversation to anyone. To most parents, the possibility that their own child might be gay is unthinkable. Before the girl moved on to high school, I told her about John Anderson, an openly gay teacher at her future school. She had the good fortune to have him for her first year English class. She came out as a lesbian early in her high school career.

As assistant principal, I am often called on to resolve disciplinary problems. One afternoon, a bus driver asked for help with an 8th grade boy, who the driver claimed had been unruly and had called him a derogatory sexual name. The boy protested his innocence. His mother called me later that afternoon to say that she strongly doubted that her son would ever call anyone that word because someone very close to him is gay and he knows the epithet is hurtful. I let the boy know, that I believed him. When the ABC network ran a broadcast on same-sex marriage, he stopped in my office to see if I had seen the show. I had, and we talked about it for a while. My point of view validated same-gender unions, and this seemed to somehow validate this student. He later sought me out to help him after school to get his grades back on track.

## Providing an Accepting School Environment

Other teachers and administrators at Stratford schools also have provided support to gay and lesbian students and parents. For example, Garrett Stack, the openly gay principal of Franklin Elementary School, and John Anderson of Bunnell High School make no secret of their domestic partnership. Parents feel comfortable talking with both Stack and Anderson of their concerns about their children's sexual orientations and other sex-equity issues.

At Franklin, there is less abusive language directed to perceived gay and lesbian children; when kids use such language, teachers and other staff members do not tolerate it. At Bunnell, Anderson reports that the gay and lesbian topic is no longer taboo, and a teacher even brought harassment charges against a student for using derogatory language. Many of the student's peers at the high school supported the teacher.

Educators and parents can follow some simple guidelines to provide equity to the gay and lesbian students in our schools and homes:

1. Use the words *gay, lesbian, bisexual.* Use inclusive language, such as *partner* or *spouse* instead of *husband* or *wife.*
2. Provide classroom speakers. Dan Kelly, a young actor, spoke at Flood Middle School in Stratford, Connecticut, as part of a diversity celebration. He spoke on what it felt like to be gay in middle and high school and had a positive effect on reducing name calling.
3. Display or wear a gay-positive symbol. When teachers in Brookfield, Connecticut, put pink triangles outside their room as a symbol of safety for gay and lesbian students, some people in the community objected. With administrative support, however, the teachers prevailed. Gay, lesbian, and questioning youth feel part of a safe school.
4. Challenge homophobic remarks everywhere and all the time. Establish an anti-slur policy.
5. Provide positive role models, both historic and current. English teachers can mention that Walt Whitman was gay; art teachers can tell students that Michelangelo was gay.
6. Demand inservice training for all staff. Fran Evans in Danbury, Connecticut, brought in speakers from PFLAG. A group now meets weekly to discuss sexual orientation issues.
7. Include discussions of gay, lesbian, or bisexual issues in the class, as with any other minority issue.
8. Create social situations for both gay and straight friends. Flood Middle School hosted a "teen night" instead of a dance and encouraged all kids—both gay and straight—to attend.

Our experiences in Stratford have convinced us that as parents and educators, we can provide a healthy, supportive climate for all—including the minorities of gay and lesbian students, parents, and staff members.

## References

Alyson, S. (1980). *Young, Gay, and Proud.* Boston: Alyson Publications.

Hershberger, S. (January 1995). "The Impact of Victimization on the Mental Health and Suicidality of Lesbian, Gay Male and Bisexual Youths." *Developmental Psychology* 31, 1: 65–74.

Hunter, J. (September 1990). "Violence Against Lesbian and Gay Male Youths." *Journal of Interpersonal Violence* 5, 3: 295–300.

Savin-Williams, R. (April 1994). "Verbal and Physical Abuse as Stressors in the Lives of Lesbian, Gay Male and Bisexual Youths." *Journal of Counseling and Clinical Psychology* 62, 2: 261–269.

# What Does It Mean?
## Exploring the Myths of Multicultural Education

NATALIE G. ADAMS
*Georgia Southern University*

> This article, based on an ethnographic study in a predominantly White working-class middle school in the South, critically examines why a literature-based multicultural curriculum failed to get "students to talk about racism." According to the author, attempts to implement a multicultural curriculum fail not because of unprepared, insincere, or insensitive teachers; rather these efforts fail primarily because of dominant mainstream beliefs embedded in the institution of schooling and in the discourse of multiculturalism itself.

> What does it mean when a White female English professor is eager to include a work by Toni Morrison on the syllabus of her course but then teaches that work without ever making reference to race or ethnicity?
>
> (hooks, 1993, p. 93)

What does it mean when a high school English teacher substitutes *The Color Purple* for *A Tale of Two Cities*, but the focus of her teaching deviates little from the curriculum guide with its emphasis on plot, character, setting, and theme? What does it mean when a middle school language arts teacher decides to use Mildred Taylor's *Roll of Thunder, Hear My Cry* as a way of talking about racism but denies that racism exists at her school? What does it mean to have a multicultural English curriculum in classrooms that continue to be teacher-directed and teacher-centered? What does it mean when school officials proclaim that they have a multicultural English curriculum, but no Blacks are represented on the school's cheerleading squad although the majority of the school's football players are Black? In short, what does it mean to have a "multicultural" English curriculum that never addresses issues of racism, classism, and sexism?

The purpose of this paper is to explore the many "what does it mean" questions that highlight the ambiguities, contradictions, and tensions embedded in the discourse of multiculturalism in the English classroom.

---

Natalie G. Adams, "What Does It Mean? Explaining the Myths of Multicultural Education," *Urban Education,* April 1995, Vol. 30, No. 1, pp. 27–39. Reprinted by permission from Sage Publications, Inc. (US), Corwin Press.

## What Does It Mean To Have a Multicultural English Curriculum?

In 1993 I conducted an ethnographic study in a middle school, Centerville Middle School, located approximately 15 miles from a large city in the Deep South. The school was composed of 610 students, 85 percent were White and 15 percent were Black. Thirty percent of the families in Centerville's school community had incomes below the poverty level. Approximately 10 percent of the students were from middle to upper-middle class families. The remainder of the students were from working-class families. Sixty-four percent of the students' parents did not complete high school.

For six months, I visited weekly one eighth-grade language arts classroom taught by a White female teacher, Ms. Lafitte. The class was composed of 23 students, of whom eight were White males, two were Black males, nine were White females, and four were Black females. As both a participant and an observer of this class, I had the opportunity to informally interview the students as well as the teacher on numerous occasions. I also formally interviewed the teacher and 10 students who volunteered to participate in the research project (four White males, two Black males, one Black female, and three White females). Additionally, all 23 students completed a questionnaire designed to ascertain the reading and writing habits of their parents as well as the students' understandings of literacy and multiculturalism in today's society.

In 1992, the English teachers at Centerville Middle School adopted what they called "a multicultural literature-based English curriculum." As explained by Lafitte, who was instrumental in implementing this curriculum, "we wanted to quit relying so much on the basal readers and use novels that the kids could relate to. You know, teach kids grammar, writing, and reading skills within the context of a novel." A central component of this "multicultural" curriculum was the incorporation of several novels written by non-White authors. According to Lafitte, the teachers believed that their students would gain an appreciation and tolerance for cultural and ethnic diversity through the reading and discussing of novels that centralized the experiences of non-Whites.

This move to use multicultural literature to teach an appreciation of cultural diversity reflects one approach of multicultural education, described by McCarthy (1990) as the model of cultural understanding. Central to the cultural understanding model of multicultural education is the belief that attitudinal changes will occur if individuals are exposed to programs that foster respect for ethnic and racial differences (McCarthy, 1990). Embedded in Centerville's philosophy of multicultural education is the assumption that if students understand cultural differences (accomplished through the reading of the "Other's" experiences as portrayed in literature) and if they accept and respect those differences (accomplished through meaningful class discussions), then students will become more tolerant of racial and cultural diversity.

A useful way to "deal with prejudice," according to Lafitte, was to use the novel *Roll of Thunder, Hear My Cry*, which was written by Mildred Taylor, a Black female author. The novel explores racism in the South in the 1930s through the eyes of the main character Cassie Logan, a 10-year-old Black female. Lafitte had not read the novel prior to assigning it to the students; however, it had been recommended to her by a colleague who had read it in a "Multicultural Adolescent Literature" class taught at a nearby university. Prior to beginning the novel, Lafitte stated that the novel would be an "excellent tool to get students to talk about racism." During the three-week unit, students were assigned a chapter to read each night; class time was spent discussing the chapter and periodic quizzes were given; upon completion of the novel, a unit test was given.

Despite the intentions of Lafitte to use the novel as an impetus for "talking about" racism, classroom discussions were void of any meaningful and honest discussions about racism in today's society. Disappointed and puzzled after teaching the novel for three weeks, Lafitte asserted, "I thought the kids would relate to the book—you know, really talk about it. But they didn't, and I don't know why." Like many well-intentioned teachers, Lafitte had simplified racism, thinking it to be something that could be "dealt with" by using literature written by non-White writers. Her assumption was that White students would become more knowledgeable, and, thus, more understanding, of the oppression that Blacks encounter in society by reading about it in fiction. However, Lafitte (as well as the other Centerville teachers) had implemented a model of multicultural education without critically analyzing the complex dynamics of race, power, and structured inequality within the institution of schooling itself. In the following three sections, I discuss three problematics inherent in Centerville's approach to multicultural education that make "talking about" racism virtually impossible.

## "Why Do We Rave To Read This Book?": Implementing a Multicultural Curriculum in a Traditional Teacher-Centered Classroom

One of the central obstacles in getting students to talk openly and honestly about racism centers on the unequal power relations embedded in traditional classroom practices. L. J., one of the Black male participants in the study, asserted during an interview, "I don't know why we have to read this book. My mamma told me the only place I should talk about racism is with my family or with the counselor." L. J.'s comments illustrate quite poignantly one problem in implementing a multicultural curriculum in classrooms steeped in traditional understandings of the roles of teacher and student.

First and foremost, the decision to read *Roll of Thunder, Hear My Cry* was made by Lafitte and the other English teachers with no input from the students. In short, reading the novel was not a choice. Because a large portion of their grade for that nine-week period was based on the weekly quizzes and the unit test related to the book, students were compelled to read the novel to meet course requirements. Although every student read at least a portion of the novel, many of the students demonstrated their resistance by remaining silent or by not choosing voluntarily to participate in classroom discussions (e.g., of the four Black females in class, one chose never to participate and was, in fact, chastised for doing math homework during one of the discussions, while another Black female spoke so quietly when called upon by the teacher that I could not hear her response).

Secondly, classroom discussions were teacher-centered and teacher-directed. Lafitte determined the nature of classroom discussions by controlling how time would be used, who would talk and when, what questions would be asked, and what answers were right or acceptable. Most of the three weeks devoted to the novel were spent in whole-class discussions centered on questions that Lafitte had reproduced from a teacher's manual written by one of the leading textbook publishers in the United States (for a discussion of the political implications of using commercial reading materials, see Shannon, 1992). The dominant pattern of discourse in Lafitte's class was a three-part sequence characterized by the teacher initiating a discussion with a question, followed by a student response, and ending with the teacher evaluating or summarizing the student's response (Kutz & Roskelly, 1991). Student responses were elicited by either the teacher calling on someone who had his/her hand raised (the most popular method in this class) or by the teacher calling on someone who had not raised his/her hand (often used as a disciplinary tool to get those students who were daydreaming "back on track").

These differential power relations are described by Freire (1983) as being an integral component of the "banking concept" of education, which maintains the teacher as the guardian of knowledge who dispenses that knowledge into empty-headed students. According to Freire, this kind of education serves to numb students' creative powers, keeping them passive, unquestioning members of a hegemonic society. In Lafitte's classroom, the already established and seemingly impermeable pattern of discourse and power between teacher and students functioned as a serious deterrent to honest discussions of racism.

## The "Not in My Backyard Syndrome": Distancing Multiculturalism from the Local Community

Another potential danger in an uncritical implementation of multiculturalism in the classroom is the tendency to distance and detach issues of multiculturalism from local and community concerns. Certainly the teachers and administrators at Centerville readily admitted that ethnic diversity often translates into racial unrest in the world at large; therefore, an appreciation of cultural diversity needed to be taught at school. However, these same teachers and administrators denied the existence of racism and racial tension at their particular school. According to Lafitte, "Racism is not a problem here, so we don't need to start something that so far we've managed to avoid."

In reality, Centerville Middle School is not immune to the racism that plagues the world at large. Certainly, the problems of riots and violence between Blacks and Whites are missing at Centerville, but nevertheless, institutional discrimination exists there. For example, 15 percent of the students at Centerville are Black; however, there is only one Black regular education teacher; there are no Black cheerleaders although the football team is composed of a large number of Black players, and the only bulletin board depicting great Black Americans is situated in the far corridor outside the Black teacher's room. As illuminated by the following two quotes, the students themselves were quite perceptive about the "hidden" racism evident at Centerville:

> Terrence (Black male): You got a big group of Black people who don't hang around with White people and talk all crazy and all that; they talk about they don't like White people.

> Paul (White male): There is still a lot of racism around here and a lot of prejudiced teachers. I'm a little prejudiced to some Black people here. There are still some Blacks treated as if they are slaves, mostly because so many Black people are on welfare, so people who really need it can't get it. It's a shame.

This tendency to distance multicultural issues from immediate settings and personal experiences manifested itself in the discussions about the novel. The students' personal experiences about racism in the South were not elicited. In fact, only once during my observations did the discussion of racial inequality ever extend beyond the events portrayed in the book. Terrence, a gregarious and outspoken Black male, asked Lafitte if there was racism when she was a child growing up in the South. She perfunctorily answered, "Yes, but I went to Catholic schools which were not integrated." The discussion ended.

Interestingly, three weeks earlier, the class had read *The Diary of Anne Frank*, which evoked poignant and personal comments from both students and teacher. During one discussion, Lafitte asked the students to look around the room and name those students who would have been oppressed during Hitler's reign and those who would have been part of his chosen race. As students began to point out those dark-haired and dark-eyed students as the persecuted, Lafitte quickly reminded them that indeed most of them would have been targets for persecution because of their religion, the large majority of the students being Catholic. One White female student then asked Lafitte if she would have helped Anne and her Jewish family if she had lived during their time. Lafitte answered "I think so. I would go out on a limb." Then she added, "I don't know in this society."

Because there were no Jewish students in the class and because the Holocaust is not an issue of local concern or personal significance to these students, the persecution of Jews in the 1930s and 1940s in Europe is a "safe" and, thus, appropriate topic to discuss; the persecution of the Logan family in *Roll of Thunder, Hear My Cry* and the oppression of Blacks in the South today is not a "safe" topic and, thus, should be avoided. Wong (1993) talks about this contradiction as being the "not in my backyard" syndrome:

> To learn about another people and another culture would be wonderful for one's education, but if this people or culture happens to be a domestic minority from one's backyard, and alive and kicking and making noise to boot, then the humanistic ideal be damned—let us stick to exotic peoples, preferably safely dead, in faraway lands! (p. 112)

Despite Lafitte's initial belief that her White students "would be more understanding of Blacks if they read about their experiences in a book," her comments and discussions often served to detract attention from racial conflicts between Blacks and Whites as illustrated by the following comment she made about a Black female character in the novel: "She [the Black character] says some White people have to put Blacks down to feel better about themselves. I think this is true of a lot of people, not just Whites. Before they can step up, they feel they have to step on others." Furthermore, she never asked the White students if they would have helped the Black family in the novel, a question that had been asked in the *Diary of Anne Frank* discussion. She also distanced the setting of the book, Mississippi, from their immediate setting by saying: "I know you may have difficulty reading the book, but sometimes the dialect of the Deep South is hard to understand." However, the most illuminating example of how issues of racial unrest were distanced, detached, and depersonalized is revealed in the following discussion that involved Lafitte and two Black students. Lafitte initiated a discussion by saying, "If you were Black and read the book, you might get angry." (I wrote in my field notes next to this comment, "Why didn't she just ask the Black students how they felt about the book?") A Black female, who never participated in any other discussion, quickly responded, "It don't bother me." Terrence, who was the only Black student who consistently responded in class discussions, asserted, "It bothered me when they burnt that Black man." Lafitte quickly ended the discussion by saying, "The author did not write this book to make you take negative action."

## "Schools Can't Cure the Problems of Society": Eliminating Racism and Social Action From the Multicultural Agenda

Centerville's approach to multicultural education subscribes to the belief that being a good citizen and a productive American worker requires individuals to appreciate the ethnic diversity that comprises America, the "melting pot." However, according to McCarthy (1990), the problem with this kind of approach to multicultural education is that discussions of racism get erased from the multicultural agenda: "By focusing on sensitivity training and on individual differences, multicultural proponents typically skirt the very problem which multicultural education seeks to address: WHITE RACISM" (pp. 34–35). Such a depoliticized view of schooling and multiculturalism dismisses the belief that schools should be about educating for social action; therefore, discussions of racism have no place in the schools' curriculum.

Inevitably, honest discussions of racism will make people uncomfortable, angry, and even fearful as they are forced to acknowledge the institutional discriminatory practices in their own school (e.g., tracking, special education classifications, criteria for gifted education, selection of cheerleaders), in their own community, and in the larger society. In the classroom, racism is viewed as a volatile subject that can be explosive and divisive. In short, discussions of racism translate into discipline problems and discipline problems should be avoided at all costs—even if the cost is truthfulness. Thus, talking about racism in many classrooms, even ones that purport to have a multicultural curriculum like those in Centerville, actually means talking around racism, talking in generalities and half-truths, talking in ways that do not make people feel uncomfortable or angry or afraid—talking in "safe" ways.

Most of the students interviewed were well aware that the silences and half-truths evident in the discussions of *Roll of Thunder, Hear My Cry* were conscientious actions on the students' part to maintain a "safe" level of discussion. As evident in the following conversations between the author and two Black male students, open and honest discussions about racism were considered grounds for "getting in trouble":

Author: Are there things that you would say about that book that you don't?

Terence: Yes ma'am, like how they treated them people. It gets me real upset to think how someone was treated that was just as much a man, just as much a woman, just as much a boy, just as much a girl that you are and get treated that bad. I think that's real sad.

Author: Have you gotten to talk about that much in class?

Terrence: No ma'am.

Author: Why do you think?

Terrence: I guess Ms. Lafitte don't want any racial stuff, cause, you know, some people might get upset and fight cause of that, stupid stuff like that.

Author: How come the other Blacks don't say anything?

Terrence: Cause they're scared of what might happen to them; they might get in trouble, but I risk getting in trouble to say what I think is right.

Author: Did that book make you angry?

L. J.: Some parts, when I come to the racist parts, I get angry sometimes.

Author: And so you don't say anything?

L. J.: No ma'am. I just be quiet.

Author: Why?

L. J.: So I won't get in trouble.

Lisa, a White female participant, rationalized the silence of her Black classmates as follows: "They don't talk about the book cause they know they're going to get in trouble if they do and get beat up by the White people if they said anything."

Although most of the Black students were reticent to discuss issues of racism in the class, according to LaTanga, a Black female, some of the Black students did express anger outside of class about the instances of racism in the book. However, LaTanga felt their anger was unjustified:

> Some of them in class think that the way they treated the Blacks in those days was wrong, and some of the Blacks in our class are getting mad. But, I don't see reason to get mad, you know, because they're not doing it these days, [lowers voice] so I don't see why they're getting mad.

Also, while discussions of the novel were dominated by the White students, many White students were also quiet during class discussions. They felt that honest discussions about racism would lead to further conflicts between Blacks and Whites as evident in the following comment made by David, a White male:

> The Whites in the class didn't want to talk about it [the novel] cause they didn't want to offend the Blacks. Personally, I don't like to talk about it [racism] cause I know I might offend them.

Although Lafitte believed the novel would encourage students to talk about racism, she expressed on numerous occasions that schools "could not and should not cure the problems of society." This belief that schools should not involve themselves in political and social issues, such as establishing equality in the world at large, was quite implicit in the multicultural curriculum at Centerville. Consequently, finding solutions to racial inequality, encouraging students and teachers to challenge racism in their school and community, and educating students for social action were not goals of Centerville's multicultural literature-based English curriculum.

As discussed in the above sections, Lafitte's attempts to talk about racism failed because her school implemented a multicultural curriculum without examining the discontinuities, ambiguities, and tensions between the discourse of schooling and the discourse of multiculturalism. In short, the teachers and administrators failed to seriously consider the many "what does it mean" questions embedded in their own philosophy of multicultural education. A critical analysis of multicultural education would have forced the teachers at Centerville Middle School to ask: What *should* it mean to have a multicultural English curriculum?

## Conclusion: What Should It Mean To Have a Multicultural English Curriculum?

First, having a multicultural English curriculum means getting beyond the "talkinbout" kind of classroom discourse to a kind of classroom language that fosters truthfulness in our students' voices as well as our own. Brown (1991) describes this "talkinbout" form of discourse as:

> ... a peculiarly stiff, jargon-ridden language of expression or reflection. It is a language of work and technique, oriented toward achieving some narrowly (and often trivially) defined success, rather than toward achieving deeper understanding. It is about effectiveness, not truthfulness or rightness in the moral sense. (p. 234)

For students to talk honestly, an atmosphere of trust and community must be evident in the classroom so that students can learn to "talk back" without negative consequences. Talking back, as described by hooks (1989), means expressing one's self honestly with an insistence upon being heard; it means making one's presence count. Terrence succinctly explained the significance of his talking back about issues of racism: "My mama told me if you can't express yourself it makes no sense to talk."

Secondly, having a multicultural English curriculum means that the texts we use, the instructional strategies we employ, the discussions we initiate, the classroom climate we provide, and the student talk we encourage reflect the belief that students have a voice in the classroom and their lived experiences are valued and seen as an important springboard for future learning. Furthermore, as teachers and students, we must collectively challenge many long held assumptions about schooling that are, in fact, antithetical to the philosophy of multiculturalism. We must also critically examine the political implications of multiculturalism itself. Whose purpose does it serve? Does it further perpetuate the "not in my backyard syndrome"? Adding literature written by non-White authors to the English curriculum without challenging and changing how we teach or why we teach is but one more form of tokenism.

In short, having a multicultural English curriculum means a commitment to reading texts reflective of a diversity of cultural experiences for the purpose of exploring how difference is constructed and used to marginalize those considered to be the "Other" (Giroux, 1992). It means helping our students and ourselves develop a language for talking back—for challenging oppression and domination. Only then can we become critical citizens, intent on changing the power structures that sustain racism in our society.

## References

Brown, R. (1991). *Schools of thought*. San Francisco: Jossey-Bass.

Freire, P. (1983). *Pedagogy of the oppressed*. New York: Continuum.

Giroux, H. (1992). Textual authority and the role of teachers as public intellectuals. In C. M. Hurlbert & S. Totten (Eds.), *Social issues in the English classroom* (pp. 304–321). Urbana, IL: National Council of Teachers in English.

hooks, b. (1989). *Talking back: Thinking feminist, thinking black*. Boston: South End Press.

hooks, b. (1993). Transformative pedagogy and multiculturalism. In T. Perry & J. Fraser (Eds.), *Freedom's plow: Teaching in the multicultural classroom* (pp. 91–97). New York: Routledge.

Kutz, E., & Roskelly, H. (1991). *An unquiet pedigogy: Transforming practice in the English classroom*. Portsmouth, NH: Heinemann.

McCarthy, C. (1990). *Race and curriculum*. New York: The Falmer Press.

Rose, M. (1989). *Lives on the boundary*. New York: Free Press.

Shannon, P. (1992). Commercial reading materials, a technological ideology, and the deskilling of teachers. In P. Shannon (Ed.), *Becoming political* (pp. 182–207). Portsmouth, NH: Heinemann.

Wong, S. (1993). Promises, pitfalls, and principles of text selection in curricular diversification: The Asian-American case. In T. Perry & J. Fraser (Eds.), *Freedom's plow: Teaching in the multicultural classroom* (pp. 109–120). New York: Routledge.

# Multicultural Education

## S. REX MORROW, ED.D.
*Associate Professor of Education, Old Dominion University*

Multicultural education is a complex and controversial curriculum which was born in the years immediately following the societal challenges of the American Civil Rights movement in the 1960s and 1970s. The range of discussion and interpretation in defining multicultural education includes the identification and inclusion of significant individuals of diversity in American history to a more social active role of students as problem solvers of social injustice and prejudice in modern American society. Due to the continuing discussion as to what should be the role of multicultural education in the public school curriculum, I provide the formal guidelines for multicultural education as prescribed by the National Council for the Social Studies (NCSS) in 1976. The guidelines were designed to provide teachers and school leaders with a basic understanding and sense of curriculum direction necessary to infuse multicultural education into preexisting curricula.

According to NCSS (1976), multicultural/multiethnic education is effective ethnic and cultural studies instruction that can best take place within a school that is sensitive to ethnic and cultural diversity. The focus of multicultural education is ethnic and cultural pluralism, a concept celebrating the cultural contributions of all groups in a society. Teachers from preschool on are asked to modify their curricula and teaching strategies so they reflect the ethnic and cultural diversity characteristic of American society. Additionally, multicultural education is needed by all students, not just minority studies.

Paralleling the movement toward more culturally responsive instruction is the emergence of global education. Global education focuses on world studies and integrating an international perspective in the K–12 schooling. Educational leadership for both multicultural education and global education largely emerged from the social science educational community. Both educational groups share many unique goals and objectives in seeking curriculum reform and school improvement. Since multicultural education targets school learning about diverse ethnic and cultural groups within one's country or nation, many intercultural educators contend multicultural education is a vital movement within the broader context of what is global education. As global education seeks to provide for improved international learning of cultures and societies—past, present, and future—curricula should focus not only on learning about Latin American nations and their cultures, but also about Hispanic or Latin cultures that exist within the United States. Just as teaching about modern issues of society and culture in Mexico is important, so is understanding needed by U.S. school students about issues of culture and society in Hispanic and other diverse communities across the United States. This is in keeping with the primary goal of global education which states that "developing cultural sensitivity and a sense of global interdependence is a challenge to all educators" (Bruce, Anderson, and Podemski, 1991).

---

S. Rex Morrow, Ed.D., Associate Professor of Education, Old Dominion University.

As early as 1973, the National Council for the Social Studies demonstrated interest in the issue of diversity education with the publication of *Teaching Ethnic Studies: Concepts and Strategies* (1973). The text, edited by James A. Banks, is a compilation of essays by leading social educators on how and why teachers should teach about ethnic issues. The essays are grouped into three sections: 1) Racism, Cultural Pluralism, and Social Injustice; 2) Teaching about Ethnic Minorities; and 3) Teaching about White Ethnic Groups and Women's Rights. One of Banks' leading contentions was that one way to assist in White dominant culture's awareness in teaching about ethnic diversity was to demonstrate that White Americans also have ethnic identity and encourage White majority students to discover their ethnic heritages. To this end, the text discussed several instructional practices to encourage students to learn about their English, Irish, German, Italian, Polish, and other Euroethnic backgrounds. An example activity is located in Appendix A at the end of this chapter.

In 1975, Banks wrote *Teaching Strategies for Ethnic Studies*, which is now in its fifth edition (1991). The text is a much more complete and teacher-centered text than his 1973 publication. In the 1991 edition, he provides many multi-ethnic teaching strategies for public school teachers, including: American Indians, Native Hawaiians, African American, European Americans/Jewish Americans, Hispanic Americans, and Asian Americans. It was in the 1975 edition of this text that Banks began systematically using the term "multicultural education."

One of the most noted cases which promoted multicultural awareness in the ability of students to learn in school was made famous in the movie, "Stand and Deliver." The film narrates the success of a Los Angeles mathematics teacher, Jaime Escalante, who worked with at-risk inner-city Hispanic youth. In 1978, four out of seven of Escalante's advanced placement pre-calculus students had passed the A.P. test. By 1987 Escalante and his school colleague, Benjamin Jimenez, had 89 passing students in A.P. pre-calculus out of a total of 129. This represented a 66% passage rate, a figure attained nowhere in the United States at that time (Davidman and Davidman, p. 14.). Escalante broke the mythology that at-risk students from the barrios of Los Angeles could not successfully master advanced mathematical principals, which were designed for America's college preparatory school curriculum.

In 1981, Banks wrote *Multiethnic Education: Theory and Practice*. In this text, Banks discusses how multicultural education has become international in scope. The book discusses how other nations of the world are coping and engaging in multi-ethnic and multicultural education as part of their national curricula for schools. In the 1988 edition, Banks adds several other national profiles, including Canada, the United Kingdom, and Australia. He further notes that some European countries, such as France, Switzerland, Germany, and the Netherlands, prefer the term "Intercultural Education." However, the educational objectives are largely the same and seek similar societal outcomes.

The 1980s began an era of significant growth and awareness in the need for educational planning for multicultural education. As a result, many publications were written and a school of educational leaders for multicultural education appeared. It was in the mid 1980s that U.S. multiculturalists intensified their efforts and created their own professional organization, the National Association of Multicultural Educators, or "NAME" as it is commonly referred. The organization first appeared as an affiliate of the Association of Teacher Educators, a major professional organization of professors in colleges of education. However, as the National Association of Multicultural Education grew and defined its own mission, it dissolved its linkages with the Association of Teacher Educators in the early 1990s. Today, NAME is a large professional organization that has a large annual meeting and several organizational publications. For more information, educators may contact the *National Association for Multicultural Education* at Southern University, College of Education, P.O. Box 9983, Baton Rouge, Louisiana 70613 or by telephone at (504) 771-2290.

In one of the first texts to use multicultural education in the title, Banks describes multicultural education as being three distinct operations: 1) an idea or concept; 2) an educational reform movement; and 3) a process. As a concept, multicultural education is an opportunity for all students to learn about their ethnic, racial, and cultural characteristics within the school setting.

A significant number of schools systematically deny equal opportunity for students to learn about their culture because the existing curriculum places little value on their cultural heritage. This situation exists regardless of socioeconomic status and physical and mental ability. Multicultural education is a reform movement which is attempting to change schools and educational agencies, so that students regardless of social class, gender, race, and culture will have an equal opportunity to learn. Lastly, multicultural education is a process with goals that will never be fully realized. Educational equality, like liberty and justice, is an ideal toward which human beings work but never fully attain. Racism, sexism, and handicapism will exist to some extent no matter how hard we work to eliminate these problems (Banks, 1989, p. 3).

Banks argues that multicultural education has become an international movement with educators examining reform in such countries as the United Kingdom, Germany, France, the Netherlands, Switzerland, Australia, and Canada. Furthermore, he states that other victimized groups stimulated by social ferment and the quest for human rights during the 1970s articulated their grievances and demanded that institutions be reformed so they would face less discrimination and acquire more human rights. Disabled persons, senior citizens, and gay rights advocates were among the groups that organized politically during this period and made significant inroads in changing institutions and laws. The Education for All Handicapped Children Act of 1975 (P. L. 94-172), which requires that handicapped students be educated in the least restricted environment and which institutionalized the word mainstreaming in education, is perhaps the most significant legal victory of the movement for the rights of the handicapped in education (Banks, 1989, p. 5). In 1995, the Americans with Disabilities Act furthered the rights of handicapped citizens to equal opportunity and access to both public and private facilities, businesses, and institutions throughout the United States.

In *Turning on Learning: Five Approaches for Multicultural Teaching Plans for Race, Class, Gender, and Disability*, Grant and Sleeter (1989) provide lesson planning strategies for classroom usage in promoting multicultural understanding. In "American Indians in Our State," the authors provide a lesson plan for students in grades 5–8, which: 1) identifies American Indian reservations on a state map and names tribes in the state; 2) lists towns, cities, rivers, and other geographical features in the states that have American Indian names; and 3) fosters appreciation for local Indian art and literature. This lesson plan is provided in Appendix B.

Grant and Sleeter (1989) also describe the importance of social reconstructionism in teaching for multicultural understanding. One of the classroom activities provided in this section is entitled "The Court System." This social studies lesson is designed for grades 8–12 and takes approximately one week to complete. The objectives include: 1) describing how the court system works and how a trial takes place; 2) explaining the roles of the main persons involved in the court system; and 3) distinguishing between the different kinds of courts and identifying the kinds of legal issues that each court deals with. The authors suggest that legal recourse is often the most successful approach in confronting human rights, as is well documented during the later half of this century in the Civil Rights Movement. This lesson plan is outlined in Appendix C.

It was not until the early 1990s that the issue of gay rights and sexual orientation emerged as another dimension of multicultural education. With the determination by the American Psychological Association that homosexuality is a condition of human experience and individual choice—not a social disease or mental illness which requires treatment—and a growing gay awareness by the U.S. public, gay and lesbian rights has become associated with the multicultural education movement. Gollnick and Chinn (1990) in *Multicultural Education in a Pluralistic Society*, fourth edition, state that adolescent homosexuals are either treated as though they do not exist or as objects of hate and bigotry. Professional educators have the responsibility for eradicating sexism in the classroom and school. Their roles require that they not limit the potential of any student because of gender or sexual orientation. However, in a study of school counselors and teacher candidates, it was found that educators, in general, lack the sensitivity, knowledge, and skills to address effectively the needs of students with same-sex feelings. Classroom interactions, resources, extracurricular activities, and counseling practices must be evaluated to ensure that students are not being discriminated against because of their sexual orientation. Many teachers have found that well planned and structured role play scenarios and student-interactive simulations focusing on controversial issues have been useful in aiding the majority culture in better understanding and recognizing the human rights and civil liberties of those of minority cultures.

In *Racial and Ethnic Relations in America*, S. Dale McLemore (1994) provides fourteen concise examinations for the reader to test his or her ethnic and multicultural knowledge of America. The following abridged test is designed to assist the reader in understanding how well he or she understands issues in Black American history and culture from the years of slavery to desegregation. The answers to these questions are found in Appendix D.

Black Americans: From Slavery to Segregation
*A Multicultural Awareness Inventory*

1. The United States Supreme Court ruling that school desegregation must proceed "with all deliberate speed" occurred in the _____ case.

   A. *Dred Scott*

   B. *Bakke*

   C. *Brown*

   D. *Griggs*

2. A notable armed revolt in Virginia by Black slaves against White domination was led by:

   A. Richard Frethorne

   B. Sojourner Truth

   C. Nat Turner

   D. Dred Scott

3. The United States Supreme Court decision that paved the way for the legalization of segregation in America (called Jim Crow) was

   A. *Dred Scott v. Sanford*, 1857; *Milkin v. Bradley*, 1864

   B. *Fredrick Douglass v. Jefferson Davis*, 1866; *Brown II*, 1965

   C. *Nat Turner v. Richmond*, 1833; *Green v. County School Board of New Kent County*, 1968

   D. *Plessy v. Ferguson*, 1896; *Brown v. Board of Education of Topeka*, 1954

4. The "hand" metaphor describing a type of pluralism in which groups are limited on matters essential to mutual progress but separate in social matters was proposed by:

   A. Booker T. Washington

   B. W. E. B. DuBois

   C. W. D. Fard

   D. Elijah Muhammad

5. Stanley Elkins argued that African slaves did not retaliate more openly and frequently than they did because:

   A. Even though the slave system of the U.S. was harsh, it was better than that of Latin America

   B. The plantation system was psychologically destructive for the slaves, reducing them to a child-like, irresponsible condition

   C. The masters used the carrot as well as the stick to motivate their slaves to be better workers

   D. The typical slave was a highly motivated and productive worker

6. The Jim Crow system of segregation that existed until the 1960s was:

   A. A custom of the South dating back to the time before the Civil War

   B. Only applied to transportation facilities in the South

   C. Created by the enactment of laws around the beginning of the present century

   D. Supported by W. E. B. DuBois

7. The twentieth century has seen an enormous immigration of Blacks to the North and of Mexicans to the U.S. The theoretical importance of these facts is stressed in:

   A. The doctrine of White supremacy

   B. The ideology of cultural pluralism

   C. The colonial model

   D. The immigrant model

8. In the famous Supreme Court case *Plessy v. Ferguson,* the U.S. Supreme Court:

   A. Cleared the way for the destruction of Jim Crowism

   B. Ruled that separate facilities were inherently unequal and violated the Fourteenth Amendment

   C. Legally approved the "separate but equal" doctrine

   D. Ruled that separate facilities are inherently unequal

9. The sharecropping system that developed in the South after the Civil War:

   A. Permitted the landowners significantly to improve their standard of living

   B. Permitted the tenants significantly to improve their standard of living

   C. Created a beneficent circle of lending and sales that reconstructed the southern agricultural economy

   D. Kept Black tenants in a condition of peonage that was hardly better than slavery

10. Beginning in 1890, the legal effort of White Southerners to force Black Americans back into a position of complete subordination focused on segregation in:

   A. Education

   B. Public transportation

   C. Housing

   D. Jobs

11. A number of Black protest organizations like the New Negro Alliance, CORE, and the March on Washington Movement were formed during the depression years 1930–1940. Their common strategy was:

   A. To persuade Black Americans to join the Communist Party

   B. The use of the economic boycott

   C. The use of mass and, if necessary, violent confrontations

   D. An effort to elect officials who would favor Black Americans and to defeat those who would not

12. White resistance to the *Brown* decision was immediately met by a change of tactics by Black Americans. The change was primarily to a greatly expanded use of:

   A. Violent protest

   B. Criminal litigation

   C. Slowdowns and strikes

   D. Nonviolent direct action

13. T or F: President Lincoln favored deportation as a solution to the "Negro problem."

14. T or F: The historical record indicates that the Blacks' resistance to slavery was minimal, with only a handful of insurrections being planned.

15. T or F: The campaign of the old southern planter to end Radical Republican Reconstruction and capture political control of the South was successful.

The issues of multicultural education in the twenty-first century will continue to present challenges to the classroom teacher attempting to provide a positive and nurturing learning climate that exemplifies the world of diversity in which our students live. In the 1990s the term, "at-risk" has been used to denote any group that presents a challenge for successful achievement in U.S. schools. To a large degree, "at-risk" is defined as any student at risk of failure in a White dominant school culture. It will be more constructive for educators to identify the dominant culture of the individual student and use this awareness to motivate and scholastically challenge students to new heights in learning and individual achievement. Ronald Takaki (1994) in *From Different Shores* summarizes the current context of racism and prejudice in America by stating to diminish the significance of racial oppression in America's past and to define racial inequality as a problem of prejudice and limit the solution as the outlawing of individual acts of discrimination is effectively to leave intact the very structures of racial inequality (p. 34). Takaki concludes that until the institutions of American society are reformed so as to reflect the multicultural nature of this nation, prejudice and racism will continue as a major social issue in the U.S.

# References

Banks, James A. *Teaching Ethnic Studies: Concepts and Strategies.* National Council for the Social Studies: Washington D.C., 1973.
—*Teaching Strategies For Ethnic Studies* (5th edition). Allyn and Bacon: Needham Heights, MA, 1991.
—*Multicultural Education: Theory and Practice.* Allyn and Bacon: Newton, MA, 1988.
Banks, James A., and Cherry A. Banks. *Multicultural Education: Issues and Perspectives.* Allyn and Bacon: Needham Heights, MA, 1989.
Bruce, Michael G., Podemski, Richard S., and Carrel M. Anderson. "Developing a Global Perspective: Strategies for Teacher Education Programs." *Journal of Teacher Education.* Washington D.C.: American Association of Colleges for Teacher Education, vol. 42, no. 1, 1991, pps. 21–27.
Davidman, Leonard and Pamela Davidman. *Teaching with a Multicultural Perspective: A Practical Guide.* Longman: New York, 1994.
Gollnick Donna M. and Philip C. Chinn. *Multicultural Education in a Pluralistic Society* (4th\tab edition). Merrill Publishing: New York, 1990.
Grant, Carl A. and Christine E. Sleeter. *Turning on Learning: Five Approaches for Multicultural Teaching Plans for Race, Class, Gender, and Disability.* Merrill Publishing: New York, 1989.
McLemore S. Dale. *Racial and Ethnic Relations in America: Test Items* (4th edition). Allyn and Bacon: Needham Heights, MA, 1994.
National Council for the Social Studies. *Curriculum Guidelines for Multiethnic Education: Position Statement.* National Council for the Social Studies: Washington D.C., 1976.
Takaki, Ronald. *From Different Shores: Perspective on Race and Ethnicity in America.* Oxford University Press, New York, 1994.

# Appendix A

**Concept:** Immigration
**Generalization:** Europeans immigrated to the United States for various economic, political, and social reasons. Their experiences in the United States were both similar and different.

1. To help the students gain the needed content background to study American immigration, assign the appropriate readings that will enable them to answer the following questions about the first or old immigrants to America.

    A. What European nations did the first immigrants come from during the colonial period?

    B. Why did they come?

    C. Was America like they expected? If so, in what ways? If not, why not? Explain.

2. After the students have completed their reading assignment, discuss the three questions above with the class. During the discussion, list on the board the reasons that various groups immigrated to the U.S. When the reasons have been listed, group them with the class, into three or four categories, such as "economic," "political," and "social."

3. Ask the students to read about the new immigrants who came to the U.S. from Europe in the late 1800s and early 1900s. The students should discuss the same three questions as previously listed.

4. After the students have read and discussed the old and new immigrants, they should compare and contrast these groups. The following questions can be helpful in guiding the discussion.

    A. Did the old and new immigrants come to the U.S. for similar or different reasons? Explain.

    B. How did the new immigrants differ from the old? Why?

    C. How did American life differ at times when the old and new immigrants came to America?

    D. Both the old and new immigrants experienced problems on the trip across the Atlantic. How were these problems similar and different?

    E. How were the problems of settlement and finding jobs in America similar and different for the two groups of immigrants?

    F. Ethnic conflict developed in the settlement of European nationality groups in America. What problems of prejudice and discrimination were experienced by the various groups? Which immigrant groups were discriminated against the most? The least? Why?

5. European immigrants in the United States often wrote to their friends and relatives in Europe describing the wonders of America and occasionally their problems. Three such letters are reprinted in *America's Immigrants: Adventures in Eyewitness History* by Rhoda Hoff. "A letter to a friend at home" by a Norwegian immigrant tells about the greatness of America. "Letter to mother and daughter in Norway" by Guri Endersen talks about tragic encounters her family had with American Indians. Read and discuss these letters with the class. Ask the students to pretend that they are new European immigrants

in America in the 1800s. They should write to a friend or relative in Europe telling about their experiences.

6. Have students role play the situation below, which involves a poor Italian farmer and an agent of a steamship company who tries to persuade the farmer to immigrate to the U.S. After the role play, ask students the questions that follow.

## Mr. Parieto, a poor Italian farmer in southern Italy in the 1800s.

Mr. Parieto is in his thirties. He is a hard worker and is close to his family, which includes his wife, eight children, and both of his parents. For the past three years, Mr. Parieto has been unable to feed and clothe his children well because of severe crop failures. He has heard about the greatness of America and has often thought about going there. However, he knows that his father feels he should stay in Italy so that he can depend on him in his old age. He also realizes that if he goes to America he will have to leave his wife and children in Italy.

## Mr. Ross, an agent for a steamship company that makes trips to America.

Mr. Ross tries to persuade Italian men to immigrate to America. The more men that he can persuade to go to America on his company's ship, the more money he makes. He goes up to Mr. Parieto at the village market and tells him about the wonders and opportunities in America. He tells Mr. Parieto that if he goes to America he will not be able to carry his family. However, Mr. Ross tells Mr. Parieto that he will be able to send for his family within two or three months after he arrives in America.

## Questions:

1. Did Mr. Ross persuade Mr. Parieto to go to America? Why or why not?

2. If Mr. Parieto goes to America what do you think will happen to his family?

3. If Mr. Parieto stays in America, how do you think he will take care of his family?

4. How do you think his wife, parents, and children will react if Mr. Parieto goes to America?

5. What else can Mr. Parieto do besides stay in his village or immigrate to America?

6. If you were Mr. Parieto, would you immigrate to America? Why or why not?

7. Conclude this unit by viewing a film or videotape about European immigration.

(Abridged activity from *Teaching Strategies for Ethnic Studies,* pages 258–262.)

# Appendix B
# American Indians in Our State

**Subject: Interdisciplinary Lesson**

**Time: Three or four class periods**

## Objectives:

1. Students will identify American Indian reservations on a state map, and identify and name the tribes in the state.

2. Students will list cities, rivers, and other geographical features in the state that have American Indian names.

3. Students will appreciate local American Indian art and literature.

## Suggested Procedures:

1. Ask students if they know what American Indian tribes live in the state. List the tribes on the board, and teach students how to pronounce each name.

2. On a wall map, show where American Indian reservations are located in the state, and where the majority of the tribes live.

3. Ask students if they are aware of any American Indian names of places in the state. Start a board list. Help students see several cultural legacies left by American Indians in the state, and if possible, locate them on a state map.

4. Share pictures and examples of American Indian artwork and crafts. Have students engage in creating crafts and artwork indicative of Indian groups in the state.

5. Select folktales, poems, myths, and songs produced by American Indians and have students enact or recreate them and discuss their significance.

## Evaluation:

1. Evaluate students' appreciation of American Indian art by 1) their reactions and discussion, or 2) by evaluating their replica crafts and artwork.

2. Evaluate students' knowledge of American Indian reservations (yesterday and today), tribal names, Indian vocabulary and place names, and modern day perspectives on Indians in the state, etc.

(Abridged activity is from *Turning On Learning*, pages 100–111.)

# Appendix C
# The Court System

**Subject: Social Studies**

**Grade Level: 8–12**

**Time: Ongoing**

## Objectives:

1. Students will use democratic procedures to establish classroom rules.

2. Students will use courtroom procedures to enforce classroom rules.

3. Students will describe how the court system works and how a trial takes place.

4. Students will describe local agencies that help low-income and minority families with legal problems.

5. Students will analyze local legal conflicts involving race, class, gender, or disability issues.

## Suggested Procedures:

1. Ask students to brainstorm a list of classroom conditions that they need to study and learn (e.g., lighting, heat, room, noise, and so on); write the list on the board. Ask them to discuss the list and to suggest any additions or corrections.

2. Explain that students will select the rules that they will follow daily in the classroom. Take a secret ballot vote to give each student the chance to express his or her opinion. If the rules receive a majority vote, they are passed. If they do not pass, return to the list and discuss changes. This may be time-consuming, but students' time is not wasted if the teacher explains that this is the way laws are passed, and that discussion is important.

3. Ask students if they have ever seen a courtroom, either in person or on television. Explain that in a democratic society, courts are considered a fair way to decide an individual's guilt. Ask why it is better for a court and judge/jury to decide a person's guilt rather than an individual person.

4. Arrange a visit to a local courthouse to watch a real trial in action.

5. Ask students to write an essay showing how they think the legal system works and possible problems in the system.

6. Invite a member of the legal community to come to the classroom as a learning resource about the legal system.

7. Have students role-play a local trial involving a race, social-class, disability, or gender issue. Help students gather information on the case. Assign the following roles: judge, bailiff, prosecutor, plaintiff, defendant, defense counsel, witnesses, jury, and court reporter. Conduct the court using the following steps:

   a. Opening of the court
   b. Selection of the jury
   c. Opening instructions of jury
   d. Opening statements
   e. Direct examination and cross examination of the defendant and witnesses
   f. Closing statements
   g. Jury deliberation
   h. The verdict

Have the class discuss or debrief the case discussing issues of fairness and equality.

## Evaluation:

1. Assess students' skills in using democratic and legal procedures through their development and enforcement of the classroom rules.
2. Assess students' understanding of how the court system operates and trials take place through a quiz.
3. Assess students' analyses of race, class, gender, and disability issues in legal conflict and their knowledge of local agencies that deal specifically with these issues through class discussion and a quiz/test.

(Abridged activity is from *Turning On Learning*, pages 250–252.)

# Appendix D
# Multicultural Awareness Test—Answer Key

Black Americans: From Slavery to Segregation
*A Multicultural Awareness Inventory*

1. The United States Supreme Court ruling that school desegregation must proceed "with all deliberate speed" occurred in the _____ case.

    A. *Dred Scott*

    B. *Bakke*

    **C. *Brown***

    D. *Griggs*

2. A notable armed revolt in Virginia by Black slaves against White domination was led by:

    A. Richard Frethorne

    B. Sojourner Truth

    **C. Nat Turner**

    D. Dred Scott

3. The United States Supreme Court decision that paved the way for the legalization of segregation in America (called Jim Crow) was

    A. *Dred Scott v. Sanford*, 1857; *Milkin v. Bradley*, 1864

    B. *Fredrick Douglass v. Jefferson Davis*, 1866; *Brown II*, 1965

    C. *Nat Turner v. Richmond*, 1833; *Green v. County School Board of New Kent County*, 1968

    **D. *Plessy v. Ferguson,* 1896; *Brown v. Board of Education of Topeka,* 1954**

4. The "hand" metaphor describing a type of pluralism in which groups are limited on matters essential to mutual progress but separate in social matters was proposed by:

    **A. Booker T. Washington**

    B. W. E. B. DuBois

    C. W. D. Fard

    D. Elijah Muhammad

5. Stanley Elkins argued that African slaves did not retaliate more openly and frequently than they did because:

   A. Even though the slave system of the U.S. was harsh, it was better than that of Latin America

   **B. The plantation system was psychologically destructive for the slaves, reducing them to a childlike, irresponsible condition**

   C. The masters used the carrot as well as the stick to motivate their slaves to be better workers

   D. The typical slave was a highly motivated and productive worker

6. The Jim Crow system of segregation that existed until the 1960s was:

   A. A custom of the South dating back to the time before the Civil War

   B. Only applied to transportation facilities in the South

   **C. Created by the enactment of laws around the beginning of the present century**

   D. Supported by W. E. B. DuBois

7. The twentieth century has seen an enormous immigration of Blacks to the North and of Mexicans to the U.S. The theoretical importance of these facts is stressed in:

   A. The doctrine of White supremacy

   B. The ideology of cultural pluralism

   C. The colonial model

   **D. The immigrant model**

8. In the famous Supreme Court case *Plessy v. Ferguson,* the U.S. Supreme Court:

   A. Cleared the way for the destruction of Jim Crowism

   B. Ruled that separate facilities were inherently unequal and violated the Fourteenth Amendment

   **C. Legally approved the "separate but equal" doctrine**

   D. Ruled that separate facilities are inherently unequal

9. The sharecropping system that developed in the South after the Civil War:

   A. Permitted the landowners significantly to improve their standard of living

   B. Permitted the tenants significantly to improve their standard of living

   C. Created a beneficent circle of lending and sales that reconstructed the southern agricultural economy

   **D. Kept Black tenants in a condition of peonage that was hardly better than slavery**

10. Beginning in 1890, the legal effort of White Southerners to force Black Americans back into a position of complete subordination focused on segregation in:

    A. Education

    **B. Public transportation**

    C. Housing

    D. Jobs

11. A number of Black protest organizations like the New Negro Alliance, CORE, and the March on Washington Movement were formed during the depression years 1930–1940. Their common strategy was:

    A. To persuade Black Americans to join the Communist Party

    **B. The use of the economic boycott**

    C. The use of mass and, if necessary, violent confrontations

    D. An effort to elect officials who would favor Black Americans and to defeat those who would not

12. White resistance to the *Brown* decision was immediately met by a change of tactics by Black Americans. The change was primarily to a greatly expanded use of:

    A. Violent protest

    B. Criminal litigation

    C. Slowdowns and strikes

    **D. Nonviolent direct action**

13. **T** or F: President Lincoln favored deportation as a solution to the "Negro problem."

14. T or **F**: The historical record indicates that the Blacks' resistance to slavery was minimal, with only a handful of insurrections being planned.

15. **T** or F: The campaign of the old southern planter to end Radical Republican Reconstruction and capture political control of the South was successful.

(Abridged from *Racial and Ethnic Relations in America: Test Items.*)

# MI and Curriculum Development

## THOMAS ARMSTRONG

> We do not see in our descriptions [of classroom activity] . . . much opportunity for students to become engaged with knowledge so as to employ their full range of intellectual abilities. And one wonders about the meaningfulness of whatever is acquired by students who sit listening or performing relatively repetitive exercises, year after year. Part of the brain, known as Magoun's brain, is stimulated by novelty. It appears to me that students spending twelve years in the schools we studied would be unlikely to experience much novelty. Does part of the brain just sleep, then?
>
> —John I. Goodlad (1984, p. 231)

MI theory makes its greatest contribution to education by suggesting that teachers need to expand their repertoire of techniques, tools, and strategies beyond the typical linguistic and logical ones predominantly used in American classrooms. According to John Goodlad's pioneering "A Study of Schooling" project, which involved researchers in observing over 1,000 classrooms nationwide, nearly 70 percent of classroom time was consumed by "teacher" talk—mainly teachers talking "at" students (giving instructions, lecturing). The next most widely observed activity was students doing written assignments, and according to Goodlad (1984, 230), "much of this work was in the form of responding to directives in workbooks or on worksheets." In this context, the theory of multiple intelligences functions not only as a specific remedy to one-sidedness in teaching, but also as a "metamodel" for organizing and synthesizing all the educational innovations that have sought to break out of this narrowly confined approach to learning. In doing so, it provides a broad range of stimulating curricula to "awaken" the slumbering brains that Goodlad fears populate our nation's schools.

## The Historical Background of Multimodal Teaching

Multiple intelligences as a philosophy guiding instruction is hardly a new concept. Even Plato, in a manner of speaking, seemed aware of the importance of multimodal teaching when he wrote: ". . . do not use compulsion, but let early education be a sort of amusement; you will then be better able to find out the natural bent" (Plato, 1952, p. 399). More recently, virtually all the pioneers of modern education developed systems of teaching based upon more than verbal pedagogy. The 18th century philosopher Jean Jacques Rousseau declared in his classic treatise on education, *Emile*, that the child must learn not through words, but through experience; not

---

Thomas Armstrong, "MI and Curriculum Development," *Multiple Intelligences in the Classroom,* 1994, pp. 48–64. Reprinted by permission from Association for Supervision and Curriculum Development.

through books, but through "The book of life." The Swiss reformer Johann Heinrich Pestalozzi emphasized an integrated curriculum that regarded physical, moral, and intellectual training based solidly on concrete experiences. And the founder of the modern-day kindergarten, Friedrich Froebel, developed a curriculum consisting of hands-on experiences with manipulatives ("gifts"), playing games, singing songs, gardening, and caring for animals. In the 20th century, innovators like Maria Montessori and John Dewey evolved systems of instruction based upon multiple-intelligence-like techniques, including Montessori's tactile letters and other self-paced materials, and Dewey's vision of the classroom as a microcosm of society.

By the same token, many current alternative educational models essentially are multiple-intelligence systems using different terminologies (and with varying levels of emphasis upon the different intelligences). Cooperative learning, for example, seems to place its greatest emphasis upon interpersonal intelligence, yet specific activities can involve students in each of the other intelligences as well. Similarly, whole language instruction has at its core the cultivation of linguistic intelligence, yet it uses music, hands-on activities, introspection (through journal keeping), and group work to carry out its fundamental goals. Suggestopedia, a pedagogical approach developed by the Bulgarian psychiatrist Georgi Lozanov, uses drama and visual aids as keys to unlocking a student's learning potential, yet it seems that in this approach music plays the greatest role in facilitating learning, for students listen to music as an integral part of their instruction.

MI theory essentially encompasses what good teachers have always done in their teaching: reaching beyond the text and the blackboard to awaken students' minds. Two recent movies about great teachers, *Stand and Deliver* (1987) and *Dead Poets Society* (1989), underline this point. In *Stand and Deliver*, Jaime Escalante (played by Edward James Olmos), a Hispanic high school mathematics teacher, uses apples to introduce fractions, fingers to teach multiplication, and imagery and metaphor to clarify negative numbers (if one digs a hole in the ground, the hole represents negative numbers, the pile of dirt next to it signifies positive numbers). John Keating (played by Robin Williams), the prep school instructor in *Dead Poets Society*, has students reading literary passages while kicking soccer balls and listening to classical music. MI theory provides a way for *all* teachers to reflect upon their best teaching methods and to understand why these methods work (or why they work well for some students but not for others). It also helps teachers expand their current teaching repertoire to include a broader range of methods, materials, and techniques for reaching an ever wider and more diverse range of learners.

## The MI Teacher

A teacher in an MI classroom contrasts sharply with a teacher in a traditional classroom. In the traditional classroom, the teacher lectures while standing at the front of the classroom, writes on the blackboard, asks students questions about the assigned reading or handouts, and waits while students finish their written work. In the MI classroom, the teacher continually shifts her method of presentation from linguistic to spatial to musical and so on, often combining intelligences in creative ways.

The MI teacher may spend part of the time lecturing and writing on the blackboard at the front of the room. This, after all, is a legitimate teaching technique. Teachers have simply been doing too much of it. The MI teacher, however, also draws pictures on the blackboard or shows a videotape to illustrate an idea. She often plays music at some time during the day, either to set the stage for an objective, to make a point, or to provide an environment for study. The MI teacher provides hands-on experiences, whether this involves getting students up and moving about, or passing an artifact around to bring to life the material studied, or having students build something tangible to reveal their understanding. The MI teacher also has students interacting with each other in different ways (e.g., in pairs, small groups, or large groups), and she plans time for students to engage in self-reflection, undertake self-paced work, or link their personal experiences and feelings to the material being studied.

Such characterizations of what the MI teacher does and does not do, however, should not serve to rigidify the instructional dimensions of MI theory. The theory can be implemented in a wide range of instructional contexts, from highly traditional settings where teachers spend much of their time directly teaching students to open environments where students regulate most of their own learning. Even traditional teaching can take place in a variety of ways designed to stimulate the seven intelligences. The teacher who lectures with rhythmic emphasis (musical), draws pictures on the board to illustrate points (spatial), makes dramatic gestures as she talks (bodily-kinesthetic), pauses to give students time to reflect (intrapersonal), and asks questions that invite spirited interaction (interpersonal) is using MI principles within a teacher-centered perspective.

## Key Materials and Methods of MI Teaching

There are a number of teaching tools in MI theory that go far beyond the traditional teacher-as-lecturer mode of instruction. Figure 1 provides a quick summary of MI teaching methods. The following list provides a broader, but still incomplete, survey of the techniques and materials that can be employed in teaching through the multiple intelligences.

## Linguistic Intelligence

- lectures
- large- and small-group discussions
- books
- worksheets
- manuals
- BRAINSTORMING
- writing activities
- word games
- sharing time
- student speeches
- STORYTELLING
- talking books and cassettes
- extemporaneous speaking
- debates
- JOURNAL KEEPING
- choral reading
- individualized reading
- reading to the class
- memorizing linguistic facts

## FIGURE 1
### Summary of the "Seven Ways of Teaching"

| Intelligence | Teaching Activities (examples) | Teaching Materials (examples) | Instructional Strategies |
|---|---|---|---|
| Linguistic | lectures, discussions, word games, storytelling, choral reading, journal writing, etc. | books, tape recorders, typewriters, stamp sets, books on tape, etc. | read about it, write about it, talk about it, listen to it |
| Logical-Mathematical | brain teasers; problem solving, science experiments, mental calculation, number games, critical thinking, etc. | calculators, math manipulatives, science equipment, math games, etc. | quantify it, think critically about it, conceptualize it |
| Spatial | visual presentations, art activities, imagination games, mind-mapping, metaphor, visualization, etc. | graphs, maps, video, LEGO sets, art materials, optical illusions, cameras, picture library, etc. | see it, draw it, visualize it, color it, mind-map it |
| Bodily-Kinesthetic | hands-on learning, drama, dance, sports that teach, tactile activities, relaxation exercises, etc. | building tools, clay, sports equipment, manipulatives, tactile learning resources, etc. | build it, act it out, touch it, get a "gut feeling" of it, dance it |
| Musical | superlearning, rapping, songs that teach | tape recorder, tape collection, musical instruments | sing it, rap it, listen to it |
| Interpersonal | cooperative learning, peer tutoring, community involvement, social gatherings, simulations, etc. | board games, party supplies, props for role plays, etc. | teach it, collaborate on it, interact with respect to it |
| Intrapersonal | individualized instruction, independent study, options in course of study, self-esteem building, etc. | self-checking materials, journals, materials for projects, etc. | connect it to your personal life, make choices with regard to it |

| Intelligence | Sample Educational Movement, (primary intelligence) | Sample Teacher Presentation Skill | Sample Activity to Begin a Lesson |
|---|---|---|---|
| Linguistic | Whole Language | teaching through storytelling | long word on the blackboard |
| Logical-Mathematical | Critical Thinking | Socratic questioning | posing a logical paradox |
| Spatial | Integrated Arts Instruction | drawing/mind-mapping concepts | unusual picture on the overhead |
| Bodily-Kinesthetic | Hands-On Learning | using gestures/dramatic expression | mysterious artifact passed around the class |
| Musical | Suggestopedia | using voice rhythmically | piece of music played as students come into class |
| Interpersonal | Cooperative Learning | dynamically interacting with students | "Turn to a neighbor and share . . ." |
| Intrapersonal | Individualized Instruction | bringing *feeling* into presentation | "Close your eyes and think of a time in your life when . . ." |

- TAPE RECORDING ONE'S WORDS
- using word processors
- PUBLISHING (e.g., creating class newspapers)

## Logical-Mathematical Intelligence

- mathematical problems on the board
- SOCRATIC QUESTIONING
- scientific demonstrations
- logical problem-solving exercises
- CLASSIFICATIONS AND CATEGORIZATIONS
- creating codes
- logic puzzles and games
- QUANTIFICATIONS AND CALCULATIONS
- computer programming languages
- SCIENCE THINKING
- logical-sequential presentation of subject matter
- Piagetian cognitive stretching exercises
- HEURISTICS

## Spatial Intelligence

- charts, graphs, diagrams, and maps
- VISUALIZATION
- photography
- videos, slides, and movies
- visual puzzles and mazes
- 3-D construction kits
- art appreciation
- imaginative storytelling
- PICTURE METAPHORS
- creative daydreaming
- painting, collage, and other visual arts
- IDEA SKETCHING
- visual thinking exercises
- GRAPHIC SYMBOLS

- using mind-maps and other visual organizers
- computer graphics software
- Visual pattern seeking
- optical illusions
- COLOR CUES
- telescopes, microscopes, and binoculars
- visual awareness activities
- draw-and-paint/computer-assisted-design software
- picture literacy experiences

## Bodily-Kinesthetic Intelligence

- creative movement
- HANDS-ON THINKING
- field trips
- mime
- THE CLASSROOM THEATER
- competitive and cooperative games
- physical awareness exercises
- hands-on activities of all kinds
- crafts
- BODY MAPS
- use of kinesthetic imagery
- cooking, gardening, and other "messy" activities
- manipulatives
- virtual reality software
- KINESTHETIC CONCEPTS
- physical education activities
- using body language/hand signals to communicate
- tactile materials and experiences
- physical relaxation exercises
- BODY ANSWERS

## Musical Intelligence

- MUSICAL CONCEPTS
- singing, humming, or whistling
- playing recorded music
- playing live music on piano, guitar, or other instruments
- group singing
- MOOD MUSIC
- music appreciation
- playing percussion instruments
- RHYTHMS, SONGS, RAPS, AND CHANTS
- using background music
- linking old tunes with concepts
- DISCOGRAPHIES
- creating new melodies for concepts
- listening to inner musical imagery
- music software
- SUPERMEMORY MUSIC

## Interpersonal Intelligence

- COOPERATIVE GROUPS
- interpersonal interaction
- conflict mediation
- peer teaching
- BOARD GAMES
- cross-age tutoring
- group brainstorming sessions
- PEER SHARING
- community involvement
- apprenticeships
- SIMULATIONS
- academic clubs
- interactive software
- parties or social gatherings as context for learning
- PEOPLE SCULPTING

## Intrapersonal Intelligence

- independent study
- FEELING-TONED MOMENTS
- self-paced instruction
- individualized projects and games
- private spaces for study
- ONE-MINUTE REFLECTION PERIODS
- interest centers
- PERSONAL CONNECTIONS
- options for homework
- CHOICE TIME
- self-teaching programmed instruction
- exposure to inspirational/motivational curricula
- self-esteem activities
- journal keeping
- GOAL-SETTING SESSIONS

## How to Create MI Lesson Plans

On one level, MI theory applied to the curriculum might best be represented by a loose and diverse collection of teaching strategies such as those listed above. In this sense, MI theory represents a model of instruction that has no distinct rules other than the demands imposed by the cognitive components of the intelligences themselves. Teachers can pick and choose from the above activities, implementing the theory in a way suited to their own unique teaching style and congruent with their educational philosophy (as long as that philosophy does not declare that all children learn in the same way).

On a deeper level, however, MI theory suggests a set of parameters within which educators can create new curricula. In fact, the theory provides a context within which educators can address any skill, content area, theme, or instructional objective, and develop at least seven ways to teach it. Essentially, MI theory offers a means of building daily lesson plans, weekly units, or monthly or year-long themes and programs in such a way that all students can have their strongest intelligences addressed at least some of the time.

The best way to approach curriculum development using the theory of multiple intelligences is by thinking about how we *can translate* the material to be taught from one intelligence to another. In other words, how can we take a linguistic symbol system, such as the English language, and translate it—not into other linguistic languages, such as Spanish or French, but into the languages of other intelligences, namely, pictures, physical or musical expansion, logical symbols or concepts, social interactions, and intrapersonal connections?

The following seven-step procedure suggests one way to create lesson plans or curriculum units using MI theory as an organizing framework:

1. **Focus on a Specific Objective or Topic.** You might want to develop curricula on a large scale (e.g., for a year-long theme) or create a program for reaching a specific instructional

## Figure 2

### MI Planning Questions

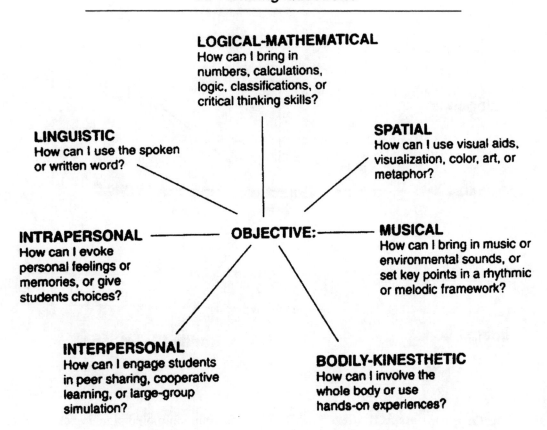

objective (e.g., for a student's individualized education plan). Whether you have chosen "ecology" or "the schwa sound" as a focus, however, make sure you have clearly and concisely stated the objective. Place the objective or topic in the center of a sheet of paper, as shown above in Figure 2.

2. **Ask Key MI Questions.** Figure 2 shows the kinds of questions to ask when developing a curriculum for a specific objective or topic. These questions can help prime the creative pump for the next steps.

3. **Consider the Possibilities.** Look over the questions in Figure 2, the list of MI techniques and materials in Figure 1. Which of the methods and materials seem most appropriate? Think of other possibilities not listed that might be appropriate.

4. **Brainstorm.** Using an MI Planning Sheet like the one shown below in Figure 3, begin listing as many teaching approaches as possible for each intelligence. You should end up with something like the sheet shown in Figure 4. When listing approaches, be specific about the topic you want to address (e.g., "videotape of rain forest" rather than simply "videotape"). The rule of thumb for brainstorming is "list *everything* that comes to mind." Aim for at least twenty or thirty ideas and at least one idea for each intelligence. Brainstorming with colleagues may help stimulate your thinking.

# Figure 3

*MI Planning Sheet*

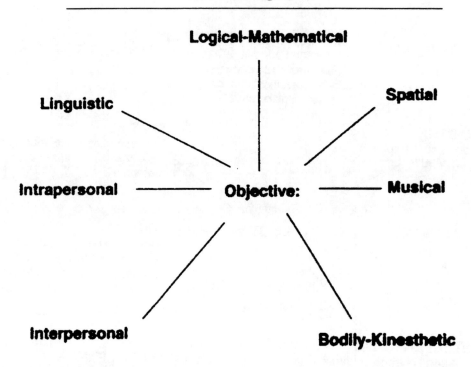

5. **Select Appropriate Activities.** From the ideas on your completed planning sheet, circle the approaches that seem most workable in your educational setting.

6. **Set Up a Sequential Plan.** Using the approaches you've selected, design a lesson plan or unit around the specific topic or objective chosen. Figure 5 shows what a seven-day lesson plan might look like when perhaps thirty-five to forty minutes of class time each day are allotted to the objective.

7. **Implement the Plan.** Gather the materials needed, select an appropriate time frame, and then carry out the lesson plan. Modify the lesson as needed to incorporate changes that occur during implementation.

**Figure 4**

*Completed MI Planning Sheet*

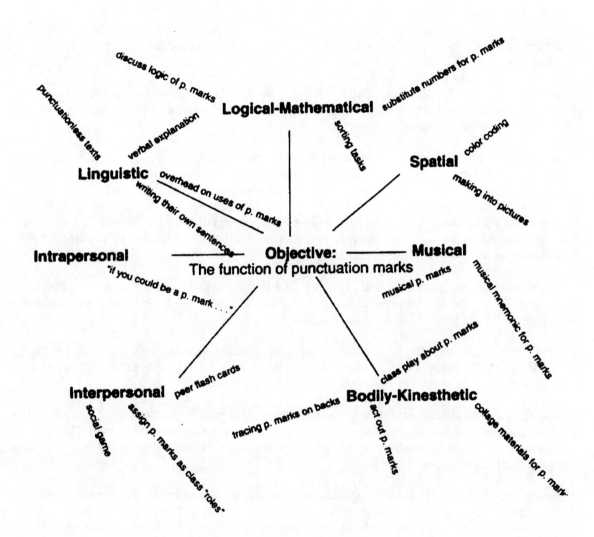

## Figure 5

### Sample Seven-Day MI Lesson Plan

**Level:** 4th grade

**Subject:** Language arts

**Objective:** To understand the function of, and differences between, four punctuation marks: the question mark, period, comma, and exclamation mark.

**Monday** (Linguistic Intelligence): Students listen to a verbal explanation of the function of punctuation marks, read sentences having examples of each mark, and complete a worksheet requiring them to fill in their own marks.

**Tuesday** (Spatial Intelligence): The teacher draws on the board graphic images that correspond in meaning and form to each mark (Question mark = a hook, since questions "hook" us into requiring an answer; exclamation point = a staff that you pound on the floor when you want to exclaim something; a period = a point, since you've just made your point, plain and simple; and a comma = a brake pedal, since it requires you to temporarily stop in the middle of a sentence. Students can make up their own images and then place them as pictures in sentences (with different colors assigned to different marks).

**Wednesday** (Bodily-Kinesthetic Intelligence): The teacher asks students to use their bodies to form the shapes of the different punctuation marks as she reads sentences requiring these marks (e.g., a curved body posture for question mark).

**Thursday** (Musical Intelligence): Students make up different sounds for the punctuation marks (as Victor Borge did in his comedy routines), and then make these sounds in unison as different students read sample sentences requiring the use of the four marks.

**Friday** (Logical-Mathematical Intelligence): Students form groups of four to six. Each group has a box divided into four compartments, each of which is assigned a punctuation mark. The groups sort sentence stubs with missing punctuation marks (one per sentence stub) into the four compartments according to the punctuation needed.

**Monday** (Interpersonal Intelligence): Students form groups of four to six. Each student has four cards, and each card has a different punctuation mark written on it. The teacher places a sentence requiring a given punctuation mark on the overhead projector. As soon as students see the sentence, they toss the relevant card in the center of their group's circle. The first student in the group to throw in a correct card gets five points, the second four, and so on.

**Tuesday** (Intrapersonal Intelligence): Students are asked to create their own sentences using each of the punctuation marks; the sentences should relate to their personal lives (e.g., a question they'd like somebody to answer, a statement they feel strongly about, a fact they know that they'd like others to know about).

## MI and Thematic Instruction

More and more educators are recognizing the importance of teaching students from an interdisciplinary point of view. Although academic skill teaching or the teaching of isolated chunks of knowledge may provide students with competencies or background information that can prove useful to them in their further education, such instruction often fails to connect students to the real world—a world that they will have to function in as citizens a few years hence. Consequently, educators are turning toward models of instruction that more closely imitate or mirror life in some significant way. Such instruction is frequently *thematic* in nature. Themes cut through traditional curricular boundaries, weave together subjects and skills that are found naturally in life, and provide students with opportunities to use their multiple intelligences in practical ways. As Susan Kovalik (1993, p. 5), developer of the Integrated Thematic Instruction (ITI) model, puts it:

> A key feature of *here and now* curriculum is that it is immediately recognized (by the student) as being relevant and meaningful. . . . Furthermore, it purports to teach our young about their world and the skills necessary to act within and upon it, thus preparing themselves for living the fast-paced changes of the 1990s and beyond.

Kovalik's ITI model is based on year-long themes (such as "What Makes It Tick?") that are themselves made up of month-long components (such as clocks/time, electrical power, transportation) and weekly topics (such as seasonal change and geologic time). Other curricular approaches focus on alternative time frames, such as semester units or three-month themes. Regardless of the time element involved, MI theory provides a context for structuring thematic curricula. It provides a way of making sure the activities selected for a theme will activate all seven intelligences and therefore draw upon every child's inner gifts.

Figure 6 outlines the kinds of activities that might be used for the theme "Inventions." It shows how activities can be structured to address traditional academic subjects as well as each of the seven intelligences. Significantly, this chart illustrates how science activities needn't focus only on logical-mathematical intelligence and how language activities (reading and writing) needn't focus only on linguistic intelligence; they can, in fact, span all seven intelligences.

Keep in mind that MI theory can be applied to the curriculum in a variety of ways. There are no standard guidelines to follow. The ideas in this chapter are suggestions only; I invite you to create other forms or formulas for lesson planning or thematic development, and I encourage you to incorporate other formats, including those developed by educators such as Kovalik (1993) and Hunter (see Gentile, 1988). Ultimately, you should be guided by your deepest and sincerest attempts to reach beyond the intelligences you may currently be teaching to, so that every child has the opportunity to succeed in school.

## References

Gentile, J. R. (1988). *Instructional Improvement: Summary and Analysis of Madeline Hunters's Essential Elements of Instruction and Supervision.* Oxford, Ohio: National Staff Development Council.

Goodlad, J. I. (1984). *A Place Called School: Prospects for the Future.* New York: McGraw-Hill.

Kovalik, S. (1993). *ITI: The Model—Integrated Thematic Instruction.* 2nd ed. Village of Oak Creek, AZ: Books for Educators.

Plato. (1952). *The Dialogues of Plato.* Chicago: Encyclopedia Britannica.

## Figure 6
### MI and Thematic Instruction

Sample Theme: Inventions

| | Math | Science | Reading | Writing | Social Studies |
|---|---|---|---|---|---|
| Linguistic | Read math problems involving inventions | Talk about the basic scientific principles involved in specific inventions | Read a general book about inventions | Write about what you'd like to invent | Write about the social conditions that gave rise to certain inventions |
| Logical-Mathematical | Learn a math formula that served as the basis for an invention | Create a hypothesis for the development of a new invention | Read a book about the logic and math behind inventions | Write a word problem based on a famous invention | Create a time line of famous inventions |
| Spatial | Sketch the geometry involved in specific inventions | Draw a new or existing invention showing all working parts | Read a book with lots of diagrams of the inner workings of inventions | Label the individual components of your drawing of an invention | Paint a mural showing inventions in social/historical context |
| Bodily-Kinesthetic | Create an invention to measure a specific physical activity | Build your own invention based on sound scientific principles | Read the instructions for putting together an existing invention | Write instructions for building your own invention from scrap materials | Put on a play about how a certain invention came to be |
| Musical | Study the math involved in the invention of musical instruments | Study the science behind the invention of electronic music | Read about the background to invention songs such as "John Henry" | Write the lyrics for a song promoting a new invention | Listen to music about inventions at different historical periods |
| Interpersonal | Be in a study group that looks at the mathematics involved in specific inventions | Form a discussion group to study the science behind inventions | Read about the cooperation necessary for developing an invention | Write a play about inventions that can be put on by the class | Hold a discussion group about how a certain invention came to be |
| Intrapersonal | Create your own word problems based on inventions | Develop a self-study program to examine the scientific basis for a specific invention | Read the biography of a famous inventor | Write your personal autobiography as a "famous inventor" | Think about this question: if you could invent a time machine, where would you go? |

# How Our Brain Learns, Remembers, and Forgets

## ROBERT SYLWESTER

If you want to know what water is, don't ask a fish. And if you want to know what memory is, don't ask a memory researcher—at least that's what memory researchers themselves believed until only recently. Indeed, after 30 years of intense study, the noted pioneer memory researcher Karl Lashley wrote somewhat wryly in 1950, "In reviewing the evidence on the localization of the memory trace, I sometimes feel that the necessary conclusion is that learning is just not possible." And only a dozen years ago, Neisser (1982) concluded that years of extensive psychological research studies have led to major generalizations about memory that the average middle-class American 3rd grader already knows through personal experience. Apparently, memory is so integral to our existence that we have trouble isolating it in order to learn what it is and how it works.

That situation is rapidly and dramatically changing. Brain-imaging technology can now clearly identify the brain areas that activate when a subject remembers or responds to something. For example, researchers have monitored the brain activity of subjects who have been asked to respond to nouns with the first verb that comes to mind (e.g., *cup* might suggest *drink*). A brain-imaging machine locates and reports the brain areas where the nouns and their related verbs are processed (Posner and Raichle, 1994). Such research findings will help to resolve Lashley's lament of 45 years ago.

Other researchers have studied memory at the cellular level and much of this research has been done with simple marine snails. Scientists discovered that when an animal learns something, physical changes occur in the synapses of the network of neurons that process the memory, in effect strengthening the neural connections that constitute a memory network.

This cellular research required the development of technologies that could identify small changes in the structure of a neuron during memory formation. Since thousands of neurons can fit in a space the size of a pinhead, it's not surprising that such complex technologies didn't emerge until very recently. Historically, major developments in research technologies have sparked dramatic increases in our knowledge of various phenomena. Memory scientists are riding the latest technological wave, and they are rapidly moving from centuries of speculation to spectacular discoveries.

This chapter will focus on three elements of memory: (1) the physical changes that occur in neural networks when a memory is formed or erased, (2) the functional organization and operation of the several memory systems our brain uses, and (3) the procedures we use to maintain selected important memories.

---

Robert Sylwester, "How Our Brain Learns, Remembers, and Forgets," *A Celebration of Neurons: An Educator's Guide to the Human Brain,* 1995, pp. 48–64. Reprinted by permission from Association for Supervision and Curriculum Development.

## Memory at the Cellular and Network Levels

Memory networks form their chemical connections at the synapse, the narrow gap that neurotransmitters cross when they move from the axon terminal of a presynaptic neuron to attach to receptors on the dendrites or cell body of a postsynaptic neuron.

When our sensory system focuses on an object or event, it activates a large number of neurons that are assembled into a variety of related brain networks (generally, combinations of columns). Each network processes a specific property of the object or event, such as its shape or movement pattern. This initial simultaneous activation creates synchronized response patterns in the coalition of activated neural networks—principally in their firing rates and in the amounts of neurotransmitters their neurons release. This activity somehow links the networks that process the various properties of the object or event, and we get a mental impression of spatial integration—for instance, we perceive a face as a unit—even though the various brain areas are operating independently, and no single brain site contains the total face. Scientists don't completely understand the process, but they believe that attention, thought, and memory emerge out of such synchronized patterns of neural network activity.

This activity is emergent because, for example, someone entering your classroom will perceptually cause a few key networks governing shape, texture, color, and movement to fire rapidly and rhythmically enough to quickly identify a moving human being. This limited initial activity causes other memory networks that deal with related information to fire in a similar rhythmic pattern. These networks may provide more information: the stranger is a female—about the same age as your students—with a bewildered look on her face. Networks dealing with abstractions may also activate—for instance, seeing a religious symbol on the girl's clothing may activate thoughts related to her value system. Thus, thought emerges out of attention when a continuous, quite active, synchronized firing pattern resonates between a critical mass of related neural networks in the thalamus (which processes the immediate situation) and the cortex (which contains memories related to objects and events in the immediate situation). When enough information emerges about the immediate situation and about related past experiences, a motor response often forms—in this example, perhaps to move toward the girl in greeting.

A long-term memory begins to develop out of such a temporary event when our brain determines that the event is emotionally loaded and may reoccur. Suppose that the girl is a new student assigned to your classroom. The same sets of neural networks that initially represented the shapes that make up her face and body will fire every time you see her because these shapes don't change much from day to day.

The frequent activation of such interrelated networks and the emotional realization that she'll be your student for some time will eventually spark a pattern of similar physical changes that will strengthen the synaptic connections of all the neurons in the relevant networks. One such change is the development of small spines on the dendrites of postsynaptic neurons. This change increases the number of receptors and allows more presynaptic neurotransmitters to attach to the postsynaptic neurons. Because a receiving neuron fires when the influx of information contained in neurotransmitters reaches the neuron's firing threshold, a presynaptic neuron that sends in a lot of neurotransmitter information will have more influence on the firing pattern of the receiving neuron than will a neuron that sends in less information. Adding receptors to a synapse increases the likelihood that the receiving neuron will quickly reach its threshold and fire (just as raising the thermostat setting a few degrees during the winter increases the likelihood that the furnace will turn on to heat the room).

Experience supports this theory: it doesn't take much input to trigger a strong memory, but it may take a lot of mental activity to activate a weak memory. When you enter a room containing a dozen strangers and acquaintances and a couple of close friends, your friends will *pop out* of the total group in instant recognition, because the networks that represent them activate more easily than the networks that process strangers and mere acquaintances—thanks to strengthened synaptic connections. Further, the complex changes that created the total memory of your new student activate her memory as a unit in rapid recognition of her when any part of the network is activated. For example, her face will instantly pop up in your mind when you see her from the back or when you hear her voice without seeing her.

An individual neuron can be part of many memories, much as a light bulb in an advertising reader board can light up as a part of many different letters, digits, and words. Thus, our relatively small brain can process a vast number of related memories. Further, our emotions connect memory networks to other related networks. When one network fires, it can activate other related networks. A visit to your childhood home will activate the memory of many events and objects that you hadn't thought of in years.

Memories are composed of constant and changing elements that vary in their ease of recall. For example, the shape of your student's face would remain constant over the year, but the clothing she wears would change from day to day. Your brain would thus create a much stronger memory of her constant elements than of her changing elements, and so you would easily recognize her face a year later. You might remember the *kinds* of clothing she tended to wear, but you would have difficulty remembering what she wore on any given day when she was a student in your class, unless it was something unusual that could create a strong memory after one experience.

A functional library model of our brain that located memories within interconnected cortical columns, much as a library book and its bibliography contain interconnected cultural memories. Would that things were so simple. Rose (1993) cautions us that although similarities do exist in how machines (books, tapes, films) and brains store and retrieve information, technological and biological memories aren't really the same. Technological memories are localized and static, while biological memories are distributed and dynamic. Moreover, a book requires an external reader's brain to retrieve its information; a brain retrieves its own memories to enhance both its survival and the quality of its life, two concerns irrelevant to a machine. So use the book model if you wish, but with the caution that models are simplifications of complex phenomena. Brains write books, but books don't develop brains.

Consider taped information. When musicians tape a song, their playing alters the magnetic properties of the tape, and so they've stored the musical information in a form of memory. But we need a tape deck, tuner, and loudspeaker to retrieve and use that *memory* information. The situation is similar with biological memory. The information stored via synaptic changes in neural networks (the memory engram) is only a small part of a much larger system that involves such processes as perception and emotion, synthesis and analysis, speech and writing. Memory is a function of the entire system, not just the synaptic changes. Further, the taped information sounds the same every time it's played on the recorder, but the musicians who taped the song alter their performance somewhat every time they play the song that's in their biological memory. Human memory is thus less precise than machine memory, and it is more adaptive and inventive over time as experience adds depth and breadth to the variability of our memories. Unfortunately, much of our school testing program requires our students' very inventive biological memory system to exhibit high-level technological precision.

## The Functional Organization and Operation of Our Memory Systems

Our memory system is functionally organized much like an office that must process a continual influx of information that competes for attention. We quickly label some information irrelevant and discard it. Other information we handle immediately and then forget about it. We set aside information to handle later, store some in permanent files containing related information, and incorporate some into the existing procedures for operating the office. Our brain has several interrelated types of memory systems that correspond to these general categories of information processing:

## Short-Term Memory

Short-term memory is an initial memory buffer that allows us to hold a few units of information that we're attending to for a short period of time while we determine their importance. We must decide quickly because the continual influx of new information into our brain will delete anything not consciously held. This limited attentional capacity has important survival value in that it permits us to completely experience the current situation without having to remember any of it, and thus to quickly shift our attention to emerging situations that may be important. Luria (1968) described the terrible plight of a man without short-term memory, who remembered almost everything that ever happened to him. Imagine a complete lifetime memory of every boring party conversation you've ever had!

We've all experienced the fragility of short-term memory—just think of those times you've looked up a phone number and then instantly forgot it when someone interrupted with a question before you could dial it. The short-term memory process appears to function through temporary synchronized firing patterns that emerge between related networks in the thalamus (the current situation) and the cortex (related memories). The more rapidly firing, synchronized thalamus-cortex networks become foreground (attentional) information, and the less active neural networks become background (or context).

Because short-term memory space is limited to perhaps a half-dozen units of foreground information at any one time, we must rapidly combine (or chunk) related bits of foreground information into larger units by identifying similarities, differences, and patterns that can simplify and consolidate an otherwise confusing sensory field. Thus, we see a face as a unit, not as individual eyes, ears, nose, mouth, hair, and cheeks. This need to respond quickly has enhanced our ability to conceptualize and estimate, certainly major brain strengths.

Our conscious brain thus monitors the total sensory field while it simultaneously searches for and focuses on familiar, interesting, and important elements—separating foreground from background. Extensive experience in a field develops these rapid editing skills to the expert level. The curriculum enhances this remarkable brain capability when it focuses on the development of classification and language skills that force students to quickly identify the most important elements in a large unit of information.

The strong appeal of computerized video games may well lie in their lack of explicit directions to the players, who suddenly find themselves in complex electronic environments that challenge them to quickly identify and act on rapidly changing elements that may or may not be important. Failure usually sends the player back to the beginning, and success brings a more complex and attractive challenge in the next electronic environment. Contrast such an experience with the paper worksheets students receive in schools, which usually have clearly stated directions and static and uncomplicated information—but little cognitive challenge.

Students need many opportunities to develop their short-term memory capabilities through experiences such as debates and games that require them to rapidly analyze complex information and briefly hold key points in their memory. Flowing games, such as soccer and basketball, also have this continuous challenge. When textbooks and teachers highlight all the important information and when software is too user-friendly, instruction perhaps becomes more efficient, but students are deprived of the challenge and pleasure involved in continuously separating foreground from background.

## Long-Term Memory

The development of a long-term memory emerges out of an ill-understood, often conscious decision that elements of the current situation are emotionally significant and will probably reoccur. If the situation does indeed reoccur after the memory is formed, sensory and perceptual processes will represent it in the thalamus. The cortical memory of the previous experience, resonating with the current thalamic perception, will then create an attentional state, and help to determine the response—current behavior influenced by past experience. Short-term memory allows us to experience the present, but we would become a prisoner of the present without our two interrelated forms of long-term memory, declarative (or explicit) and procedural (or implicit).

**Declarative long-term memories** are factual label-and-location of memories—knowing the name of my computer and where it's located, for example. They define named categories, and so are verbal and conscious. The principal brain mechanisms involved in processing declarative memories are the hippocampus (in the limbic system) and the cortex (especially the temporal lobes). In library terms, think of the hippocampus as the card catalog that organizes the collection, and the cortex as the library's collection of books.

Episodic declarative memories are very personal—intimately tied to a specific episode or context (my first attempt to run my computer, my joy at discovering how it simplified writing and editing).

Semantic declarative memories are more abstract—context-free and often represented by symbols, such as those used in language and mathematics. They can be used in many different settings and so are important in teaching for transfer (knowing how to use function keys and software). My semantic understanding of the typewriter and its keyboard simplified my moves from manual typewriter to electric typewriter to word processor.

Initial skill learning, such as learning to type, is often episodic—the memories contain both foreground and background elements of the experience. When I learned to type, my teacher, classroom, and typewriter provided an important, easily remembered emotional context during the initial learning period. It would have been inefficient for me to continue to recall all these background elements whenever I typed, however, and so my teacher used class and home drills and different typewriters to help me eliminate the context of the learning (background) from the execution of the skill (foreground).

My typing knowledge and skill had thus become more semantic—more abstract, but also more useful in a wide variety of keyboard settings and tasks (background). In effect, my brain erased the background information from my memory by reducing its frequency and significance, and strengthened the foreground information (actual typing) by focusing on it.

My typing speed was limited, however, by my simultaneous conscious spelling of words and activation of keys, and so I also had to eliminate this conscious behavior through a transfer of skills from semantic declarative memory to procedural memory—to master automatic touch typing.

**Procedural long-term memories** are automatic skill sequences—knowing how to touch type, for example. Because they don't rely on conscious verbal recall (except to initiate, monitor, and stop the extended movement sequence), they are fast and efficient. They are also difficult to master and to forget (we can roller skate after years of not doing it) and are best developed through the observation of experts, frequent practice, and continual feedback. As a skill develops, the number of actions processed as a behavioral unit increases, and prerequisite skills are integrated into advanced skills. For example, learning to navigate a bicycle around the neighborhood during childhood enhances car-driving skills during our adult years.

The principal brain mechanisms involved in processing procedural memories are the amygdala (our brain's emotional center, located in the limbic system), the cerebellum (located in the lower back of our brain), and the autonomic nervous system (which regulates circulation and respiration)—but procedural memories also involve altered muscle systems.

At the survival level, the procedural system needs an emotional trigger (the amygdala) that can quickly activate the automatic motor system (the cerebellum) when enough threatening factors emerge, such as the sight and sounds of a rapidly approaching predator. It's advantageous for our running mechanisms to fire automatically, sequentially, and in loops (right foot, left foot, right foot, etc.) so that our conscious cognitive systems can focus on determining the best escape route. Thus, the procedural system is designed so that the motor neurons that process each action automatically trigger the next action in the sequence (think of a line of falling dominoes). What we've done over eons of time is to adapt the automatic neural machinery initially designed for running down food and escaping from predators to such contemporary activities as typing, speaking, driving cars, and playing piano passages.

When I was learning how to type I knew where all the keys were, and I was slow because my typing actions weren't connected into a complex automatic sequence. Today I don't consciously know where any of the keys are and I'm a fast, efficient typist. My fingers have now become an automatic extension of my brain's language mechanisms.

We can easily move between conscious and unconscious activation of these processes. For example, suppose I decide to walk to a meeting. My initial steps will be conscious, but the process will quickly become automatic (procedural)—and I can then simultaneously walk and think about items on the meeting agenda. Were I to confront an obstruction, however, my walking would become more conscious as I moved around the obstruction, and I would temporarily stop thinking about the meeting because the two activities could have overloaded my conscious attentional brain processes.

Our brain is most efficient at recalling and using episodic memories that have important personal meanings. It is much less efficient at mastering the important context-free semantic and procedural memories. That's why schools have to spend so much time and energy on worksheet-type facts and skills that are isolated on specific contexts (but that generally have the important value of being useful in a wide variety of contexts). Conversely, computers, reference books, and so on are very reliable with facts and procedures, but they lack the emotional contexts that make our value-laden episodic memories so rich.

A memory is a neural representation of an object or event that occurs in a specific context, and emotionally important contexts can create powerful memories. When objects and events are registered by several senses (e.g., seeing, hearing, touching, tasting), they can be stored in several interrelated memory networks. A memory stored in this way becomes more accessible and powerful than a memory stored in just one sensory area, because each sensory memory checks and extends the others.

Recognition is easier than recall because recognition occurs in the original context of the memory, or in one quite similar. If the emotional setting in which a memory originally occurred is tied to the memory, re-creating the original emotional setting enhances the recall of that memory and related memories. Thus, such emotional, multisensory school activities as games, role playing, simulations, and arts experiences can create powerful memories.

Procedural memories tend to govern the behavior of animals, whereas the much more flexible (easily learned, easily forgotten) declarative memory system is very important to humans. The combination of the two types of memory creates a powerful, integrated human memory system: thoughtful action.

The two systems appear to collaborate in such memory issues as the repressed memories of sexual abuse. Kandel (1994) suggests that the fearfulness of the abusive experience can lead to the release of norepinephrine (adrenaline), and this strengthens the amygdala-cerebellum connections that process the emotional memory of the event. Conversely, the painfulness of the experience can lead to the release of opiate endorphins that weaken the hippocampus-cortex connections that process the conscious memory of the factual circumstances surrounding the event.[1] Subsequently, the victim tends to avoid anything that triggers the fearful emotion, but doesn't consciously know why. This behavior makes sense. It would be important to avoid the person and the specific setting involved in the abuse, but if the person is someone whose support you need, and if the setting is your home, not remembering the details of the abuse may actually be a form of survival—a case of psychological escape when physical escape is difficult.

Years later, a chance combination of similar characters, location, actions, and emotions may be enough for the strong emotional memory to trigger the recall of the weak factual memory of the original circumstances of the abuse. Think of how the strong emotional and contextual overtones of a class reunion suddenly trigger the weak factual memories of many decades-old and seemingly forgotten school experiences.

Kandel cautions that this explanation of repressed memories doesn't ensure the truth of such a memory because the same neural structures that process perception also process our imagination. Thus, we can incorrectly believe that an imagined event is a real, perceived event.[2] A nightmare, which can often leave us trembling and sweating, is a common example of this kind of error.

## Maintaining Our Memories

Why do we devote a third of our life to sleeping and dreaming, several hours a day to mass media, and much of the rest of our waking day to informal conversation and storytelling? Why do remembrances of previous experiences suddenly pop into our mind while we're listening to stories that other people tell? Why do parables and songs provide us such a rich lore of moral and ethical information? Why do our brightest students take the leading roles in classroom discussions?

A growing number of memory researchers believe that such activities are fundamental to the development, maintenance, editing, and retrieving of our long-term memories (Schank, 1990, 1991). Although only limited research evidence supports this thesis, the basic idea has a ring of rightness about it that further research would confirm. For educators, the theories provide a fascinating new view of the role school activities play in the maturation of a child's brain.

Because the development of a long-term memory requires the physical reconstruction of the synapses in the affected neural networks, it is necessary to shut down their activity during the rebuilding process, much as a paving crew must detour traffic during the reconstruction of a road. Our brain does this through sleep (and perhaps during daytime periods of reduced activity in the relevant brain area). It reduces sensorimotor activity for about eight hours every night while it reconstructs and resets the memory networks that have emerged out of the day's events. Thus, sleep enhances the creation, editing, and erasing of memories.

### How Dreaming Maintains Survival Memories

Memory networks must be constantly stimulated or else the neural synapses that were rebuilt to create an easily activated network will revert to their original state, and the memory network will disintegrate. We call this disintegration forgetting. Some researchers believe that dreaming helps to extend the life of important survival memories that aren't sufficiently activated during normal daytime activities (Hobson, 1994) and to erase obsolete memories (Crick and Mitchison, 1983).

Hobson (1989) discovered that every 90 minutes during a sleep period called REM sleep (rapid eye movement), certain brainstem structures begin to fire randomly into the cortex, where declarative memories are stored; it's as if sensory information is entering our brain when it isn't.[3] Whenever this random firing activates a memory network, however, what occurs in our mind is much like what occurs when we unexpectedly hear the word *banana*. We instantly see a banana in our mind, even though no banana is present. The sound memory of banana activated the sight memory of banana. Thus, during these four to five nightly dream periods (which range from 10 to 40 minutes in length), our brainstem constantly and randomly activates memory networks. It's somewhat like a circuit-testing and updating process for maintaining key survival programs so that they'll be functional when we need to use them.

Our mind assembles this random activity into the unconscious stories we call dreams—trying to make sense out of random nonsense. If the dream gets too emotionally intense (often around 3 a.m.), we wake up and end the dream. This two hours of nightly dreaming (more than 700 hours a year) maintains many of the important memories of things that don't occur often in real life (e.g., escaping from danger, remembering relatives who don't live nearby). The more connections a memory has to other memories, the more apt it is to be activated by this nightly random firing. The sexual and violent content of dreams tends to be high because our sexuality and survival are so closely tied to many aspects of our self—and so to many of our memories. Hobson (1994) reports that about two-thirds of our dreams are about unpleasant events—and so the parting expression "Pleasant dreams" has a hopeful meaning.

During REM sleep, our motor system is inhibited so that we don't act out our dreams. We commonly experience this phenomenon as running in sand while trying to escape danger. The cortex has sent a message to the motor system to run, but the motor inhibition stops the message prior to action.

Although dream periods allow animals to maintain the relatively few memories of survival strategies they must develop, we humans live in a much more complex social environment. We can't depend on random dreaming to maintain all the important memories that we develop.

What we appear to have done is to adapt the skill-sequencing efficiency of procedural memory to our need to recall the combinations and sequences of facts that form declarative memory. We've created a variety of *storytelling* formats that efficiently string together the related factual events of an experience. Recall that in a procedural memory, each action in a skill sequence automatically triggers the next action, and so simplifies recall. Stringing the separate events of a remembered extended factual experience into a logical sequence (such as the order in which vacation events occurred) places each subsequent event into the context of the previous event, and context enhances recall. Thus, it's much easier to use a narrative format to remember several events that occurred during an experience than to try to remember the events in isolation from each other. Just begin the story and let it logically and automatically unfold. The discussion below explores the basic storytelling formats we humans have developed to enhance the recall of declarative memories.

## How the Mass Media Maintain Broad Cultural Memories

Metaphoric forms of mass media provide us with many narrative opportunities to consciously stimulate the memories we want to maintain. We go to a theater to see a film we believe will focus on something that's important to us. We get comfortable, the lights go down, conversation stops, and film images and sound magically appear. We soon find ourselves caught up in the events depicted on the screen, and personal memories related to the story suddenly pop into our mind.

So it is with novels, TV programs, songs, games, and pageants. We consciously seek out those that we hope will stimulate our memories of broad cultural issues that we consider important. It's like we're seeking out dream possibilities while we're awake. We tend to be unhappy with a mass media experience if we can't identify with any of the characters, locations, or events. Many people who watch professional sporting events are probably reliving their own adolescent participation in these games. They probably weren't nearly as good as the athletes they watch, but they can imagine that they were.

The most powerful metaphoric experiences are those that focus on important cultural issues, but define the story's characters and locations somewhat loosely, so that many people can easily identify with the issues the story explores. The parables that Jesus told are a good example. Many people learn these stories during childhood, but discover their deeper meanings as adults. Seeing a person in need or someone committing an altruistic act could spark the recall of the story of the good Samaritan, and over time many increasingly deeper connections could evolve between the parable itself and our life experiences.

Similarly, the songs of our adolescent years often become the beacons of our adult life because they help us to recall the important developments of our adolescence. We often listen to this music and flock to nostalgic concerts to relive the memories in the extended reverie of song. The song slows the simple message so that we can savor all the emotions of the experience.

## How Conversation Maintains Informal Personal Memories

The mass media tend to focus on broad social concerns, but many of the memories we want to maintain are highly personal, and not apt to be turned into a novel or film. We maintain such memory networks informally through conversations with friends who shared the experience or who had similar experiences. Families get together on celebratory occasions that spark the recall of earlier celebrations. We'll converse with anyone who is willing to talk about a topic of mutual interest—each person's story or comment in turn sparking a contribution from the other person.

When we listen to another person's story and a related experience suddenly pops into our mind, we tend to quit listening to the other person's story and focus our thoughts on ours, and then we insert our story into the conversation at the first opportunity. Schank (1990) argues that we don't listen to the stories of others to learn about their experiences, but rather in the expectation that the interaction will enhance the maintenance of our own related memories. Conversely, we may opt out of an event, such as a class reunion, because of the high probability that it will spark painful memories. Further, a friend may decline to respond to our query about a recent negative event because she is trying not to create a story memory of it.

The value of a long-term relationship may reside in the constant availability of a storytelling partner who is willing to help recall mutually important events. The furnishings and art in a couple's home are historical artifacts that spark memories that define the relationship. The tragedy of the dissolution of such a relationship may well lie in the irreplaceable loss of the common history. The memories become vulnerable when the conversations cease. Conversely, a partnership may deliberately dissolve when the partners no longer want to listen to the other's version of events that were perceived and remembered differently.

## How Schools Maintain Formal Societal Memories

Much of our culturally important information doesn't come up in dreams, in mass media, or during meal conversations—for example, the multiplication tables, how to spell *accommodate*, the names of countries and their capitals, how to compute the area of a circle. Our society has created schools to ensure that students master such culturally important information and then have opportunities to test their memories in simulated and real-life settings.

A school functions somewhat like daytime sleeping and dreaming. Recall that six hours (75 percent) of our nighttime focuses on the creation and editing of memory networks, and two hours (25 percent) involves dream periods in which the networks are randomly activated and organized into dream stories. This set-up sounds a lot like school. Teachers tend to focus more of their time and energy on teaching new information (i.e., creating memory networks) and less on using that knowledge in such social problem situations as discussions, games, simulations, role playing, storytelling, music, and art (i.e., circuit testing).

Further, students tend to view school as a somewhat surreal, random, dreamlike experience. Spelling follows arithmetic or vice versa; the teacher suddenly lashes out at a student; someone throws up; it's time to go to the library; it's anyone's guess whose hand will go up when the teacher asks a question, or even what question will be asked . . .

It is the student's task to make sense out of all this nonsense, to translate the seeming curricular randomness into a coherent view of our culture. Unfortunately, our culture seems to value random facts, and schools tend to reinforce this bias. For example, the students who score the highest on objective tests containing many discriminating items (items that elicit many incorrect responses) tend to be those with the best command of the most obscure facts—and obscure facts are almost by definition *unimportant* facts. The same holds true for the winners of TV quiz shows and games like Trivial Pursuit—some of us love to bounce random facts around in our mind.

Our task as educators, however, is to help students begin to find *relationships* between the somewhat random, often trivial fact-filled experiences of everyday life and the fewer enduring principles that define life—and then to help them create and constantly test the memory networks that solidify those relationships.

The memory theory and research presented in this chapter suggest that the best school vehicle for this search for relationships is storytelling as a broad concept that includes such elements as conversations, debates, role playing, simulations, songs, games, films, and novels. The brightest students are the ones who always have their hand in the air to expand the discussion through stories about their own experiences—to maintain and extend their own memory networks through active recall, as it were. It's as if their brains know how important it is for them to act on their knowledge and beliefs, to not sit passively and let their classmates make the mental connections.

We certainly can't rely on the random nature of dreams and the mass media to maintain our students' memories of socially important information. And it isn't enough for us to forcefully forge new memory networks in our students. We must constantly help students test their memories in real and metaphoric life settings that encourage stimulating interaction, or else all our efforts to create the memories are for naught. Watching a film doesn't help us maintain memory networks unless we become emotionally involved in its story; likewise, watching a teacher teach doesn't help students maintain their memory networks unless they, too, become emotionally involved in the exploration of the "story" at hand.

Bright students tend to have the mental ability to take a simple (even random) event and find relationships between it and some curricular issue. They thus can become a very useful classroom resource for beginning story lines and moving them along. But this can happen only in a classroom in which the teacher encourages the students to tell their stories to each other and to themselves.

## How Our Brain Processes Our Stories

Schank (1990) suggests that the formation of a story about an experience is a memory process involving five stages that can reduce something like a two-week vacation trip to a relatively short, basic story that we effortlessly tell when someone asks us about the experience, and that we can adapt to the specific interests of different listeners:

1. **Define the Gist.** As we recall the total experience to create the memory, we distill all the separate events into the key elements that constitute the gist (or essence) of the experience. Over time, these elements may change as the experience becomes a more important event in our life, perhaps in ways that we didn't anticipate when the experience occurred.
2. **Sequence the Activities.** We arrange the events into a sequence that enhances the memory and the meaning, such as chronologically or in terms of importance. Each event triggers the memory of the next event in the logical sequence.
3. **Index the Story.** As we formulate our story, we index key concepts and terms in our memory that will allow us to easily recall the events that we plan to include in our story—and that will trigger memories when we listen to the stories that others tell. This is also what we do externally when we jot down notes for a lesson presentation.
4. **Tell the Story.** We tend to use natural conversational language when we tell our story, and so we actually make up much of the story during its telling.
5. **Amplify the Story.** When we tell our story, we insert credible details that we didn't specifically store in our long-term memory of the event, but that we know occurred from our general knowledge of how the world normally functions (e.g., that the plane landed when we reached our destination).

Storytelling is a natural process that we generally don't consciously carry out (except perhaps when we formally plan a lesson—write out a story). It allows us to combine a complex combination of objects and events from our declarative memory storehouse to a sequential format that, in some ill-understood manner, enhances the memory of all of them. As suggested earlier, it's probably a declarative memory adaptation of the ancient skill sequences that constitute procedural memory. Our brain tends to adapt and recycle any efficient process.

Many things can't be communicated easily outside the context of a story—for instance, telling a family member what we did during the day requires us to at least list a sequence of events, which is the very framework of a story. And sometimes a story clarifies meaning for us in a way that facts or concise definitions cannot—when defining a word such as *chutzpah, weird,* or *insensitive,* for example, storytelling seems to work better than a dictionary definition because it creates many possible connections for remembering the meaning of the word.

Part of our brain's love of storytelling certainly lies in the challenges that stories often present to us. We enjoy figuring out the gist of a story, independent of the storyteller's explaining it to us. We tend to get angry if someone tells us how a film we plan to see turned out. We prefer to get the point of a joke on our own. We enjoy mystery stories, even though they follow a formula.

To be human is to be a storyteller. A computer can tell us how many words are in a story, correct some spelling errors, and execute other mechanical tasks—but it doesn't have a clue to what the story is about. Conversely, although children will miscount the words and miss many spelling errors, they can easily tell us the gist of the story—and even imaginatively recount the story in their own words. Perhaps our brain's affinity for storytelling explains why the humanities are so basic to education.

## Notes

1. Norepinephrine (or adrenaline, when it's a hormone) is associated with both alertness and the stress response. Sylwester (1995) describes in *A Celebration of Neurons: An Educators' Guide to the Human Brain,* the actions of the endorphins, a class of peptides that reduce awareness of pain in dangerous situations in which the pain might immobilize a person who might need to escape. The release of endorphins in an emotionally painful or hurtful situation, as described by Sylwester (1995), might thus interfere with the development of a memory. (See Appendix A in *A Celebration of Neurons* for information on the neurotransmitters.)
2. Elizabeth Loftus is a major memory researcher who questions the validity of repressed memories. In *The Myth of Repressed Memories* (New York: St. Martin's Press, 1994), she cites research studies in which subjects were led to fabricate entire memories of traumatic events from their childhood that hadn't occurred.
3. The norepinephrine and serotonin neurotransmitter systems that play an important role in maintaining attention when we're awake reduce their activity during non-REM sleep and shut down during REM sleep, when we dream. The acetylcholine neurotransmitter system, which is active when we're awake, begins to dominate brain activity during sleep, especially during dream periods. Acetylcholine plays an important role in the operation of the hippocampus, which codes long-term memories. Hobson's *The Chemistry of Conscious States* (Boston: Little, Brown, 1994) provides a clear explanation of the process.

## References

Crick, F., and G. Mitchison. (1983). "The Function of Dream Sleep." *Nature* 304: 111–114.

Hobson, J. (1989). *Sleep.* New York: Scientific American Library. Distributed by W. H. Freeman.

Hobson, J. A. (1994). *The Chemistry of Conscious States: How the Brain Changes Its Mind.* Boston: Little, Brown.

Kandel, M., and E. Kandel. (May 1994). "Flights of Memory." *Discover* 15, 6: 32–38.

Luria, A. R. (1968). *The Mind of a Mnemonomist.* New York: Basic Books.

Neisser, U. (1982). *Memory Observed: Remembering in Natural Contexts.* San Francisco: W. H. Freeman.

Posner, M. I., and M. E. Raichle. (1994). *Images of Mind.* New York: Scientific American Library.

Rose, S. (1993). *The Making of Memory: From Molecules to Mind.* New York: Anchor Books/Doubleday.

Schank, R. C. (1990). *Tell Me A Story: A New Look at Real and Artificial Memory.* New York: Scribner.

Schank, R. C. (1991). *The Connoisseur's Guide to the Mind: How We Think, How We Learn, and What It Means to Be Intelligent.* New York: Summit Books.

# Promising Practices That Foster Inclusive Education

ALICE UDVARI-SOLNER and JACQUELINE S. THOUSAND

The inclusive education movement has often been viewed as a separate initiative running parallel or even counter to other curricular and instructional reform efforts (Block and Haring, 1992). We take a holistic rather than separatist viewpoint and propose that the innovative changes occurring in general education are the same kinds of changes required for effective inclusion.

A number of established and emerging general education practices emulate the principles of inclusive education. When these practices are used, educators may be better equipped to facilitate meaningful and effective inclusive education for students perceived as disabled, at risk, or gifted, as well as those considered "average." Among the initiatives that have great promise for building inclusive schools are outcome-based education, multicultural education, multiple intelligence theory, constructivist learning, interdisciplinary curriculum, community-referenced instruction, authentic assessment of student performance, multi-age grouping, use of technology in the classroom, peer-mediated instruction, teaching responsibility and peacemaking, and collaborative teaming among adults and students. The remainder of this chapter examines each practice in the light of inclusive education.

## Outcome-Based Education

Outcome-based education (OBE) is not a new concept to educators; it has evolved over the past 40 years to its current conceptualization with three central premises (Spady and Marshall, 1991):

- All children can learn and succeed, although not in the same way or on the same day.
- Success breeds success.
- Schools determine the conditions of success.

In addition, OBE is guided by four principles (Brandt, 1992–1993). The first, *clarity of focus,* implies that all aspects of education (curriculum, instruction, assessment) are centered on what we want children to demonstrate by the end of their schooling career. Everyone is clear at all times about the goals of education.

The second principle, *expanding opportunity,* recognizes that students learn in different ways and at different rates, and that various methods and contexts (perhaps out of the school building) are needed to optimize learning. Outcome demonstration is not tied to the calendar.

---

Alice Udvari-Solner and Jacqueline S. Thousand, "Promising Practices That Foster Inclusive Education," *Creating an Inclusive School,* 1995, pp. 87–109. Reprinted by permission from Association for Supervision and Curriculum Development.

The third principle, *high expectations,* is rooted in the assumption that every student is "able to do significant things well" (Bill Spady, quoted by Brandt, 1992–1993, p. 66). All students are expected to demonstrate success in their own way.

Finally, the fourth principle, *designing down,* turns the traditional method of designing curriculum upside down. Long-range outcomes are established first, and then curriculum is designed, always with an eye on where students ultimately are expected to end up.

Why institute OBE for students with disabilities? Clearly, OBE is consistent with and supportive of an inclusive education philosophy. OBE professes to encompass all students and focus on success for all. Additionally, many community members and education leaders are attracted to the autonomy schools are given to establish the means—the curriculum—for achieving significant outcomes. Teachers are encouraged to be flexible and to provide educational experiences in a variety of ways for a diverse student body. Students are not required to do the same things in the same ways in the same amount of time as same-aged peers.

Some question how students with severe disabilities can be included in OBE. McLaughlin and Warren (1992) argue that students with intensive challenges can be a part of the OBE model if the curriculum is defined in broad and balanced areas of knowledge and skill rather than narrow subject areas. To illustrate, 4,000 adults and students provided input into the development of Vermont's Common Core of Learning (Vermont State Department of Education, 1993). The Common Core identified these skills as vital results: communication, reasoning and problem solving, personal development, social responsibility, and fields of knowledge including technology and new disciplines that "may be only just coming into existence" (p. 12).

Clearly, students with severe disabilities can achieve in many of these vital domains, although the performance criteria and method of assessing success may be quite different from that of their classmates. Central to this question is the notion of "personal best" (Shriner, Ysseldyke, Thurlow, and Honetschlager, 1994, p. 41). For example, literacy may be an expectation for all graduates. One student demonstrates his personal best by writing a persuasive speech, whereas another demonstrates her personal best by effectively using her assistive communication device to express her wants and interests.

## Multicultural Education

The term *multicultural education* has been used to describe various policies related to educational equity and practices that foster understanding of human' differences and similarities (Banks and McGee Banks, 1989; Sleeter and Grant, 1994). The principles of multiculturalism were formulated in the 1960s and '70s as issues of culture and diversity rose to the forefront of political and educational arenas. Initially, these principles were most prominently associated with gender, ethnicity, and class distinctions. Only recently have the issues at disability and sexual orientation made their way into the multicultural literature (Tiedt and Tiedt, 1990). As a result multicultural education has rarely been linked effectively with inclusive education.

When the underlying goals of a multicultural approach are examined, they fit well with the ideological framework of inclusive education. The goals and outcomes of multicultural education are to:

- foster human rights and respect for difference,
- acknowledge the value of cultural diversity,
- promote an understanding of alternative life choices,
- establish social justice and equal opportunity, and
- facilitate equitable power distribution among individuals and groups (Gollnick, 1980).

When school communities employ a multicultural approach, they make a commitment to empower students and to attempt to increase academic achievement by redesigning the entire educational agenda to make learning environments responsive to students' cultures, behavior, and learning styles (Banks and McGee Banks, 1989).

Grant and Sleeter (1989) have extended the concept of multicultural education using a *reconstructionist* viewpoint. Simply stated, reconstructionism requires a critique of contemporary culture and a reconceptualization of what it can and should be to realize a more humane society (Brameld, 1956). Students are encouraged to critically evaluate inequities and instances of discrimination or bias and to identify strategies for change. By engaging in a meta-analysis of existing conditions and establishing visions that reflect a value system, even the youngest members of school communities are encouraged to make a personal commitment to change. A reconstructionist orientation holds promise for accelerating educational reform by embedding reformation in teachers' and students' day-to-day discourse.

Although multicultural education and inclusion are not synonymous, administrators, educators, and community members need to recognize the commonalities between them so they may coordinate reform activities within schools to maximize the use of resources and optimize the number of children who will benefit.

## Multiple Intelligences Theory

The theory of multiple intelligences (MI-theory) proposed by Howard Gardner (1983) questions the adequacy and efficacy of the traditional conceptualization of knowledge, aptitude, and intellect. As defined in the western world, intelligence has long been equated with logical and linguistic abilities. The underlying assumptions of this view are that the processes of the mind are quantifiable and can be translated into a singular construct. Furthermore, all children can be compared and rank ordered by intellectual prowess (Goldman and Gardner, 1989), hence our reliance on I.Q. scores as essential descriptors of students' abilities and predictors of academic success.

MI-theory is based on the supposition that several distinct forms or families of intelligence exist—or, more accurately, co-exist—to create a constellation of ability for any one individual. Gardener (1983) has recommended consideration of at least seven types of intelligence: linguistic, logical-mathematical, musical, spatial, bodily-kinesthetic, interpersonal, and intrapersonal.

These categories are constructed to promote the valuing of skills beyond the conventional representations of verbal ability, written expression, and mathematical reasoning. Gardner's valued capacities include: the ability to depict and manipulate spatial representations; to think in and produce musical forms; to use kinesthetic action to perform, produce, and problem solve, and to use effective communication and interaction skills to understand others or reflect on one's own behavior.

The notion of multiple intelligences has important implications for inclusive education. Gardner based his theory in part on observations and studies of the capacities of children with disabilities and on the meaning of intelligence in varied cultures (Gardner, 1983), thus validating a broader perspective. Teachers equipped with this perspective are in the position to appreciate students' "unconventional" behavior and seek productive applications of these skills within a learning context. They will arrange learning activities to allow expression of knowledge through multiple modes and the use of different intelligences. Teachers may use the student's strongest modalities or intelligences as vehicles to promote skill acquisition in weaker areas of performance.

MI-theory does not allow the student to be viewed only through the constricted lens of logic or language. Instead, learning and memory are seen as multifaceted and not completely understood. Consequently, our mechanisms for assessing intelligence are at the very least far too narrow and perhaps misguided. This calls into question the current systems used to identify and label any child as disabled. Embracing the tenets of MI-theory could interrupt the vicious cycle of labeling and the social construction of disability based on one or two aspects of ability. The use of a multiple-intelligence orientation liberates educators to see the idiosyncrasies in learning styles and differentiate curriculum for all students, thus making "difference" usual in the classroom.

## Constructivist Learning

From a constructivist perspective, learning is the creation of meaning when an individual makes linkages between new knowledge and the context of existing knowledge (Poplin and Stone, 1992). A key characteristic of this view, then, is that learners "construct' their own knowledge (Peterson, Fennema, and Carpenter, 1988–1989). Generally speaking, the ideas of Brownwell, Vygotsky, Dewey, and Piaget are constructivist (Resnick and Klopfer, 1989); underpinning their theories is the idea that knowledge is not quantitative but interpretive and must develop in social contexts of communities and communicative interchanges (Peterson and Knapp, 1993).

Constructivism challenges the assumptions and practices of reductionism that have pervaded our educational practices for generations. In a deficit-driven reductionist framework, effective learning takes place in a rigid, hierarchical progression. Each concept or skill is broken into small segments or steps, and students learn each one in sequence (Poplin and Stone, 1992). A supposition exists that children are unable to learn higher-order skills before mastering those of lower order (Peterson et al., 1988–1989). Learning, then, is an accumulation of isolated facts. It is presumed that through this accumulation process, learners will build skills and generate new knowledge.

Conceptualizing curriculum and instruction from a constructivist vantage point intersects productively with the practices of inclusive education, Constructivism fosters the idea that all people are always learning, and the process cannot be stopped (Poplin and Stone, 1992). "No human being understands everything; every human being understands some thing. Education should strive to improve understanding as much as possible, whatever the student's proclivities might be" (Siegel and Shaughnessy, 1994, p. 564). Both of these statements imply that there are few, if any, prerequisites for learning and that children must be met at their current level of performance without undue focus on remediation. It is acknowledged that all students enter school with different knowledge that is influenced by background, experiences, and cultural practice. Consequently, teachers must take into account these factors and ensure that new information is related in meaningful ways to each learner's existing knowledge.

## Interdisciplinary Curriculum

An interdisciplinary approach is a curricular orientation that expressly employs methodology and language from more than one discipline to examine a central theme, issue, problem, topic, or experience (Jacobs, 1989). Teachers and students are encouraged in a learning partnership to examine one area in depth from complex and multiple perspectives. Interdisciplinary curriculum may be implemented in several ways. At the elementary level, a single teacher can interface the content of assorted disciplines throughout one unit of study. Instructors of art, music, and physical education can further infuse the theme across the instructional day. Most true to the interdisciplinary philosophy is the practice at middle and high school levels of uniting teachers of separate disciplines to team teach around a selected set of issues.

Interdisciplinary/thematic approaches have grown out of dissatisfaction with discipline-based or subject-driven methods of curriculum organization. Discipline-based models are premised on the teaching of content knowledge. However, knowledge in all areas of study is growing exponentially each day. Essentially, there are not enough hours in the day to teach all that is new. Jacobs (1989) believes this indicates a need to rethink the way we select areas of study, deciding not only what should be taught, but what should be eliminated. Fragmentation of schedules and subject matter into allotted time periods is common practice in discipline-based approaches. With thematic methods, students are not forced to create bridges among seemingly unrelated splinters of information but instead can view issues in a holistic manner.

If discipline-based approaches pose drawbacks for the typical population, the impact on students with disabilities is likely to be more significant. One reason students with disabilities failed in the past in general education classes was that the subject matter presented was unrelated, out of context, and practiced only a few minutes per day without consideration for generalization and transfer. A thematic orientation offers a way to show how different subject areas relate and influence students' lives, thereby affirming the relevance of the curriculum (Ackerman and Perkins, 1989; Bean, 1990).

## Community-Referenced Instruction

Community-referenced or community-based instruction is characterized by students applying skills in nonschool settings that have some relationship, relevance, and purpose to their lives now or in the future (Falvey, 1989). Instruction takes place regularly in community environments where age-appropriate vocational, domestic, community, or recreational skills can be acquired (Brown, Branston, Baumgart, Vincent, Falvey, and Schroeder, 1979). The premise behind community-referenced instruction is that all students need an education that prepares them with the skills to live and work as part of the adult community—in other words, to achieve functional outcomes (Clark, 1994).

A community-based approach to instruction evolved as a best practice for students with moderate and severe disabilities (Brown et al., 1979). However, it is now recognized as a valuable tool in the education of all students (Peterson, LeRoy, Field, and Wood, 1992). Given the complexities of adult life in the 21st century, educators are realizing that all the skills that are relevant, critical, and enriching cannot be taught effectively within the confines of the classroom. For students with significant disabilities who may experience problems generalizing skills acquired in one setting to another, the need for systematic instruction in the actual environment of concern is evident. For students without disabilities, there is a need to connect with the larger community, work in concert with community members to engage in problem-solving skills, and integrate themselves as participants in businesses and organizations long before graduation comes.

Accessing the community for instruction can provide a student with disabilities the context to learn or maintain a new skill. The same environment can be used as a "fidelity check" for the application of math, science, or language skills for typical students. For example, each week a group of three students from a 7th grade home economics class goes to the local grocery store. For the member of the group who has disabilities, the weekly trip serves as an opportunity to learn to travel by city bus, cross streets, and select and pay for groceries. The other students in the group must use mathematics and nutrition skills by comparative shopping to select items that are most economical and contain the least grams of fat, thus employing skills emphasized within the classroom.

## Authentic Assessment of Student Performance

The need for better alignment between assessment and instruction is even more evident (Chittenden, 1991) as schools have begun to shift their curriculum to include multicultural, constructivist, interdisciplinary, and community-referenced approaches, and as teachers have placed more emphasis on the meaning of learning with attention to children's interest and proclivities. Traditional measures of performance that do not provide information about students' understanding and quality of thinking are out of step with dynamic, student-centered instructional practices.

Perrone (1994) noted that typical assessment techniques relying primarily on the recall of knowledge provide an artificial, decontextualized view of the learner. Assessment has been equated with the possession of information rather than the acquisition of global constructs (e.g., learning the process of writing) (Zessoules and Gardner, 1991). In most cases, data acquired from these assessments are unrelated to the ways students naturally learn or will need to use the knowledge. This problem is amplified for students with disabilities. When traditional measures such as standardized, non-referenced tests are used for evaluation, their performances predictably fall below those of their nondisabled peers. Thus, a unidimensional and deficit-oriented profile of the learner is maintained.

Based on the need for more realistic and responsive outcome measures, a number of alternative evaluation techniques—authentic assessments—have evolved. Authentic assessment occurs when students are expected to perform, produce, or otherwise demonstrate skills that represent realistic learning demands (Choate and Evans, 1992; Diez and Moon, 1992). According to Meyer (1992), the contexts of the assessments are real-life settings in and out of the classroom without contrived and standardized conditions. Authentic assessments can be considered exhibitions of learning that are gathered over time to show evidence of progress, acquisition, and application. For example, written expression may be assessed through the use of a portfolio that includes several samples of writing representing conceptual ideas, rough drafts, self-edited papers, and final versions. Included in this assessment may be products such as poems, letters, or research papers that illustrate ability to use other forms of written expression. The student also is encouraged to include self-evaluations and personal goals for progress.

Common features of authentic assessments are:

- Students must integrate and apply skills to accomplish a larger task (Choate and Evans, 1992).
- The processes of learning, higher-level thinking, or problem-solving are emphasized, as well as the product of these actions (Diez and Moon, 1992).
- Assessment tasks must help students make judgments about their own performance. Through self-appraisal, children set goals for progress and provoke further learning (Perrone, 1994; Zessoules and Gardner, 1991).
- The criteria for performance are negotiated and made explicit to students in advance (Wiggins, 1989).

The use of authentic assessments is an important component in creating inclusive classrooms. This form of evaluation is closely linked to the individualized, performance-based assessment that has been the preferred mode of assessment in special education. These techniques are less likely to be culturally biased for students who are limited in English proficiency or in any other intellectual, physical, or emotional capacity.

Students with unique learning characteristics and their peers are allowed to express knowledge through multiple modes and in nontraditional ways (Perrone, 1991). Instruction and assessments are provided with relevant tasks so students who have difficulty generalizing skills or using them out of context are not required to transfer learnings to demonstrate understanding (Choate and Evans, 1992). Functional expressions of competence more readily enable teachers to identify skills that are discrepant or mastered, thus giving direction to instruction of highest priority.

## Multi-Age Grouping

Kasten and Clarke (1993) define multi-aged grouping as "any deliberate grouping of children that includes more than one traditional grade level in a single classroom community" (p. 3). Also referred to as nongraded, family, or vertical grouping, multi-age classrooms are considered single learning communities made up of a balanced collection of students from the school population with consideration for heterogeneity in gender, ability, ethnicity, interests, and age levels. It is not unlikely to have siblings and members of the extended family within one vertical grouping.

The multi-age classroom, a well-established practice in countries such as Canada, New Zealand, Britain, and some parts of the United States (Elkind, 1987), is based on several underlying assumptions that directly oppose the traditional practice of grade-level grouping. Grade-level grouping presumes that students who are the same age have like learning needs and abilities, thus benefiting from similar instruction. Placement of the child is based solely on age or physical time (Elkind, 1987). Learning by grade level is viewed as a predictable, sequential, and orderly procedure; and one year of schooling is a product that can be judged and rated by a standard of performance (Kasten and Clark, 1993).

In contrast, a multi-age approach is based on the assumption that learning is a continuous and dynamic process. Student diversity is essential. Children are expected and, in fact, encouraged to learn at different rates and levels. The growth of the child is viewed in both biological and psychological time, rather than merely physical time, so that learning experiences are designed as developmentally appropriate.

Many elements of multi-age grouping that work for students without disabilities are also advantageous for students with disabilities. The emphasis on *heterogeneity* requires a classroom organization flexible enough to accommodate children at different levels of maturity and with different levels of intellectual ability. In fact, diversity impedes the use of lock-step instructional methods aimed at the whole class or a specific grade level.

The sense of *community* created over time among teachers and students is advantageous to promote long-term networks of support for students with disabilities. Transitions from setting to setting and teacher to teacher are associated with recoupment, generalization, and social adjustment difficulties for some students with disabilities. These "passages" are reduced in nongraded groupings, and teachers have time to get to know a particular student. Teachers can use information gained about the child in one year to plan learning experiences for the next year without the risk of losing that knowledge in a transition to new staff.

## Use of Technology in the Classroom

Technology is proving to be a catalyst for transforming schooling by fostering excitement in learning for all children. As Peck and Dorricott of the Institute for the Reinvention of Education (1994) observed:

> To see students so engaged in learning that they lose track of time, to see a level of excitement that causes students to come to school early and stay late, and to have time to develop strong relationships with students and to meet their individual needs allows educators to fulfill age-old dreams (p. 14).

What is technology? It is more than computers and software packages, and it reinvents itself almost daily. Technology in education includes calculators, video cameras, VCRs, portable personal computers, printers, general-purpose software such as word processing programs and *HyperCard*, computer-assisted instruction for drill and practice, laser videodisks, telecommunication networks such as electronic mail, distance education, interactive multimedia, scanners, text-to-speech and speech-to-text software, pen-based notepads such as the Apple *Newton,* and more. Given the current and expanding access to technology inside and outside of the classroom, the climate is conducive to including students with disabilities who need technology to access the curriculum, express their knowledge, communicate, or control their environment.

In the past, technology was only in the possession of a few experts such as the computer lab teacher or those who designed or programmed augmentative communication systems for students with communication limitations. Today, technology has become "user friendly." Educators are joining the ranks of adults and children who rarely go a day without interacting with their laptop computer for desktop publishing, data management, game playing, or instantly communicating and socializing via the Internet.

Technological tools of a student with disabilities that once seemed too complex, cumbersome, or expensive (for example, massive computers bolted to a table or voice synthesizers) have become very portable, affordable, and standard hardware and software features of schools. Technology that used to be unusual for a single student now is usual within the classroom. Never before have educators been in such an ideal position to capitalize on technological advances in order to readily educate students who have different learning styles and rates or who rely on technological support to learn, communicate, and control their world.

In an interview with Frank Betts (1994), David Thornburg laid out a scenario that would equip every 2nd or 3rd grade child with a "loaded" computer for $100 per child. For $200, the system could be upgraded in 8th grade. But he cautions that even with the feasibility for advanced technology at every child's fingertips at home and school, equipment and software does not guarantee an excellent educational program (Thornburg, 1992). Teachers still need to get to know each child and base decisions on how each child learns. Commenting on the use of technology to support the inclusion and learning of students with disabilities in school, Dutton and Dutton (1990) state it this way:

> Remember that technology is not a "cure" for a disability; rather, it is a tool for everyone in society. Focus should not be placed on how the equipment itself will work, but efforts should be placed toward developing strategies, utilizing effective teaching practices, and working with the strengths of all students in the class. Technology can help move barriers, but it is people, working together, who learn and succeed (p. 182).

## Peer-Mediated Instruction

"Peer mediated instruction" (Harper, Maheady, and Malletter, 1994, p. 229) refers to any teaching arrangement in which students serve as instructional agents for other students. Cooperative group-learning model and peer tutoring or partner learning strategies are two forms of peer-mediated instruction that support inclusive education.

As Johnson and Johnson (1994) point out, students may interact in three ways during learning. They may compete to see who is the best, they may work alone and individually toward their goals without attending to other students, or they may have a stake in one another's success by working cooperatively. Competitive learning interferes with community building, which is one objective of inclusive education. Yet, "research indicates that a vast majority of students in the United States view school as a competitive enterprise where one tries to do better than other students" (Johnson and Johnson, 1994, pp. 32–33).

It is critical to note the dramatic difference between simply asking students to sit together and work in a group, and careful structuring so students work in cooperative learning groups. A group of children chatting together at a table as they do their own work is not a cooperative group, because no sense of positive interdependence exists, and no need for mutual support is arranged for them. It is only under particular conditions that groups will have healthy and productive relationships and may be expected to be more productive than in individualistic or competitive learning arrangements. Common to the diverse approaches of cooperative learning are five conditions or attributes:

- a joint task or learning activity suitable for group work,
- small-group learning in teams of five or fewer members,
- a focus on cooperative interpersonal behaviors,
- positive interdependence through team members' encouragement of one another's learning, and
- individual responsibility and accountability for the participation and learning of each team member (Davidson, 1994).

A rich research base supports the use of cooperative learning to facilitate successful learning in heterogeneous groupings of students with varying abilities, interests, and backgrounds. Within the context of inclusive education, cooperative learning makes great sense as an instructional strategy as it enables students "to learn and work in environments where their individual strengths are recognized and individual needs are addressed" (Sapon-Shevin, Ayres, and Duncan, 1994, p. 46). In other words, cooperative learning allows the classroom to be transformed into a microcosm of the diverse society and work world into which students will enter and a place for acquiring the skills to appreciate and cope with people who initially might be perceived as different or even difficult. Within this context, students learn what a society in which each person is valued would be like.

Partner learning or peer tutoring systems are not new; teachers of one-room school houses relied heavily on students as instructors. Children are continually teaching one another informally when they play games and engage in sports. Partner learning systems build relationships among students and offer a cost-effective way of enhancing engaged learning time on the part of children. Peer tutor systems can be same-age or cross-age and can be established within a single classroom, across classes, or across an entire school. Evidence of the social, instructional, and cost effectiveness of tutoring continues to mount (e.g., Fuchs, Fuchs, Hamlett, Phillips, and Bentz, 1994; Thousand, Villa, and Nevin, 1994). Benefits to students receiving this type of instruction include learning gains, interpersonal skill development, and heightened self-esteem. Good and Brophy (1987) suggest the quality of instruction provided by trained tutors may be superior to that of adults for at least three reasons:

- Children use more age-appropriate and meaningful language and examples.
- Having recently learned what they are to teach, they are familiar with their partner's potential frustrations.
- They tend to be more direct than adults.

Tutors also experience benefits similar to those of their partners. Namely, they develop interpersonal skills and may enhance self-esteem. Further, tutors report that they understand the concepts, procedures, and operations they teach at a much deeper level than they did before instructing. This likely is due to their meta-cognitive activity when preparing to teach.

## Teaching Responsibility and Peacemaking

Among the children who are perceived as the most challenging to educate within current school organizational structures are those who demonstrate high rates of rule-violating behavior, children who have acquired maladaptive ways of relating, and children who are perceived as troubled or troubling. Adversity at home and in the community negatively affects an increasing number of children's ability and motivation to learn. The educator's job has broadened from providing effective and personalized learning opportunities to addressing the stressors in children's lives by offering a variety of school-based social supports (for example, breakfast program, free lunch, mental health, and other human services on campus). Personal responsibility and peacemaking have risen to the top as curriculum priorities (Villa, Udis, and Thousand, 1994).

Educators long have recognized that for students to master a content area such as mathematics or science, they need continuous and complex instruction throughout their elementary, middle, and high school years. When a child does not grasp concept or skill, we react with a "teaching response" and attempt to reteach the material with additional or different supports and accommodations. The content area of responsibility, however, has not received the same immediate treatment. The explicit teaching of patterns of behavior and habits representative of responsible behavior often never occurs. Instead, instruction is relegated to reactive, add on, quick-fix methods such as seeing a guidance counselor, going to a six-week social skills group, or instituting a written behavior change plan. To teach responsibility is as demanding as teaching any other content; it requires careful thought and complex, ongoing instruction from the day a child enters school.

Requisite to students learning responsible values, attitudes, and behaviors is the perception that somebody in the school community genuinely cares about them. Thus, teachers, above all, must demonstrate caring and concern by validating students' efforts and achievements. They must also directly teach responsibility by setting limits to ensure safety; establishing a school-wide discipline system that promotes the learning of responsibility; and directly instructing students in pro-social communication skills, anger management, and impulse control techniques (Villa et al., 1994).

Models of discipline that are responsibility based (Curwin and Mendler, 1988; Glasser, 1986) acknowledge conflict as a natural part of life. They consider behavior to be contextual and transform the educator's role from cop to facilitator. There are no "if-then" consequences (e.g., three tardies equals a detention; 10 absences results in a grade of "F" in the missed class). Instead, responses to rule-violating behavior depend on all kinds of factors such as the time of day, the frequency and intensity of the behavior, and the number of other people exhibiting the behavior. These responses range from reminders, warnings, re-directions, cues, and self-monitoring techniques to behavioral contracts and direct teaching of alternative responses. Most important is to recognize that the development of student responsibility should be part of the curriculum and considered as important as any other curriculum domain. It should be concerned with teaching young people how to get their needs met in socially acceptable ways and should include modeling, coaching, and ongoing thought and reflection on the part of school personnel.

One way to incorporate the development of student responsibility into the curriculum and culture of a school is to turn conflict management back to the students by using them as peer mediators. In an increasing number of North American schools, students are trained to mediate conflicts and are available during school hours to conduct mediations at student, teacher, or administrator request. A small number of students may be selected and trained in mediation processes or all students may receive training in conflict resolutions skills. Students who serve as mediators sometimes are called peacemakers or conflict managers.

Schrumpf (1994) outlined a structured process for establishing a peer mediation program within a school. He emphasizes that peer mediation must be made highly visible so students and teachers actually use the program as an alternative to adult intervention. Mediators need ongoing adult support through regular meetings in which they discuss issues and receive advanced training. Data collection in the form of Peer Mediation Requests and Peer Mediation Agreements are analyzed to determine the nature, frequency, and outcomes of mediation request.

Emerging data suggest that peer mediation programs correlate with improved school attendance and decreases in fights, student suspensions, and vandalism. For example, of the 130 teachers in a New York City school in which students practiced peer mediation, 71 percent reported reductions in physical violence, 66 percent heard less verbal harassment, and 69 percent observed increased student willingness to cooperate with one another (Meek, 1992). Learning to resolve conflict, with peers is an empowering action consistent with the principles of inclusive education, with the potential of generalizing from peace between two people to peacemaking in community and global contexts.

## Collaborative Teaming Among Adults and Students

Schools showing great success in responding to student diversity have redefined the role of general and special educators and other support personnel to that of collaborative team members who assemble to jointly problem solve the daily challenges of heterogeneous schooling. Among the benefits of collaborative planning and teaching team arrangements is the increased instructor/learner ratio and the resulting immediacy in diagnosing and responding to individual student's needs. Teaming arrangements capitalize on the diverse knowledge and instructional strengths of each team member and, when special educators are included on the team, eliminate the need to refer students to special education in order to access special educators' support. Although is not yet the norm in North American schools, when the term is discussed, it generally conjures up images of adults (usually professional educators, sometimes community members) sharing planning, teaching, and evaluation responsibilities. Until recently, the students themselves were missing from the teaming concept.

Villa and Thousand (1992) have identified multiple rationales for including students in collaborative educational roles with adults. First, students represent a wealthy pool of expertise, refreshing creativity, and enthusiasm at no cost to schools. Second, educational reform recommendations call for students to exercise higher-level thinking skills to determine what, where, when, and how they will learn. Third, collaborating with adults in advocacy efforts for other learners help students develop the ethic and practice of contributing to and caring for a greater community and society. Fourth, given the information explosion and the complexity of our networked global community, collaborative teaming skills are necessary for survival in the workplace. Educators, then, have a responsibility to model collaboration by sharing their decisionmaking and instructional power with students and arranging for and inviting students to join in at least the following collaborative endeavors (Villa and Thousand, 1992):

- Students as members of teaching teams and as instructors in cooperative learning and partner learning arrangements.
- Students as members of planning teams, determining accommodations for themselves or classmates with and without disabilities.
- Students functioning as advocates for themselves and for classmates during meetings (e.g., individual educational plan meeting for a student with a disability) and other major events that determine a student's future educational and post-school choices.
- Students as mediators of conflict.
- Students providing social and logistical support to a classmate as peer partner or a member of a Circle of Friends (Falvey, Forest, Pearpoint, and Rosenberg, 1994).
- Students as coaches of their teachers, offering feedback regarding the effectiveness of their instructional and discipline procedures and decisions.
- Students as members of inservice, curriculum, discipline, and other school governance committees such as the school board.

All of these options for collaboration facilitate meaningful inclusion and participation of students with disabilities in school. Asking students to join with adults in collaborative action is a critical strategy for fostering the spirit of community, and equity that is foundational to quality heterogeneous schooling experiences and the desired educational outcomes of active student participation and critical thinking.

The exemplary and promising practices identified in this chapter establish the infrastructure within which the principles of inclusive education can be realized. Collectively, the initiatives have the potential to a create a unified philosophy and revolutionary standards of educational practice. The success of any change, however, always relies on the courage and determination of practicing educators to translate and put in place principles of these contemporary reform initiatives. "You can't mandate what matters," Fullan writes (1993, p. 125); instead, the complex goals of change require knowledge, skills, collaboration, creative thinking, and committed and passionate action. If widespread progress is ever to occur in education, inclusive education must not be treated as an "add on" to other pressing initiatives; it must be the central discussion, and teachers must be the central participants in a scholarly discourse on education's future. The interface between inclusive education and other exemplary practices must become clearly and publicly self-evident (Peterson and Knapp, 1993).

## References

Ackerman, D., and D. N. Perkins. (1989). "Integrating Thinking and Learning Skills Across the Curriculum." In *Interdisciplinary Curriculum: Design and Implementation,* edited by H. H. Jacobs. Alexandria Va: Association for Supervision and Curriculum Development.

Banks, J., and C. McGee Banks. (1989). *Multicultural Education: Issues and Perspectives.* Boston: Allyn and Bacon.

Bean, J. (May 1990). "Rethinking the Middle School Curriculum." *Middle School Journal* 21, 5: 1–5.

Betts, F. (1994). "On the Birth of the Communication Age: A Conversation with David Thornburg." *Educational Leadership* 51, 7: 20–23.

Block, J. H., and T. G. Haring. (1992). "On Swamps, Bogs, Alligators and Special Education Reform." In *Restructuring for a Caring and Effective Education: An Administrative Guide to Creating Heterogeneous Education,* edited by R. Villa, J. Thousand, W. Stainback, and S. Stainback. Baltimore: Paul H. Brookes.

Brameld, T. (1956). *Toward a Reconstructed Philosophy of Education.* New York: Holt, Rinehart, and Winston.

Brandt, R. (1992–1993). "A Conversation with Bill Spady." *Educational Leadership* 50, 4:66–70.

Brown, L., M. Branston, D. Baumgart, L. Vincent, M. Falvey, and J. Schroeder. (1979). "Utilizing the Characteristics of Current and Subsequent Environments as Factors in the Development of Curricular Content for Severely Handicapped Students." *AAESPH Review* 4, 4: 407–424.

Chittenden, E. (1991). "Authentic Assessment, Evaluation, and Documentation." In *Expanding Student Assessment,* edited by V. Perrone. Alexandria, Va.: Association for Supervision and Curriculum Development.

Choate, J. S., and S. Evans. (1992). "Authentic Assessment of Special Learners: Problem or Promise?" *Preventing School Failure* 37, 1: 6–9.

Clark, G. (1994). "Is Functional Curriculum Approach Compatible with an Inclusive Education Model?" *Teaching Exceptional Children* 26, 2: 36–37.

Curwin, R., and A. Mendler. (1988). *Discipline with Dignity.* Alexandria, Va.: Association for Supervision and Curriculum Development.

Davidson, N. (1994). "Cooperative and Collaborative Learning: An Integrated Perspective." In *Creativity and Collaborative Learning: A Practical Guide to Empowering Students and Teachers,* edited by J. Thousand, R. Villa, and A. Nevin. Baltimore: Paul H. Brookes.

Diez, M., and J. Moon. (1992). "What Do We Want Students to Know? . . . And Other Important Questions." *Educational Leadership* 49, 8: 38–41.

Dutton, D. H., and D. L. Dutton. (1990). "Technology to Support Diverse Needs in Regular Classes." In *Support Networks for Inclusive Schooling: Interdependent Integrated Education,* edited by W. Stainback and S. Stainback. Baltimore: Paul H. Brookes.

Elkind, D. (1987). "Multiage Grouping." *Young Children* 43, 11: 2.

Falvey, M. A. (1989). *Community Based Curriculum: Instructional Strategies for Students with Severe Handicaps.* Baltimore: Paul H. Brookes.

Falvey, M. A., M. Forest, J. Pearpoint, and R. L. Rosenberg. (1994). "Building Connections." In *Creativity and Collaborative Learning: A Practical Guide to Empowering Students and Teachers,* edited by J. Thousand, R. Villa, and A. Nevin. Baltimore: Paul H. Brookes.

Ford, A., R. Schnorr, L. Meyer, L. Davern, J. Black, and P. Dempsey. (1989). *The Syracuse Community-Referenced Curriculum Guide for Students with Moderate and Severe Disabilities.* Baltimore: Paul H. Brookes.

Fuchs, L. S., D. Fuchs, C. L. Hamlett, N. B. Phillips, and J. Bentz. (1994). "Classwide Curriculum-Based Measurement: Helping General Educators Meet the Challenge of Student Diversity." *Exceptional Children* 60: 518–537.

Fullan, M. (1993). "Innovative Reform and Restructuring Strategies." In *Challenges and Achievements of American Education,* edited by G. Cawelti. 1993 ASCD Yearbook. Alexandria, Va.: Association for Supervision and Curriculum Development.

Gardner, H. (1983). *Frames of Mind: The Theory of Multiple Intelligences.* New York: Harper Collins Publishers.

Glasser, W. (1986). *Control Theory in the Classroom.* New York: Harper and Row.

Goldman, J., and H. Gardner. (1989). "Multiple Paths to Educational Effectiveness." In *Beyond Separate Education: Quality Education for All,* edited by D. K. Lipskey and A. Gartner. Baltimore: Paul H. Brookes.

Gollnick, D. M. (1980). "Multicultural Education." *Viewpoints in Teaching and Learning* 56:1–17.

Good, T. L., and J. G. Brophy. (1987). *Looking into Classrooms,* 4th ed. New York: Harper & Row.

Grant, C., and C. Sleeter. (1989). "Race, Class, Gender, Exceptionality, and Educational Reform." In *Multicultural Education: Issues and Perspectives,* edited by J. Banks and C. McGee Banks. Boston: Allyn and Bacon.

Harper, G. F., L. Maheady, and B. Mallette. (1994). "The Power of Peer-Mediated Instruction: How and Why It Promotes Academic Success for All Students." In *Creativity and Collaborative Learning: A Practical Guide to Empowering Students and Teachers,* edited by J. Thousand, R. Villa, and A. Nevin. Baltimore: Paul H. Brookes.

Jacobs, H. H. (1989). "The Growing Need for Interdisciplinary Curriculum Content." In *Interdisciplinary Curriculum: Design and Implementation,* edited by H. H. Jacobs. Alexandria Va.: Association for Supervision and Curriculum Development.

Johnson, R. T., and D. W. Johnson. (1994). "An Overview of Cooperative Learning." In *Creativity and Collaborative Learning: A Practical Guide to Empowering Students and Teachers,* edited by J. Thousand, R. Villa, and A. Nevin. Baltimore: Paul H. Brookes.

Kasten, W., and B. Clarke. (1993). *The Multi-age Classroom: A Family of Learners.* Katonah, N.Y: Richard C. Owen Publishers.

McLaughlin, M., and S. Warren. (1992). *Issues and Options in Restructuring Schools and Special Education Programs.* College Park: University of Maryland, The Center for Policy Options in Special Education, and the Institute for the Study of Exceptional Children and Youth.

Meek, M. (Fall 1992). "The Peacekeepers." *Teaching Tolerance,* pp. 46–52.

Meyer, C. (1992). "What's the Difference Between Authentic and Performance Assessment?" *Educational Leadership* 49, 8: 39–40.

Peck, K., and D. Dorricott. (1994). "Why Use Technology?" *Educational Leadership* 15, 7: 11–14.

Perrone, V. (1991). *Expanding Student Assessment.* Alexandria, Va.: Association for Supervision and Curriculum Development.

Perrone, V. (1994). "How to Engage Students in Learning." *Educational Leadership* 51, 5: 11–13.

Peterson, M., B. LeRoy, S. Field, and P. Wood. (1992). "Community-Referenced Learning in Inclusive Schools: Effective Curriculum for All Students." In *Curriculum Considerations in Inclusive Classrooms: Facilitating Learning for All* (pp. 207–227), edited by S. Stainback and W. Stainback. Baltimore: Paul H. Brookes.

Peterson, P., E. Fennema, and T. Carpenter. (1988–1989). "Using Knowledge of How Students Think About Math." *Educational Leadership* 46, 4: 42–46.

Peterson, P., and N. Knapp. (1993). "Inventing and Reinventing Ideas: Constructivist Teaching and Learning in Mathematics." In *Challenges and Achievements of American Education,* edited by G. Cawelti. 1993 ASCD Yearbook. Alexandria, Va.: Association for Supervision and Curriculum Development.

Poplin, M. S., and S. Stone. (1992). "Paradigm Shifts in Instructional Strategies: From Reductionism to Holistic/Constructivism." In *Controversial Issues Confronting Special Education: Divergent Perspectives,* edited by W. Stainback and S. Stainback. Boston: Allyn and Bacon.

Resnick, L. B., and L. E. Klopfer. (1989). *Toward the Thinking Curriculum: Current Cognitive Research.* Alexandria, Va.: Association for Supervision and Curriculum Development.

Sapon-Shevin, M., B. J. Ayres and J. Duncan. (1994). "Cooperative Learning and Inclusion." In *Creativity and Collaborative Learning: A Practical Guide to Empowering Students and Teachers,* edited by J. Thousand, R. Villa, and A. Nevin. Baltimore: Paul H. Brookes.

Schrumpf, F. (1994). "The Role of Students in Resolving Conflicts in Schools." In *Creativity and Collaborative Learning: A Practical Guide to Empowering Students and Teachers,* edited by J. Thousand, R. Villa, and A. Nevin. Baltimore: Paul H. Brookes.

Shriner, J. G., J. E. Ysseldyke, M. L. Thurlow, and D. Honetschlager. (1994). "'All' Means 'All': Including Students with Disabilities." *Educational Leadership* 51, 6: 38–42.

Siegel, J., and M. Shaughnessy. (March 1994). "An Interview with Howard Gardner: Educating for Understanding." *Phi Delta Kappan* 75, 7: 563–566.

Sleeter, C., and C. Grant. (1994). "Education That Is Multicultural and Social Reconstructionist." In *Making Choices for Multicultural Education: Five Approaches to Race, Class, and Gender* (2nd ed.), edited by C. Sleeter and C. Grant. New York: Merrill.

Spady, W., and K. Marshall. (1991). "Beyond Traditional Outcome-Based Education." *Educational Leadership* 49, 2: 67–72.

Shornburg, D. (1992). *Edutrends 2010.* San Carlos, Calif.: Starsong Publications.

Thousand, J., R. Villa, and A. Nevin. (1994). *Creativity and Collaborative Learning: A Practical Guide to Empowering Students and Teachers.* Baltimore: Paul H. Brookes.

Tiedt, P., and I. Tiedt. (1990). "Education for Multicultural Understanding." In *Multicultural Teaching: A Handbook of Activities, Information, and Resources* (3rd ed.), edited by P. Tiedt and I. Tiedt. Boston: Allyn and Bacon.

Vermont State Department of Education. (1993). *Vermont's Common Core of Learning: The Results We Need from Education.* Montpelier, Vt.: Vermont State Department of Education.

Villa, R., and J. Thousand. (1992) "Student Collaboration: An Essential for Curriculum Delivery in the 21st Century." In *Curriculum Considerations in Inclusive Classrooms: Facilitating Learning for All Students,* edited by S. Stainback and W. Stainback. Baltimore: Paul H. Brookes.

Villa, R., J. Udis, and J. Thousand. (1994). "Responses for Children Experiencing Behavioral and Emotional Challenges." In *Creativity and Collaborative Learning: A Practical Guide to Empowering Students and Teachers,* edited by J. Thousand, R. Villa, and A. Nevin. Baltimore: Paul H. Brookes.

Wiggins, G. (1989). "Teaching to the (Authentic) Test." *Educational Leadership* 46, 7:41–47.

Zessoules, R., and H. Gardner. (1991). "Authentic Assessment: Beyond the Buzzword and Into the Classroom." In *Expanding Student Assessment,* edited by V. Perrone. Alexandria, Va.: Association for Supervision and Curriculum Development.

# Section IV

*Promising Practices in Contemporary American Education*

# Technological Literacy for Educators

RICHARD C. OVERBAUGH

*Old Dominion University*

From the workplace to everyday living and entertainment, the infusion of technology is ubiquitous. As a reflection of society, public school systems are responsible for preparing students to lead effective and responsible lives (Harrington, 1991) which includes technological proficiency. Adequate equipment and teachers who are sufficiently knowledgeable for integrating computers into curricula are essential, therefore, for preparing students for today's technologically advanced society. In spite of this need, and evidence that computers are effective across all grade levels (Khalili & Shashaani, 1994; Kulick & Kulick, 1991; Roblyer, Castine, & King, 1988) and in all settings—urban, suburban, and rural (Christmann, Lucking, & Badgett, 1997)—many of today's teachers lack the necessary skill and knowledge to use technology effectively in the classroom (Hasselbring, 1991; Maddux, 1997; Miller, 1992; Nicklin, 1992; Ray, 1991) and schools typically do not possess the equipment and resources necessary for effective instruction.

The Office of Technology Assessment (1995) anticipated that there would be 5.8 million computers in K–12 schools by the spring of 1995—a 9:1 student/computer ratio. However, such figures can be, and usually are, misleading. For example, in 1996, Virginia reported a potentially impressive average of 65.5 computers per school. This average seems impressive until one considers the 55.5 standard deviation existing among different schools (i.e., a situation wherein an average school could have anywhere from 10 to 110 computers (Virginia 6-year plan, 1996).

Equipment procurement, the misguided sole focus of technology efforts in many public schools and universities, however, is only part of the solution. The physical presence of technology certainly does not automatically lead to technologically savvy educators, pedagogically sound curricular use, or any use at all!

Because technology infusion has been hit and (largely) miss, federal and state mandates and initiatives are beginning to appear. The first national effort was spearheaded by the International Society for Technology in Education (ISTE) in 1993 with a published guide for various levels of technological literacy ranging from a bare minimum for all classroom teachers to various technology specialties. The guidelines were endorsed by the National Council for the Accreditation of Teacher Educators (NCATE) and later adopted as a basis for their own evaluative efforts. ISTE's guidelines have also begun to be adopted/adapted by many states. The Advisory Board for Teacher Education and Licensure (ABTEL) in Virginia, for example, has recommended an adaptation of ISTE's technology guidelines be accepted by the state beginning in December 1998.

In spite of the efforts being made to develop and implement technology guidelines, many of these initiatives use wording that can be interpreted quite loosely. The reason for this is twofold.

---

Richard C. Overbaugh, Assistant Professor, Old Dominion University.

First, leaving wording non-specific allows the competency item to remain viable as technology continuously and rapidly changes. Second, and much less defensible, the wording of competency guidelines are often ambiguous to allow individual school systems or individual schools to interpret the competencies in a manner that suits widely varied situations. The latter allows weak interpretations of the original "spirit" of the competency, undermining the purpose for establishing competencies. The problem of diluted interpretations of competencies is compounded by additional caveats that schools can meet the standards by showing they are working towards meeting the goals or simply having a plan to begin working toward the goals.

Although many groups are working towards defining literacy standards, many believe that a definition is elusive. Beliefs differ about what constitutes computer literacy, ranging from programming ability and hardware expertise to the use of content-specific applications. While technology is a fast moving target, technological literacy is not an elusive term and multiple interpretations should not be blamed for problems with computer literacy among educators.

This paper has two foci: First, the major components of technology literacy, which should be standard in all educators' knowledge base, will be discussed briefly. This "list" is based on fairly typical computer literacy readings and concurs with the aforementioned guidelines and current standards. However, this paper goes beyond a simple listing by offering brief descriptions and education-based examples of each technological item. Second, factors that hinder or help foster adequate knowledge, skill acquisition, and curricular infusion are explored. This writing is intended for anyone with at least a modicum of working computer knowledge in the education field who wishes to understand what technology skills today's education professionals are expected to know.

## Essential Computer Literacy Domains

Those responsible for preparing teachers to use computers often ask: "What constitutes technological literacy?" and "How can teachers best *become* technologically literate?" Perhaps a more succinct questions is: "What knowledge do teachers need to effectively utilize technology in their classrooms, and what is the best way to provide that knowledge?"

Discussions on computer literacy components typically result in two computer-use domains: (1) learning to use productivity tools (e.g., word processors and databases) for managerial and administrative tasks; and (2) learning to use technology to help teach specific content (see Bozeman & Spuck, 1991; Scheingold, Martin, & Endreweit, 1986). Productivity tools can be used for administrative and secretarial chores and to help teach content. Therefore, four domains will be discussed: (a) productivity tools; (b) computer assisted instruction; (c) productivity tools for computer assisted instruction; and (d) telecommunications.

### Productivity Tools

Productivity tools are computer applications that help reduce the amount of time consumed by administrative and secretarial chores—administrivia in current vernacular—which are inherent in managing a classroom. Administrivia includes such tasks as writing lesson plans, keeping grades and room inventories, and sending home announcements and letters. Schools typically have *integrated software* or *software suites* for these duties. Choices range from the elegant and efficient *ClarisWorks* program to feature-bloated, hardware-intensive suites such as *Microsoft Office*. Most "works" packages and suites include word processors, databases, spreadsheets, and presentation programs.

## Word Processors

Word processing is perhaps the most well known, and is typically the first, application learned. Foremost to word processors are "formatting" and "editing features." Nearly unlimited "looks" can be applied to typed documents through: (a) available fonts (the appearance of characters, e.g., Old English or Pica); (b) various styles (e.g., bold, underline, superscript, size, and color); and (c) variable spacing (e.g., margins, indents, and outlining). Format flexibility permits efficient creation of attractive documents from letters and lesson plans to flyers and announcements.

The editing features of word processors allow easy modification of existing text that, beyond document revision for higher quality, is an impressive time saver for teachers. For instance, as teachers gain new insights, old lesson plans can be reworked without being completely rewritten. Teachers can include a variety of instructional techniques, and change resources or student audiences. Also, "standard" letters which are used for parent contacts can be stored on disk and modified for each use. Imagine a letter sent to Mrs. Smith:

> Dear Mrs. Smith, Johnny threw a rock at Susy on the playground today. Would you please. . . .
>
> Sincerely,
>
> Mr. Thomas

Two weeks later the same letter might be sent to Mr. Franklin after Karen threw a rock at Sam. If the letter is kept on disk, the teacher need only retrieve the file, change the appropriate text, and print the new copy. Paper can also be conserved if the copies of various versions of the "standard" letter are saved with different names via the "save as" feature.

## Database

The database is an extremely useful tool for teachers; albeit it is usually limited to keeping student records or room inventories. Most teachers keep track of student data including (but certainly not limited to) names, addresses, phone numbers, parent names, and emergency contacts. Many teachers use index cards because they can be alphabetized by student name and retrieved quickly when needed. A database can be thought of as an "electronic" card catalog that stores the same information electronically. The same *fields* (e.g, First Name, Last Name, and Birthday) are defined and each student has a "card" or *record* (See Figure 1). Beyond the obvious formatting and editing ability, databases provide various data manipulation features grouped into three basic functions: (a) sorts, (b) finds/queries, and (c) layouts/reports.

### Sorts

Databases can sort all records alphabetically or numerically on a specified *field*. For example, after entering all student names, the computer will alphabetize the list. Some may not consider sorting by name a compelling reason to use an electronic database as sorting a set of index cards does not take long. However, teachers may wish to organize, or *sort*, the database according to other criteria (e.g., grade level, birthday, or lunch status). If the information is on index cards, the time needed to manually reorder suddenly becomes significant, not including the time needed to return the cards to alphabetical order. A database performs sorts easily and instantaneously.

### Figure 1

*Database of student information*

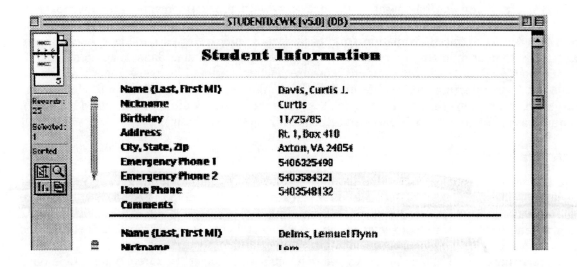

*Finds/Queries*

A database can search through all the *records* in a database to find matching information. Imagine a music teacher finding uniform hat #56 under the bus seat. The hat owner must be identified. If student records are on index cards, each card must be examined until the matching number is found—a potentially lengthy process if the match is on index card #127. In contrast, an electronic database will match the hat number and display the entire student record with a few keystrokes. This easy retrieval can be applied any time a student record needs to be identified by matching a *field* entry. The same process can find groups of students with one or more similarities. Perhaps a list of all *homeroom* students who qualify for *free* and *reduced* lunch is requested; the database can be *queried* to *find* the records of those with *homeroom* in the class field, and *free* or *reduced* in the lunch field. Or, perhaps, willing parent volunteers are needed to network a school. If a *Parent Occupation* field is included in the student database, appropriate candidates can be found.

*Layouts/Reports*

Rich student databases contain many fields but a reduced set is often desirable. For example, student lists are often required for field trips but the list should only contain students' names, addresses, phone numbers, parent/guardian name(s), and emergency contact information. An alphabetized "bus list" layout/report with only those fields can be generated with an appropriate format. Similar reports can be generated for students with October birthdays or 10th grade chemistry students who have not paid their lab fees.

A second common database use is room inventories. Many educators are responsible for maintaining current lists, such as books, globes, blinds, and desks, which include date of acquisition, condition, and original and depreciated values. A good database, once created, can be used to automatically calculate depreciated values, items' ages, and total values. Items can be added and deleted easily. Additionally, databases are superlative for recording and manipulating student data and generating specific lists and reports. Furthermore, once devised, *finds, queries, layouts,* and *reports* can be saved for future use. Ongoing database changes are, of course, reflected in subsequent manipulations.

*Database-Word Processor Merge*

A final useful feature of databases is the database-word processor merge. Like sweepstakes letters, personalized with names in the salutation and scattered throughout, teachers can create form letters that extract information from each record in a specified database. A classroom example is an interim letter to parents reporting students' grades, attendance, and comments. If the pertinent *fields* are part of the student data base, a letter is printed for every student, or *record,* in the database, inserting the specific information in the appropriate location in the letter. The merge is a tremendous time saver if midterm grade reports are required or suggested by school administration. (Address labels can also be printed from the database.)

## Spreadsheets

Like the database, the spreadsheet is a tool to organize and manipulate information but is less flexible in terms of layouts and appearance. Spreadsheets are oriented more toward numerical manipulation from math to algebra and trigonometry and include graphing/charting features. As such, spreadsheets offer more to science and math educators but any classroom teacher will find them useful for building electronic grade books or working on budgets.

Spreadsheets are organized in rows and columns. For instance, picture a paper grade book with students' names down one side and the assignments and tests listed across the top with each student's grades filled in (See Figure 2). By entering formulas, calculating final grades, midterm grades, averages, standard deviations, and minimum and maximum grades becomes automatic. Interestingly, databases are becoming more sophisticated in terms of calculation fields that do much of the same things a spreadsheet can (but without graphing/charting features) which makes doing a mid-grade-period report via a word processor-database merge very easy.

## Presentation Programs

Presentation software (such as Adobe *Persuasion,* Microsoft *Powerpoint,* the slide show feature of *ClarisWorks,* and even *Kidpix)* are programs in which attractive linear presentations are easily built for teacher and student use. Most presentation programs include: (a) a variety of backgrounds or templates; (b) animated bullet point entries, such as text flying in from the top or fading in; and (c) special effects when moving from one slide (screen) to the next.

Presentation programs are used like overhead transparencies but have additional advantages, including multimedia components such as graphics, sound, and video clips which can be inserted for clarity and heightened student interest and, like all digital media, edit options. Students enjoy using presentation programs to accompany their class presentations because, like the word processor, students find the attractive output motivating and fun.

Use of presentation software in classrooms requires computer projection devices which range from relatively crude and inexpensive scan convertors (approximately $150.00) that convert the SVGA monitor signal to NTSC television signals to display computer output on a television screen, to digital projectors (approximately $5,000.00) which project computer output crisply to a screen at a light level that permits the classroom lights to remain on. Projection equipment, which is often overlooked in technology planning, should be considered a high priority as educators will often need to project computer output when demonstrating/modeling a program or when doing group work on a single computer.

## Figure 2

### *Spreadsheet formatted as a grade book*

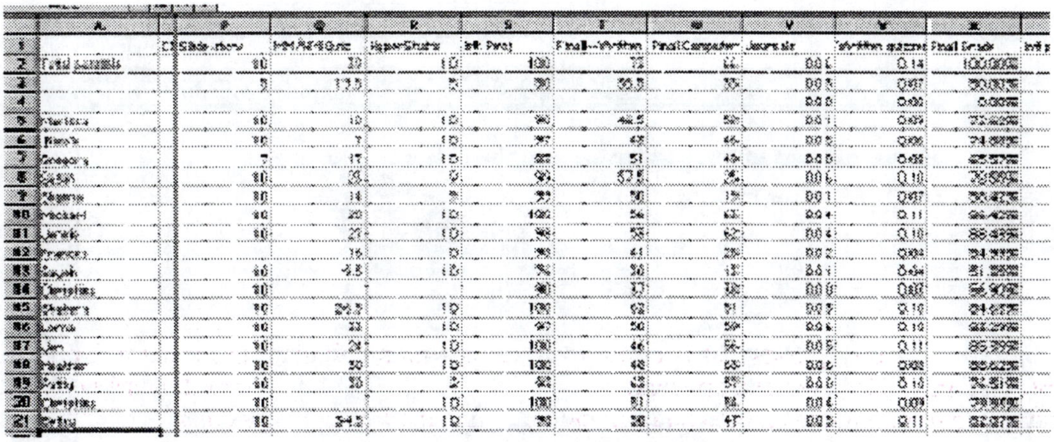

## Authoring Systems

Inexpensive hypermedia authoring systems such as *HyperCard, HyperStudio,* and *Digital Chisel* are similar to presentation programs but are much more versatile and are common in schools. Primarily, authoring programs enable teachers and students to create non-linear, or branching, programs that include: (a) navigation "buttons" that, upon a mouse click, take one from screen to screen; (b) graphics tools for drawing or painting; and (c) text fields which are specified areas of the screen that function like a word processor. Authoring programs also allow easy importing of graphics, sound, and video, as well as easy building of animations. These easy-to-use features extend software creation to any computer user from second graders to adults. (See Figure 3 for a sample screen.)

Hypermedia authoring programs should be thought of as viable media for student knowledge representation and as an alternative to more traditional formats including written reports. The advantage of creating projects/programs electronically is similar to the writing process. Students must (a) research their topic, (b) decide what is important, (c) organize the information, (d) decide how to present the information based on the audience and available resources, (e) choose *necessary* media elements, (f) construct the project, and (g) evaluate the project. By researching and making many decisions while organizing and manipulating the project chunks, students can process information many times and in many ways, supporting the tenets of constructivist and active learning.

The ease with which programs can be created via authoring programs is also a caveat to be considered. In addition to knowing how to use software, teachers and students must know how to create effective screens and how to structure projects effectively. The writing process again provides a reasonable illustration. Students usually write papers according to specifications and guidelines provided by the teacher, rather than by simply sitting down and doing an unstructured mind dump. Therefore, why should students be allowed to create computer-based projects without clearly defining a process and product?

### Figure 3

### *Screen from a simple classroom program created with HyperCard*

## Computer Assisted Instruction (CAI)

Many models of learning exist to help design effective instruction. Perhaps the most well known and still viable are the seven stages of learning identified by Gagne (Gagne & Briggs, 1979). However, the simpler four stage process proposed by Alessi and Trollip (1991) provides a reasonable framework for discussing computer assisted instruction software: (a) presentation of information or teaming experiences; (b) initial guidance as learner struggles to understand the information or execute the skill to be learned; (c) extended practice to provide fluency or speed or to ensure retention; and (d) assessment of student teaming. The first four CAI types lie on a complexity continuum from simple to complex and a cost continuum from least expensive to most expensive.

### *Drill and Practice*

Drill and Practice software is the simplest, least expensive, most popular, and most abused type of computer assisted instruction software in schools. Drill and practice software assumes previous knowledge and only fits in the third stage, extended practice, and *perhaps* the fourth, assessment, but it does not teach. Repetition is effective for practicing existing knowledge for fluency, automaticity, recall, and perhaps elaboration, or learning to apply the knowledge to different situations (e.g., learning to use foreign words in different or increasingly difficult contexts).

Drill and practice software is often abused through use as a reward or when it is assumed students will be taught something from the software. The software is easy to write and usually very inexpensive which leads to a prevalence of poor quality programs especially in financially limited schools. Quality drill and practice programs do exist but careful evaluation should be conducted before purchase.

*Tutorials*

In contrast to drill and practice programs, tutorials do *not* assume prior knowledge. Tutorials teach material and may also include a drill and practice component as well as assessment. Therefore, tutorials can play an integral part in instruction or be used as stand-alone instruction for enrichment and remediation.

*Problem Solving*

Problem-solving software is any software that is based on the traditional four-step problem-solving process: (a) defining a problem; (b) devising a solution; (c) implementing the plan; and (d) evaluating the plan. If the evaluation step shows the solution was incorrect, the learner returns to the definition step or devises a new solution plan.

*Simulations*

Computer simulations include any near-to-real experience provided by a computer. For example, many schools cannot afford field trips or expensive materials but can afford computer-based simulations. Another example is where the real experience may be too dangerous (e.g., a chemistry experiment involving caustic or volatile substances) or impractical (e.g., breeding animals in the classroom to study genetics or taking a trip to the Smithsonian Institution in Washington, D.C. or the Louvre in Paris) leaving a virtual, or simulated, experiment or trip the next best option. Additionally, computer simulations can be used for time compression. Events such as continental drift and plate tectonics that took/take place over a long period of time can be compressed into a brief time span, providing a visual representation that helps learners grasp a potentially abstract idea.

*Instructional Games*

Instructional games have potential instructional use but are often misused. When carefully integrated as part of other academic instruction, games can provide rich, varied, and interesting components. For example, the *Carmen San Diego* series is often used to teach reference skills, map reading, and geography. Many times, however, the games are used as rewards for better students which means they have little more than entertainment value.

*Software Evaluation*

Before leaving computer assisted instruction, a final, and extremely important, skill needs to be mentioned: software evaluation. Evaluation is not easy and requires a substantial amount of time. However, evaluation is critical in order to avoid wasting money. In fact, evaluation is especially important because of the technical wizardry inherent in current software. Society has become mediacentric and software developers build in many sounds, graphics, and movies that are little more than initial attention getters. Once beyond the glitz, many seemingly exciting and fun software packages have little to offer teachers and students.

Most school systems have a standard evaluation form or many are available in books (e.g., Geisert & Futrell, 1995). At the very least, software should be evaluated according to several criteria. One evaluation criterion is technical/mechanical attributes such as the ease of use, reliability, and hardware requirements. Second, whether the instructional design is agreeable to the teacher should be considered. Even if software is well designed, if the teacher's style is not agreeable with the instructional design of the software, the software will be little used. Finally, a field test should be conducted with a few students from the intended audience to identify additional or overlooked problems.

## *Productivity Tools as CAI*

Productivity tools used as CAI can be used to teach contents in many ways. Word processors, databases, spreadsheets, and presentation/authoring programs can be adapted to many situations, limited only by the imagination which can often overcome the need for expensive content specific software. The obvious appeal lies in the availability of "works" packages such as *ClarisWorks* in nearly every classroom. A few brief examples follow.

### *Word Processors*

While word processors have no inherent ability to improve student writing (beyond limited surface revisions such as spell checkers and subjective grammar checks), easy revisions help in the writing process. When students are guided through prewriting and rewriting activities, low-level and high-level editing are easy to do, enabling students to overcome negative attitudes towards research and revision. Additionally, templates, such as "standard" lab report formats, can be created for students to fill in the blanks.

### *Databases*

Databases are perhaps one of the most useful tools because of their uses for organization and data manipulation. Students in nearly any subject can use databases from research to teaming classification skills. Second graders are quite capable of researching countries or cities around the world and compiling information such as language spoken, currency and currency exchange, historic and contemporary facts and features. Additionally, databases can be used to classify data and manipulate data for problem solving.

### *Spreadsheets*

By design, spreadsheets are more suited for math and science but can also be used to teach other lessons such as budgeting, currency exchange rates, and time allocation skills. A favorite in elementary schools is to use M&Ms or dice and a spreadsheet to teach probability.

### *Presentation/Authoring/Programming*

Presentation programs (e.g., *ClarisWorks* slide show, *Kid Pix*, and *Powerpoint*) and authoring programs (e.g., *HyperCard, HyperStudio,* and *Digital Chisel*) are viable media for student knowledge representation. Students can use computer-based presentations just as they can write papers and do class presentations or unit projects. Not only does using the computer for organizing and presenting material motivate students, the multiple processing time required leads to higher levels of learning, For example, students studying a particular decade to make a multimedia presentation have to: research to find major events; research events for themselves; organize materials; decide how to present their work; create the presentation. By finding and revisiting the material many times, students are actively involved in creating meaningful knowledge.

## *Telecommunications*

Telecommunications consists of a number of tools such as email, bulletin boards, listservs, and the World Wide Web. Unlike the relatively stable productivity tools and computer assisted instruction, telecommunications is the most rapidly changing component of educational technology, with the World Wide Web seemingly changing monthly. However, it should be kept in mind that the most basic productivity suites, such as *ClarisWorks* and Microsoft *Works*, have telecommunications features that enable even those with computers as old as DOS machines and Apple IIs with 2400 baud modems to connect to text-based servers for reasonable email, bulletin board, and listserv use.

## *Email*

Electronic mail, or email, is one of the original school uses of the Internet. Students compose mail messages either online or off (and then paste the text into the mail program) and send the message to anyone in the world who has an email address. The primary advantage email has over traditional mail, or "snail mail" is speed. Mail usually reaches its destination in minutes and rarely takes more than a few hours, supporting the immediacy often needed to maintain student interest. Another email advantage is the rapid return of incorrectly addressed messages, so the sender can readdress and resend them without delay.

The speed with which students can correspond with others across grade levels, subjects, cultures, and even the globe supports the idea of providing an audience for student writers. In essence, students who have a specific audience typically attend to the quality of their writing more so than they would for traditional classroom writing which should lead to better writing skills. Moreover, regular and speedy correspondence tends to heighten this effect.

Finally, assuming a school has at least some connection to the Internet, cost for email is virtually zero. For example, Virginia still maintains a text-based public education network (VaPEN) with no subscription cost for schools which supports email for computers as old as Apple IIs and DOS machines.

## *Bulletin Boards*

Bulletin boards are on-line repositories of files related to a particular topic that can be read by anyone with an Internet connection. For example, most public education networks maintain bulletin boards for teachers to discuss classroom plans, procedures, and ideas. Interested teachers dial in or connect to the bulletin board and view a list of postings and then choose the one(s) they wish to read. Then, a reply can be posted. Readers and posters can work at anytime they wish which means this type of communication is asynchronous (as opposed to synchronous chat rooms). Bulletin boards exist for almost any topic from the teacher bulletin boards to hobbyist boards but are usually more useful to teachers than students.

## *World Wide Web*

The World Wide Web is the part of the Internet that is accessible via the point-and-click graphic user interfaces (GUI) pioneered by the Apple Macintosh computer and copied by the Microsoft *Windows* program. Until about 1990, the Internet was only accessible via cryptic, text-based, command line interfaces of which telnet, gopher, and file transfer protocol (ftp) were used to connect to remote computers, find usable files, and copy or post files. Discussion of text-based Internet tools is not only beyond the scope of this writing but unnecessary. Point-and-click browsers have virtually replaced all these tools with simple graphic tools which anyone can use. *Netscape Communicator* leads the browser market with a very full-featured and easy-to-use product which happens to be free to educators. With a browser, users connect to computers around the world by entering a URL, or uniform resource locator, which is the machine equivalent of a unique street address or phone number. Once connected to the site, users move from screen to screen or location to location via "links" (i.e., underlined hypertext or buttons) which connect the user to subsequent URLs.

Hidden amidst the myriad junk on the WWW are many excellent resources for teachers, including classroom management ideas, lesson plans, resources for student research, and sites to identify pen pals around the world. Even with the many sites dedicated to providing educational information and search engines (e.g., Alta Vista and Yahoo) which find sites based on keyword and phrase searches, finding educational resources can be difficult. However, once found, educators can "bookmark" their favorite sites and return whenever they wish.

The World Wide Web may be the major driving force behind the continually needed upgrades in hardware and operating systems which are necessary to maintain bearable connection speeds. Realistically, if WWW access were not considered essential, older computers would do a reasonable job with every other aspect of computing. Admittedly, each generation of software includes new features and claims to be easier to use, but investigations into the technology presently used for administrative and secretarial chores reveal that a significant majority of users utilize seemingly outdated machines but maintain acceptable productivity levels. In fact, many business entities (e.g., national drug store networked customer drug databases and store messaging systems) continue to use text-based computers simply because they are adequate. Additionally, instructionally sound CAI software designed for DOS and Apple II computers is still good software. Much of the demand for upgraded software and hardware is simply a result of successful marketing and to maintain reasonable connectivity to the rapidly evolving World Wide Web. Some argue, however, that because computer prices continue to drop dramatically, purchasing a new computer should not be considered a big deal. This argument ignores the basic fact that to maintain reasonable connection to the WWW, or meet the *perceived* need to stay current with software versions, requires a significant expenditure every two to three years which is simply not affordable by schools and most businesses. Imagine *any* other appliance that costs from $1,200 to $6,000 which will be considered worthless in two to four years.

## Technological Literacy Enhancements and Detractions

The education world is finally realizing initiatives and mandates from state and national levels that attempt to force educators to become technologically current with the rest of society are needed. Undeniably, many teachers lack the skills needed for efficient and effective integration of technology into the classroom. Compounding the problem, many teacher-education graduates are less technologically proficient than their future students (Nicklin, 1992) and a substantial percentage of technologically deficient practicing teachers lack the motivation to learn (Miller, 1991).

Although teachers' motivation (or lack of motivation) to become and remain technologically literate is important, some inherent obstacles merit discussion. Even a 9:1 student to computer ratio puts inadequate numbers of machines in each classroom or worse, leads to grouping machines in labs. The latter results in scheduling difficulties or classroom management issues which understandably contribute to a disinclination to use computers. Time is a major problem. Learning about computers requires substantial time and most practicing teachers simply have little to spare.

Contributing to the time problem is the ever-increasing complexity of hardware. Each "improvement" to operating systems, ostensibly to make computers easier to use, render problem-solving and configuration tasks further beyond the average user's ability. Furthermore, technology integration causes major changes in teaching methodology that can potentially diminish teachers' self-confidence (Fishman & Duffy, 1992).

However, even though the public school system is a business in which employees with archaic skills are tolerated, to postulate that all teachers are not willing to align themselves with our technologically rich society is irresponsible. A simple comparison of teachers to business employees for the purpose of blaming personnel for substandard skills ignores infrastructure which is an enormous factor.

### Required Infrastructure

Successful businesses provide the infrastructure needed for maximum employee performance. Schools typically do not. Hasselbring (1991) identified three basic conditions that must exist before educators can begin to accept technology: (a) well equipped facilities and good access; (b) on-going instruction; and (c) technical support.

For many years the common wisdom said teachers should have, at absolute minimum, their own classroom computer. Some posited that, if there is only one computer in the classroom, the computer should be on the teacher's desk (Snyder, 1993) so productivity tools could be used for administrivia chores and to completely familiarize the teacher with the machine which, in turn, allows them to more easily use the machine to teach content. Even though some classrooms still have only one (or no) computers, and software companies such as Tom Snyder Productions produce high quality software for limited hardware classrooms, current thought is a five to one ratio of students to computers which, for example, is targeted by the state of Virginia by the year 2000. A 5:1 ratio will put adequate numbers of machines in each classroom which will eliminate the pooling of computers in school labs, alleviate advance lab scheduling, and help remove the common notion that computers are a segregated topic of study rather than an integral part of everyday activities.

The second condition is on-going computer instruction delivered at regular intervals over years. Typical efforts at teaching faculty computer skills are short, disjointed workshops delivered sporadically which are largely unsuccessful. To learn new skills, especially technology skills, instruction needs to be delivered in chunks over time with ample "playtime" in between so new skills can be practiced for adequate learning.

The third condition is constantly available technical support for hardware and software. Too often, hardware and software are provided but no personnel are available for installation and maintenance and to answer the inevitable questions. For example, DOS/Windows schools trying to upgrade to *Windows 95* experience "a great deal of frustration and wasted time" (Maddux, 1997, p. 6). The complexity problem is exacerbated by the Internet. Maddux found that

> Even when telephone lines are installed, configuring hardware and software for Internet access is not a simple chore, and calls for good computer concept mastery as well as basic computer literacy skills, patience, perseverance, and the willingness to engage in trial-and-error problem solving and trouble shooting. (p. 9)

Even if teachers have the expertise needed to set up and maintain the machines, the time required to do so is unreasonable, particularly when technology fails during class activities. Constantly available technology support personnel are essential. Without them, even the most up-to-date school, staffed by superb technology experts, will flounder. Furthermore, to expect teachers to have such expertise is unreasonable in the same way those who drive cars should be expected to have the skills to install and repair fuel injection systems or transmissions. Even if faculty have the skills, the time required to use them is simply not available. Computers are tools to help content specialists share their knowledge with others and expecting teachers to have advanced technical skills is unreasonable.

Finally, infrastructure defines only the environmental conditions needed for effective technology infusion. The conditions are concrete and can be put in place by willing administrations. However, a strong infrastructure does not include the personnel and their attributes.

## Personnel Attributes

*Inservice*

Even if schools supply the basic infrastructure, teachers' personal, professional, and social beliefs regarding technology skill acquisition remain a wild card. In brief, teachers who: (a) are motivated and committed to their students' learning; (b) are dedicated to professional development; and (c) have support and collegiality in their schools and school districts are more likely to successfully integrate computers into their classrooms (Scheingold & Hadley, 1990). Essentially, when they perceive classroom computing to be useful and have adequate access to computers and technical support, teachers are willing to interrupt their routines and make time to pursue instructional technology knowledge and skills.

*Preservice*

Schools of education are responsible for preparing teachers which, of course, includes technology skills acquisition. Most schools have at least one "technology in education" course which usually begins with productivity-type skills to help students become familiar and comfortable with technology in a personally meaningful manner. Unfortunately, in many students' cases, previous beliefs are difficult to overcome (Characteristics, 1990). Preservice teachers have spent at least 12 years observing education through their own eyes and have, therefore, established very strong opinions about acceptable methodology (Hollingsworth, 1989). Students who are products of schools which have little or no technology resist learning new ideas, particularly if they don't understand them, perceive them as being forced upon them, and/or if new ideas conflict with their core beliefs (Lumsden & Norris, 1985).

## Models of Learning

Becoming computer literate clearly requires more than one or two computer-specific courses, but teacher preparation programs that have one or two basic computing classes should not add courses. A serious misconception is that the job of teaching computer-based instruction skills to education students should be given to a computer specialist. This may be true for introductory courses, but limiting technology skills acquisition to computer-specific courses perpetuates the view that computer use is something that occurs in a computer lab and is best left to the "computer people," a sentiment future teachers are likely to carry to their own classrooms.

Effective computer use needs to be modeled by content methods instructors across the curriculum. If education faculty effectively integrate computers into their methods classes, students will gain the necessary instructional computing skills without the need for additional classes.

## Models for Computer Literacy Development

Like all learning, learning about technology results in definite, predictable, stages through which learners will pass from initial exposure through mastery and application. Though a number of models have been identified to describe the stages of learning to use technology, only three that provide guides to curricula development or restructuring will be discussed.

*Two-Stage Model*

One of the most basic beliefs related to instructional computing is that computers must become personally meaningful before teachers will use them to help others (Bozeman & Spuck, 1991; Snyder, 1993). Therefore, the most important *initial* application of computers for practicing or preservice teachers and their (future) students is to learn how to use computers with productivity tools to help with administrative and secretarial tasks (Bozeman & Spuck; Woodrow, 1991). The path to computer literacy should begin with the utilization of computer management tools such as word processors, databases, and grade book programs to develop personally meaningful skills and then proceed to learning about applications that teach or help teach content (CAI).

*Four-Stage model*

Kell, Harvey, and Drexler (1990) do not separate productivity tools and CAI uses as disparate stages. Rather, they offer four stages through which new users of innovation pass: (a) awareness and preparation; (b) using the innovation mechanically and then smoothly; (c) refining the innovation's use and curricula integration; and (d) assessing the consequences of using the innovation for their students. They further state that two to three years are necessary for new users to pass through all four stages, indicating a clear need to begin early in the higher education process.

*Seven-Stage Model*

The most complex model included here identifies seven stages people generally follow when introduced to an innovation. The two-stage and four-stage models are easily subsumed by the seven ordered concerns identified by Hall, George, and Rutherford (1977): (a) awareness; (b) informational concerns; (c) personal concerns; (d) management concerns; (e) consequences; (f) collaboration concerns; and (g) refocus. These stages can be quantified with the Stages of Concern instrument which is based on the notion that, when introduced to all innovation, people tend to have very self-oriented concerns and, as they learn more, concern arises with how to manage that innovation in their classroom. Finally, explorations of how they can work with others to share their knowledge and learn new ways to use previously learned skills begin. More specifically, the first concern stage is Awareness: I am not concerned about microcomputers. The second stage is Informational: I am concerned about learning more about microcomputers. The third stage is Personal: I am concerned about how using microcomputers will affect me personally in the classroom. The fourth stage is Management: I am concerned about the time needed to learn about microcomputers. The fifth stage is Consequence: I am concerned about the effects my use of microcomputers will have on my students. The sixth stage is Collaboration: I am concerned about working with others to learn more about microcomputers. The seventh stage is Refocusing: I am concerned about learning new ways to use what I already know about microcomputers.

Regardless of how complex one wishes to be in regard to the number of distinguishable concern levels, these progressive priorities can serve as a model for computer literacy development which suggests educators should: (a) become aware of the potential uses of computers; (b) learn to use computers for managerial chores such as word processing, databasing, making presentations, and keeping grades electronically; and (c) use computers to assist with their instructional endeavors by integrating computer-assisted instruction software into regular classroom activities. This approach to computer literacy is similar to the five component program developed and implemented by Marietta College in Ohio for preservice teachers. Computer skills are introduced very early to lower-level education majors and furthered with progressive educational computing throughout the entire undergraduate career and into graduate classes (Golden, 1991). The five components are: (a) acceptance; (b) understanding; (c) application; (d) evaluation; and (e) design.

*Declarative vs. Procedural Knowledge and Active Learning*

To include a discussion of declarative and procedural knowledge in a treatment of technological literacy may strike readers as somewhat odd. However, the topic is important because understanding how technological skills should be learned is important. Also important are erroneous preconceived notions about how computing skills ought to be learned.

Whereas declarative knowledge is remembered bits of knowledge from simple rote items to deductions based on known facts and ideas, procedural knowledge is the actual demonstration, or performance, of a task. Clearly, both types of knowledge are important; but procedural knowledge is often the "Proof." I once observed an introductory computer course during an exam. No computers were on and I assumed that theory and application questions were being answered. However, I was amazed when I leaned over a student's shoulder and he was answering the question, "Describe how to move a block of text." Besides the obvious learning objective-assessment mismatch, the ability to describe how a procedure is done does not necessarily mean the student can *perform* the task. Obviously, if verification of ability to move a block of text is the objective, the student should be asked to move a block of text. This example illustrates the emphasis often placed on memorizing the steps to accomplish a computer task rather than on procedural proof the task can actually be performed.

The propensity for rote knowledge display extends into many students' and teachers' expectations of how technology skills should be taught. Most expect to be led step by step through each and every task with someone available to attend to every question or uncertainty. However, the whole notion of "step-by-step instructions" and being "walked through," often referred to as the cookbook approach, is seriously misguided. For example, almost anyone can follow the steps to assemble all the parts and end up with a finished bicycle, but doing so does not make one a bicycle mechanic or bicycle expert. Likewise, almost anyone can follow the steps on a worksheet to create a fairly complex layout in a word processor but, without thinking about what one is doing, reaching the end does not add to the existing knowledge base. The surprising propensity for the cookbook method defeats active learning, mental engagement, and concept attainment—all of which are critical to adaptation, embellishment, and continued learning.

Effective learning efforts can include step-by-step instructions but should also strongly encourage exploration and emphasize operational *concepts*. Learning by trial and error is effective so learning what technology *can do* and how to figure out *how to make the technology do a desired task* is critical. Technology changes so fast that no one can hope to learn current software and not have to worry about learning something new for five years. Successful technology users understand what technology can do and, therefore, can more easily explore existing software features, and transfer knowledge and skills to new software versions or even new software packages.

## Summary

Technology is inescapable, as is the responsibility of contemporary educators to become technologically proficient in personal and classroom management uses and curriculum integration. The need to address rampant technological illiteracy through a change in teacher attitudes and acceptance of ongoing change and continued learning is characterized by the Report to the President on the Use of Technology to Strengthen K–12 Education in the United States:

> In order to make effective use of educational technology, teachers will have to master a variety of powerful tools, redesign their lesson plans around technology-enhanced resources, solve the logistical problem of how to teach a class full of students with a smaller number of computers, and take on a complex new role in the technologically transformed classroom. (President's Committee, 1997, p. 47)

Bringing schools in line with contemporary technology is dependent on many, including administrators, who are responsible for the infrastructure needed to support technology acquisition and use, and teachers, who need to learn and continue to learn new skills to remain current.

## References

Alessi, H. & Trollip, S. (1991). *Computer-based instruction: Methods and development* (2nd ed.). Englewood Cliffs, N.J.: Prentice Hall.

Bozeman, W. D. & Spuck, D. W. (1991). Technological competence: Training educational leaders. *Journal of Research on Computing in Education*, 23(4), 514–529.

Characteristics of an effective teacher education program (1990). Morgantown, WV: West Virginia University, College of Human Resources and Education, The Benedum Project.

Christmann, E. P., Lucking, R. A. & Badgett, J. L. (1997). The effectiveness of computer-assisted instruction on the academic achievement of secondary students: A meta-analytic comparison between urban, suburban, and rural educational settings. *Computers in the Schools*, 13(3/4), 31–40.

Fishman, B. J. & Duffy, T. M. (1992). Classroom restructuring: What do teachers really need? *Educational Technology Research and Development*, 40(3), 95–111.

Gagne, R. M. (1975). *Essentials of learning for instruction.* New York: Holt, Rinehart and Winston.

Gagne, R. M., & Briggs, L. J. (1979). *Principles of instructional design* (2nd ed.). New York: Holt, Rinehart and Winston.

Geisert, P. G. & Futrell, M. K. (1995). *Teachers, computers, and curriculum, microcomputers in the classroom* (2nd ed.). Needham Heights, MA: Allyn & Bacon.

Golden, C. S. (1991). Hypermedia in teacher education: Integrating technology across the curriculum—the Marietta College model. In *Technology and teacher education annual* (88–90). Greenville, NC: School of Education, East Carolina University.

Hall, G. E., George, A. A. & Rutherford, W. L. (1977). *Measuring the stages of concern about the innovation: A manual for use of the stages of concerns questionnaire.* Austin: Research and Development Center for Teacher Education, The University of Texas.

Harrington, J. (1991). Normal style technology in teacher education: Technology and the education of teachers. *Computers in the Schools*, 8(1–3), 49–57.

Hasselbring, T. S. (1991). Improving education through technology: Barriers and recommendations. *Preventing School Failure*, 35(3), 33–37.

Hollingsworth, N. (1989). Prior beliefs and cognitive change in learning to teach. *American Educational Research Journal*, 26(2), 160–189.

International Society for Technology in Education. (1993). *Curriculum guidelines for accreditation of educational computing and technology programs.* Eugene, OR: Author.

Kell, D., Harvey, G. & Drexler, N. G. (1990, April). *Educational technology and the restructuring movement: Lessons from research on computers in the classrooms.* Report No. IR-014735). Paper presented at the Annual meeting of the American Educational Research Association, Boston, MA. (ERIC Document Reproduction Service No. ED 326 195).

Khalili, A. & Shashaani, L. (1994). The effectiveness of computer application: A meta-analysis. *Journal of Research on Computing in Education*, 27(1), 48–59.

Kulick, C. C., & Kulick, J. (1991). Learning with media. *Review of Educational Research* 7(12), 75–94.

Lumsden, D. B., & Norris, C. A. (1985). A survey of teacher attitudes and beliefs related to educational computing. *Computers in the Schools*, 2(1), 53–59.

Maddux, C. D. (1997). The newest technology crisis: Teacher expertise and how to foster it. *Computers in the Schools,* 13(3/4), 5–12.

Miller, M. J. (1992). The multimedia revolution needs teachers, too. *Newmedia*, 2(12), 12.

Nicklin, J. (1991, July 1). Teacher's use of computers stressed by education colleges. *Chronicle of Higher Education*, pp. 15–17, A 17.

President's Committee of Advisors on Science and Technology: Panel on Educational technology (1997). *Report to the president on the use of technology to strengthen K–12 education in the United States* [On-line]. Available: *http://www.whitehouse.gov/WH/EOP/OSTP/NST/PCAST/k-12ed.html*

Ray, D. (1991). Technology and restructuring: New educational directions, Part 2. *The Computing Teacher*, 18(7), 8–12.

Roblyer, M. D., Castine, W. H. & King, F. J. (1988). *Assessing the impact of computer-based instruction, a review of recent research.* New York: The Haworth Press.

Scheingold, K. & Hadley, M. (1990). *Accomplished teachers: Integrating computers into classroom practice.* New York: Bank Street College of Education.

Scheingold, K., Martin, L. M. & Endreweit, M. E. (1986). Preparing urban teachers for the technological future. In R. D. Pea and K. Sheingold (Eds.), *Mirrors of Minds: Patterns of Experience in Educational Computing*. Norwood, N.M.: Ablex Publishing Corporation.

Snyder, T. (1993). Keynote speech at the 1993 Annual Meeting of the Indiana Computer Educators (ICE) Conference, Indianapolis, IN.

U.S. Congress, Office of Technology Assessment. (1995). Teachers and technology: Making the connection. (OTA-EHR-616). Washington, DC: U.S. Government Printing Office.

Woodrow, J. E. J. (1991). Teachers' perceptions of computer needs. *Journal of Research on Computing in Education*, 23(4), 475–496.

# Integrating Children's/Young Adult Literature into the Social Studies

KATHERINE T. BUCHER
*Old Dominion University*

## Introduction

Educators are constantly searching for ways to make learning more interesting. One way to do this in the social studies classroom is to supplement the traditional textbook with quality children's and young adult (CH/YA) literature. In this chapter, the author explains how to bring subjects to life in the social studies classroom through the use of CH/YA literature. More specifically, CH/YA literature is defined; a rationale is provided for using CH/YA literature in the social studies; methods for evaluating and selecting CH/YA literature are suggested; and ways to use CH/YA literature in the social studies are explained.

## What is literature for children and young adults?

While some would argue that anything children or adolescents read can be considered CH/YA literature, a true definition of CH/YA literature is a little more complex. Not only must children or adolescents be able to read the words, they must be able to understand the concepts, feelings, and emotions presented in the writing. CH/YA literature includes both fiction and nonfiction writings that "authentically and imaginatively express the thoughts, emotions, and information about the human condition [and] relate to the experiences, developmental levels, and literary preferences of the intended audience" (Goforth, 1998, p. 3). CH/YA literature must be written in such a way that children and young adults can construct emotional, psychological, and intellectual meaning.

Literature for children and young adults consists of both fiction and nonfiction and is typically broken down into several categories or genres. A discussion of these categories can be found in major CH/YA literature textbooks such as Huck, Helper, Hickman, and Kiefer (1997) or Goforth (1998). Although different CH/YA literature textbooks refer to these genres by various names, the categories generally include the following: traditional or folk literature, including tales, legends, myths, and folklore; fantasy and science fiction; contemporary realistic fiction; mystery, suspense, and supernatural fiction; historical fiction; poetry; biography, including memoirs and diaries; and informational/nonfiction books. Jordan (1995) confines biography and informational books into a category called "nonfictional history" which is "a piece of writing

---

Katherine T. Bucher, Associate Professor and Graduate Program Director of Elementary/Middle School Education, Old Dominion University.

that attempts to accurately and fully reflect the facts relating to a particular person, event, thing, and/or era of the past" (p. 30).

While written for children and adolescents, textbooks are not considered CH/YA literature. In fact, educators distinguish between the two by using the term textbook to refer to a book that is used solely to teach, and the term trade book to refer to a book that is generally available to the public through bookstores and libraries. In this chapter, the terms trade book and CH/YA literature will be used interchangeably.

## Why should teachers use CH/YA in the social studies?

While textbooks have long been a staple in the social studies, only within the past ten years has there been a concerted effort to go beyond the offering of textbooks and to use CH/YA literature in the teaching of social studies. This change began with the call for more interdisciplinary education in schools.

In looking at the future of social studies education, the National Council of Social Studies (1989) or NCSS indicated the need to help students understand the relationships existing within the sciences and humanities by stressing the interrelationship of knowledge from the major disciplines. Jacobs (1989) also pointed out the need for teachers to develop lessons that demonstrate the major disciplines share many concepts.

Educators soon found that one way to link the disciplines is through using quality CH/YA trade books. Among the benefits of using CH/YA literature in the disciplines is high student interest in trade books and the potential for teachers to use these books to combine the learning and practice of reading skills with the learning of content area information. Additionally, while literature is often used in the language arts curriculum to develop social decision making or higher level thinking skills, the same process can be applied in other disciplines to achieve curricular goals. As Wood, Flood, and Lapp (1994) point out, literature "provides for a richer, more meaningful understanding of subject matter, and it provides a relevant way to introduce students to the pleasures and rewards of reading" (p. 67). Reading, like most skills, improves with practice; practice that should not be limited to language art, English, or reading classes.

Writing in *Social Education*, Kim and Garcia (1996) discuss the importance of CH/YA literature in the social studies curriculum as a way to present basic social studies concepts and themes. Their thoughts are echoed by McGowan, Erickson, & Neufeld (1996) who, while lamenting the lack of research on the topic, support the idea that students who are exposed to a literature-based social studies curriculum have a greater knowledge of historic events and a better understanding of social issues.

Although there is concern about the lack of research comparing textbook to literature-based social studies, there is not a total void of research in the field. For instance, from their research conducted with sixth graders, Jones, Coombs, and McKinney (1995) found that students using CH/YA books in the social studies learned more and had better attitudes about the social studies than those using only a social studies textbook.

Other educators also cite the advantages of using CH/YA trade books in the teaching of social studies. According to Kincade and Pruitt (1996), quality literature can provide information that many textbooks lack such as background information, illustrations, and global education concepts. Smith and Johnson (1995) agree and stress that CH/YA literature can be used to develop an understanding and tolerance of other cultures. Other educators such as Wunder (1995) describe the use of children's literature to teach the ten NCSS themes (National Council for the Social Studies, 1994) and emphasize the importance of using literature to support the NCSS standards.

While many educators see the importance of using nonfiction or informational books to supplement textbooks, others maintain that fiction books are important too. For instance, Danks (1995) reports that poetry can be used in social studies to provide an emotional account that is often missing in nonfictional writings. Likewise, in looking at the study of geography, Flaim and Chiodo (1994) explain that reading novels can expand students' knowledge and interest in geography. They show that elements of geography such as place names, physical features, climate, plants and animals, and economic activities can be found in CH/YA fiction. In addition to demonstrating the importance of geography in the settings of novels, Flaim and Chiodo also indicate that the five fundamental themes of geography (location, place, human/environmental interaction, movement, and region) can be taught using CH/YA novels. They explain that "good stories illustrate human behavior in various times and places and help students to form connections to those times and places. [Literature] humanizes the distant locations and events and captures the imagination" (p. 225).

In comparing social studies textbooks to trade books, Fuhler (1992) discusses the problems of textbooks. Among the problems are the lack of student involvement with reading, the controlled vocabulary, the lack of depth on a given topic, the didactic presentation which leaves "little room for critical thinking," and the lack of "memorable sights, sounds, and textures of the past" (p. 64). Conversely, CH/YA trade books contain "a wealth of historical understanding brought vividly to life through the imaginations of the readers and the gentle direction of the teacher" (p. 64).

Though there are serious limitations to textbooks, they should not be eliminated. Textbooks can be used to provide structure while trade books are used to look into the heart, mind, and soul of history—supplementing and enriching the textbook for exploring the psychological aspects of historical events. While textbooks are effective for presenting information, literature can personalize the unfamiliar, distant and, often, cold facts of history (McKay (1995).

Writing in the *Instructor* magazine, Lindquist (1995) notes her experiences as a fifth grade teacher using CH/YA historical fiction in the classroom. She finds that trade books "illuminate time periods while integrat[ing] the curriculum, and enriching social studies.... Blending stories into a study of history turns the past into a dynamic place" (p. 46). A summary of the reasons for using CH/YA literature to teach the social studies is contained in her "7 reasons I teach with historical fiction" (p. 47). She states that CH/YA literature:

> piques kids' curiosity... levels the playing field [between children with and without a background on a given topic]... hammers home everyday details... puts people back into history... presents the complexity of issues... promotes multiple perspectives... [and] connects social studies learning to the rest of... [the] school day. (p. 47–48)

## How should teachers select literature to use with social studies?

While most educators agree about the benefits of using CH/YA literature in the social studies classroom, selecting quality CH/YA literature from the over one hundred thousand CH/YA books currently in print and the almost five thousand new CH/YA titles that are published each year can be a daunting task. This point is particularly poignant given integrating CH/YA literature into the social studies curriculum necessitates the careful selection of books and thoughtful planning of instructional activities based on them (Savage and Savage, 1993). Thankfully, in addition to reading and reviewing books themselves, social studies teachers can rely on professional reviewing sources, book awards, and best books lists issued by educational and library associations when selecting trade books.

While there are many excellent CH/YA book awards given each year, such as the John Newbery Medal and the Randolph J. Caldecott Medal, several have special significance for social studies, include the following:

Carter G. Woodson Book Award, given for sensitivity and accuracy in portraying ethnic minorities.

Coretta Scott King Award, honoring both an African-American author and illustrator.

Jane Addams Book Award, given to a book that combines literary merit with themes of brotherhood, peace, and social justice.

Jefferson Cup Award, presented to a book on American history, historical fiction, or biography.

Mildred L. Batchelder Award, bestowed upon the most outstanding book published abroad in another language and translated into English.

Scott O'Dell Award given to a work of historical fiction

Because not all outstanding books can win awards, many professional associations issue annual best books lists. There are many general lists that educators can consult, including those issued by the American Library Association and the International Reading Association. One list of books for grades K–8 is selected specifically for social studies educators, the Notable Children's Trade Books in the Field of Social Studies, which is compiled by NCSS and published in the April/May issue of *Social Education.*

In addition to annual lists, there are several professional journals devoted to extensive reviews of CH/YA books and others that regularly feature book review columns. An example of the latter is *Social Studies and the Young Learner*. In contrast, *Book Links*, published by the American Library Association, is devoted to information on CH/YA books, authors/illustrators, and curriculum connections. In *Book Links*, the "Dateline USA" series of articles by Chatton and O'Laughlin (1996, 1997) feature CH/YA books on American history while other articles focus on topics from folk literature (Peterson, 1992) to biographies (Scales, 1996).

Even when using awards, best books lists, and journal reviews to make a preliminary selection of books, social studies teachers need personally to read and review the materials that they are considering using in their classrooms. In selecting CH/YA literature, social studies teachers need to keep in mind general book selection guidelines found in CH/YA literature textbooks, such as those mentioned earlier in this chapter. These selection guidelines include evaluating works based on the elements found in all good literature: plot, characterization, setting, style of writing, point of view, and theme.

Additionally, social studies educators must consider special criteria when judging books for classroom use. Teachers should use a variety of books in order to address the needs of students with differing abilities and skills. They should also consider students' background and analysis skills. Selected CH/YA books should be ones which are well-written and bring the past to life for the reader.

More specifically, in selecting books to use in the social studies, educators should look for the following:

1. well-researched, historically accurate information, including authentic portrayals of individuals and times;
2. realistic, well-rounded characters that children and young adults can relate to or care about;
3. genuine and believable settings;
4. multicultural books that avoid stereotypes;
5. supplemental materials such as prefaces and author's notes that expand the text and place the story in historical context;

6. interesting stories which make history real but are not sentimental or condescending and neither debunk nor deify historical figures;
7. authentic vocabulary; and
8. illustrations that accurately expand or supplement the text.

One sensitive issue that must be dealt with in the selection of historical fiction titles is historical context. Things such as language and social customs which were accepted or legal in the past may no longer be accepted or legal today. A skillful educator is aware of these issues in trade books and prepares the students before actually using these books.

A key individual to consult when selecting trade books for classroom use is the school library media specialist (SLMS). As they are trained in book selection and knowledgeable about recent CH/YA literature, school library media specialists should be viewed as a resource and an instructional partner in planning and implementing CH/YA literature-based social studies instruction. In addition to providing suggestions of specific titles, the SLMS is often willing to do booktalks for students and suggest resources for educators to consult when planning lessons. If an educator has identified trade books to use in the curriculum, most SLMSs will be glad to order the titles for the school library collection.

## How can CH/YA literature be used in the social studies classroom?

Selecting books to use in the social studies is the first step in providing literature-based instruction. In some cases it may be possible to use books that students are already reading for other classes. In other cases, the selection may first be made in the social studies classroom with reinforcement coming from other disciplines, primarily science and the language arts. After the initial selection of trade books is made, teachers need to plan specific ways to connect the books to the social studies curriculum.

There are many excellent professional resources that include ideas for integrating literature into the social studies. One way to locate these sources is to search *ERIC* or the *Education Abstracts* database. Doing so, teachers may wish to use one of the following expressions: social studies AND literature, or social studies AND language arts. The results from the search should be a list of reports and articles containing instructional ideas. Additionally, several authors (e.g., Concetta Ryan, 1994, and Wanda Miller, 1997) have published resource books of teaching ideas for integrating trade books into the social studies.

Many of the activities found in professional resources are built upon the *National Standards for United States History* published by the National Center for History in the Schools (1994). These standards include an emphasis on chronological thinking, historical comprehension, historical analysis and interpretations, historical research capabilities, and historical issues, analysis and decision making. While all of these ideas cannot be listed or discussed in this chapter, the following are examples of ways to integrate literature into the social studies curriculum.

One strategy for using CH/YA literature is to organize students into literature circles. Using this technique, the teacher selects trade books to accompany a specific social studies unit. Based on the number of CH/YA books that are located, the teacher can decide whether to have each student read a different book or to use only five or six books with several students reading each one. The students in the class are then divided into groups of four or five. Groups meet periodically to discuss what the group members are reading and how this contributes to their understanding of the unit topic. In addition to helping students become active readers, literature circles are an excellent way to develop oral communication skills, build self-esteem, and develop collaborative skills while fostering acceptance of diverse points of view. Other suggestions for using literature circles can be found in the writings of Jill Scott (1994) and in many language arts or children's literature textbooks.

Some of the strategies used with literature circles can be extended into individual reading to help students focus on higher level thinking skills by making comparisons, drawing contrasts, and making predictions based on what they read. A teacher can assign different books to students and ask them to keep a log or journal in which they can, using examples from their reading, record answers to the following questions:

- What effect does war or conflict have on the basic needs (food, shelter, clothing) of people?
- How are the attitudes and beliefs of individuals changed by war?
- How is family life affected by a war?
- Is this effect the same for all families? If not, what makes the difference and why?
- How does war change relationships within a community?
- What causes these changes?
- What effect does war have upon the local and national government?

Logs or journals can be turned into a whole class activity by having students share information they have found. Additionally, if students are reading books set in different countries during the war, comparisons can be made.

The use of CH/YA literature with the social studies is not limited to historical fiction and nonfiction titles. For example, the contemporary realistic fiction of Ben Mikaelsen's *Sparrow Hawk Red* can be used effectively to teach the five themes of geography (see the Joint Committee on Geographic Education, 1984). This adventure novel tells the story of young Ricky Diaz who is shocked to learn that his mother was murdered by a Mexican drug cartel in retaliation for his father's work with the U.S. Drug Enforcement Agency (DEA). When Ricky's Dad refuses to help the DEA steal an airplane with a secret radar transmitter from a Mexican cattle ranch, which is serving as a front for the drug cartel, Ricky decides to steal it himself.

Some brief suggestions for teaching geography with this book include:

- *Location:* Using an atlas to locate the places in southern Arizona and Mariposa, Mexico that are mentioned in the book.
- *Place:* Explaining how the Spanish words add to the realism of the novel. Identifying the physical details, plants (e.g., saguaro cactus and desert willow), and geographic features that distinguish the setting of this novel from the geographic features where the students live?
- *Human/environmental Interaction:* Researching the rateros, or street children, that Ricky meets. Evaluating the accuracy of the portrayal in the novel. Describing the economic and/or environmental factors having led the rateros to live on the street. Relating these factors to those leading to homelessness in the United States. Examining the causes of the chubascos. Anticipating whether a similar flood could occur where the students live.
- *Movement:* Examining the forms of transportation that Ricky uses during his journey. Evaluating the efficiency of each method. Hypothesizing about other forms of transportation Ricky could have used and their impact.
- *Region:* Discussing the differences and similarities between what Ricky sees in Mexico and Arizona. Using other resources, study about Arizona and the Mexican state of Mariposa. Identifying differences that may have resulted from government and physical geographic features.

Other teaching ideas include:

- Asking students to write about what they read. This writing may take the form of a journal, diary, or letter to the main character or to another student in class.
- Having students compare the information in the trade book to information in the textbook or other sources—determining the difference between fact and fiction.
- Assigning CH/YA books that present both sides of an issue and then having students debate that issue.
- Letting students create dramas or readers' theaters from the trade books they have read.
- Extending books through food by locating and trying recipes for the food that is mentioned in the book.
- Encouraging students to develop travel brochures for places mentioned in the books, using resources to find pictures of regions or specific locations.
- Holding an end-of-the-unit celebration and allowing students to select a method to present their books to the class, e.g, dioramas, food, illustrations, or even a computer "slide" preservation.

## Conclusion

In this chapter, the author presents a rationale for using CH/YA literature in the social studies classroom and provides suggestions for selecting and using CH/YA literature. With the careful selection of CH/YA literature and the thoughtful planning of instruction to integrate CH/YA literature into the social studies curriculum, teachers can provide students meaningful, interesting, and integrated learning experiences that yield a deeper understanding of historical events and social issues than can be found from the use of textbooks alone.

## References

Chatton, B., & O'Laughlin, M. (1996). Dateline USA: The 1950's. *Book Links* 6, 57–64.
Chatton, B., & O'Laughlin, M. (1997). Dateline USA: From the Constitution to the Civil War. *Book Links,* 7(1), 16–22.
Danks, C. (1995). Using Holocaust short stories and poetry in the social studies classroom. *Social Education, 59,* 358–361.
Flaim, M. L., & Chiodo, J. J. (1994). A novel approach to geographic education: Using literature in the social studies. *Social Studies, 85,* 225–227.
Fuhler, C. J. (1992). Integration of trade books in the social studies curriculum. *Middle School Journal, 24*(2), 63–66.
Goforth, F. S. (1998). Literature and the learner. Belmont, CA: Wadsworth Publishing.
Huck, C. S., Helper, S., Hickman, J., & Kiefer, B. (1997). *Children's literature in the elementary school* (6th ed.). Madison, WI: Brown & Benchmark.
Jacobs, H. H. (Ed.). (1989). *Interdisciplinary curriculum: Design and implementation.* Alexandria, VA: Association for Supervision and Curriculum Development.
Joint Committee on Geographic Education, Association of American Geographers and National Council for Geographic Education. (1984). *Guidelines for geographic education: Elementary and secondary schools.* Macomb, IL: Author.
Jones, H. J., Coombs, W. T., & McKinney, C. W. (1994). A themed literature unit versus a textbook: A comparison of the effects of content acquisition and attitudes in elementary social studies. *Reading Research and Instruction, 34*(2), 85–96.
Jordan, A. D. (1995). Characteristics of nonfictional history. TALL: *Teaching and Learning about Literature with Children and Young Adults, 5*(2), 30.

Kim, C. Y., & Garcia, J. (1996). Diversity and trade books: Promoting conceptual learning in social studies. *Social Education, 60*, 208–211.

Kincade, K. M., & Pruitt, N. E. (1996). Using multicultural literature as an allay to elementary social studies texts. *Reading; Research and Instruction, 36*(1), 18–32.

Lindquist, T. (1995, October). Why & how I use historical fiction. *Instructor, 105*, 46–50.

McGowan, T., Erickson, L., & Neufeld, J. A. (1996). With reason and rhetoric: Building the case for the literature-social studies connection. *Social Education, 60*, 203–207.

McKay, R. (1995, Spring). Using literature in social studies: A caution. *Canadian Social Studies, 22*, 95–96.

Mikaelsen, B. (1993). *Sparrow hawk red*. Hyperion Books for Children, 1993.

Miller, W. J. (1997). *U.S. history through children's literature*. Englewood, CO: Teacher Ideas Press.

National Center for History in the Schools. (1994). *National standards for United States history: Exploring the American experience*. Los Angeles, CA: Author.

National Council for the Social Studies. (1994). Ten thematic strands in social studies. *Social Education, 58*, 365–368.

National Council for the Social Studies. (1989). *Charting a course: Social studies for the 21st century*. Washington, D.C.: Author.

Peterson, L. (1992). American folk heroes. *Book Links, 2*(2), 41–45.

Ryan, C. D. (1994). *Learning through literature: Social studies*. Huntington Beach, CA: Teacher Created Materials.

Ryan, C. D. (1994). *Learning through literature: U.S. history*. Huntington Beach, CA: Teacher Created Materials.

Savage, M. K., & Savage, T. V. (1993). Children's literature in middle school social studies. *Social Studies, 84*, 32–36.

Scales, P. (1996). Jean Fritz's you want women to vote, Lizzie Stanton? *Book Links, 5*(6), 47–51.

Scott, J. E. (1994). Literature circles in the middle school classroom: Developing reading, responding, and responsibility. *Middle School Journal, 26*(2), 37–41.

Smith, J. L., & Johnson, H. A. (1995). Dreaming of America, weaving literature into middle-school social studies. *Social Studies, 86*, 60–68.

Wood, K. D., Flood, J., & Lapp, D. (1994). Linking the disciplines through literature. *Middle School Journal, 26*(2), 65–67.

Wunder, S. (1995). Addressing the curriculum standards for social studies with children's literature. *Social Studies and the Young Learner, 8*(2), 4–7.

Prepared by: Katherine T. Bucher, Associate Professor and Graduate Program Director of Elementary/Middle School Education, Department of Curriculum and Instruction, Old Dominion University, Norfolk, VA. 12-97.

# Characteristics, Responsibilities, and Qualities of Urban School Mentors

EDITH GUYTON
*Georgia State University*

FRANCISCO HIDALGO
*Texas A&M University, Kingsville*

In this article, the authors assume that urban schools provide a different context for beginning teachers to develop the philosophies, attitudes, knowledge, and behaviors that guide their work in the teaching profession. The context is different because urban schools tend to serve poor children of color; teachers may suffer from low morale; bureaucratic demands are high; resources are scarce; and academic processes are often implemented differently. Making these assumptions leads to the proposition that mentors for teachers in urban schools have a different role from mentors in schools serving middle-class children. They can help beginners meet the challenges of working with students whose economic situations are precarious and respond to the needs and sensitivities of diverse children. Mentors in urban schools can also help beginners negotiate the bureaucracies and idiosyncrasies of urban school systems. They need to be able to provide support for how to teach with limited resources, overcome teacher negativism, and work around requirements not in the best interests of students. Mentors in urban schools need particular characteristics to promote positive development in beginning teachers. For the purposes of this discussion, the authors assume that good urban teachers can be good urban mentors, yet we recognize the need for further exploration of variegated roles.

## Characteristics of Urban Mentors

Some studies have identified characteristics of successful teachers of children who live in poverty (Brookhart & Rusnak, 1993; Knapp & Shields, 1990; Means & Knapp, 1991). Zeichner (1993) summarized these characteristics, which include high expectations, strong identity, a variety of teaching methods, understanding of the community, and advocacy for justice. Hidalgo and Huling-Austin (1993), in a policy study of qualities needed by teachers of Latino students,

---

Edith Guyton and Francisco Hidalgo, "Characteristics, Responsibilities, and Qualities of Urban School Mentors," *Education and Urban Society,* November 1995, Vol. 28, No. 1, pp. 40–47. Reprinted by permission from Sage Publications, Inc. (US), Corwin Press.

proposed that universities prepare teachers "who (a) have the ability to produce academic results, (b) are available and willing to serve in difficult settings, and (c) possess teaching skills derived from native or community bred talents, such as language" (p. 19). Certainly, these characteristics, as well as characteristics of effective teachers in all settings, including suburban schools (Campbell, 1990), are useful in deciding which teachers would be good mentors.

Collegial support is an important factor in teacher development. Mentor programs are one way of formalizing collegial support. Many studies indicate the efficacy of mentors in working with new teachers (Fagan & Walter, 1982; Gray & Gray, 1985; Guyton, Dobbs, Gray, & Wright, 1987; Huling-Austin, 1990; Odel & Ferraro, 1992; Schaffer, Stringfield, & Wolfe, 1992). Christensen and Conway (1990–1991) extended the benefit of mentoring to situations in which mentors are self-selected. Colbert and Wolff (1992) found that a beginner teacher support system was effective in increasing the retention and reducing the isolation of teachers in urban schools.

Weiner (1993) emphasized the importance of role models in learning to teach in urban schools. She made the following statement about preparing teachers to work in urban schools:

> Urban teachers confront the greatest diversity of student needs, but the conditions in urban schools severely limit individualization, so the special demand made of urban teacher preparation is to educate teachers who can deal with students as individuals and human beings in settings that depersonalize learning, making students and teachers anonymous and powerless. (p. 110)

Cochran-Smith (1991) insisted that urban teachers need to learn to "teach against the grain" and that they learn from experienced teachers:

> Teaching against the grain is deeply embedded in the culture and history of teaching at individual schools and in the biographies of particular teachers and their individual or collaborative efforts to alter curricula, raise questions about common practices, and resist inappropriate decisions. These relationships can only be explored in the company of experienced teachers who are themselves engaged in complex, situation-specific, and sometimes losing struggles to work against the grain. (p. 280)

Haberman (1994) believes that teachers grow professionally in urban schools by being associated with "star teachers" (p. 2). One characteristic of a star teacher is the ability to deal with an irrational bureaucracy. He characterizes these teachers as people who can do "gentle teaching in a violent society" (p. 3). Haberman's contention is that often these teachers are themselves products of the urban schools.

Research on the nature of urban schools, effective mentors, and the importance of role models for beginning teachers in urban schools suggests that mentors in urban schools need to possess characteristics and skills in common with mentors in any school but that they also must have skills and characteristics peculiar to the environment. In urban settings, beginning teachers experience the complexities of teaching children beset by poverty and teaching children whose social class and/or ethnicity often is not that of the teacher, and they must do so with bureaucratic inflexibility and social isolation. As beginning teachers cope with those issues, mentors can nurture their professional growth.

## Responsibilities and Qualities

Basic responsibilities of the mentor involve helping the novice negotiate the bureaucratic environment of urban schools and modeling ways of dealing with students who live in poverty—the urban school population. Although successful urban mentors may possess characteristics similar to mentors in nonurban schools, they also must possess a special set of qualities. The characteristics categorized and listed below are a synthesis of the recommendations of Zeichner (1993), Cochran-Smith (1991), Weiner (1993), and Haberman (1987, 1991, 1993, 1994) as applied to urban mentor behaviors, attitudes, and skills. All of the elements discussed are characteristics of successful urban teachers, as these qualities are essential. Mentors must surpass the basic criteria for being an exceptional teacher and help novices in the profession develop good qualities. Achieving the latter takes time, commitment, and the special abilities described in the following categories.

## Mentor As Change Agent/Mediator of the Urban Environment

Successful urban teachers understand and operate competently within an environment often indifferent to, and even hostile to, their most basic professional needs and the needs of individual students. They are also able to articulate how and why they negotiate within this environment. They accept the environment to the extent that they do not spend all their time and energy railing against conditions, but they are powerful within that environment and are able to help students learn. Mentors are able to talk about the school and help a novice understand the ways teachers can grow and learn there.

These teachers are not passive regarding the conditions and circumstances of urban schools and are willing to question practices in the school and to take a stand. They are people who continually ask, Does it have to be this way? How can we change things? What do I need to be doing? When a situation is deemed important, these teachers will take risks to help students. They are advocates for justice, at least within the school environment. For example, one teacher who was required to fail kindergartners whose performance was considered inadequate took it upon herself to gather evidence about the harmful effects of such a practice and present it to the policymakers (Guyton, 1994). Mentors must not only ask these questions but be able to articulate their importance and help beginning teachers understand what things they can affect and what things are beyond their power.

Successful urban teachers view teaching as problematic and are willing to experiment and collaborate to solve problems. They expect problems and they believe that it is the teacher's job to solve them. They try to figure out how to get students to learn, not to set up hurdles to determine student success or failure. Mentors are able to describe their thinking processes in dealing with problems and to help beginners develop problem-solving skills.

Effective urban mentors also strive to understand the community in which they teach, as well as articulate descriptions of the community and the dynamics between community and school. They know the community. In talking with new teachers, mentors can explain, without judgment, what is going on in a given situation. They do not stereotype community members or dismiss their actions as "just things these people do," but are able to describe the complexity of school/community relations.

## Mentor As Efficacious Teacher

Successful urban teachers judge worth as a teacher on individual successes with students, not on being able to "change the world." Their sphere of influence and success is the sphere over which they have some control. They are not paralyzed by the conditions of students' lives. Instead, they are able to teach in the face of horrifying events, recognizing areas in which they can make change and those in which they cannot. Mentors are able to curb the misguided enthusiasms of new teachers who are intent on saving the world and to help novices put some distance between themselves and the problems of children in their classes.

## Mentor As Collaborator

Successful urban teachers work with other teachers to solve problems. They respect colleagues and what they contribute. They are team players when the team works for the good of the school and the students. They also respect and communicate with parents; they view parents as a resource, not as a barrier to learning. They invite parent involvement and assume that parents want what is good for their children.

Good urban teachers view anyone related to the school or to students as a potential resource: social workers, school psychologists, and central office administrators. The question is who can help get the job done, not who gets credit or who can exert power. As mentors and resource persons, they do not have power needs that they satisfy by working with students who live in poverty. They do not choose urban schools because they can feel powerful among the least powerful segment of society, the poor.

Mentors help new teachers learn how to work with different constituencies. They are willing for their mentees to accompany them when they work with others and are able to explain what and why they do certain things. They can observe new teachers' interactions with others and give constructive feedback about how to get others to work with them.

## Mentor Sense of Self

Successful urban teachers have a clear sense of their own ethnic and cultural identity. They have explored personal identity issues and are comfortable with themselves. This comfort allows them to transcend economic, racial, and ethnic differences between self and students and their families. They also meet personal needs by working with students who live in poverty. Their job is inherently satisfying because it takes advantage of an intersection between personal need and societal need. Successful urban teachers recognize feelings of hate, prejudice, and bias and strive to overcome them. They neither claim to "love everyone" nor to be "color blind." They accept negative feelings as part of being human but also as alterable or at least capable of being set aside for the good of their students. Mentors are able to describe their sense of self and how they achieved it. They also are willing to explore their mentees' feelings and attitudes in a sensitive and caring way and are not afraid to question actions exhibiting prejudice and bias.

## Mentor As Pedagogue

Effective urban teachers include diverse cultural perspectives in the curriculum. They recognize pluralism as an asset and an essential part of society. They know that students need to "see themselves" in the curriculum. They also use a variety of teaching methods, developing broad repertoires from which to choose when working to help students learn. They do not depend on "one best way" but are able to try a different method when one does not succeed. They are able to locate and use resources other than texts to provide variety and to reach more students. If a student cannot read the text, such teachers do not bemoan the situation, but look for alternatives. Mentors are able to help new teachers find resources, ask good questions about instruction, and be reflective about their teaching. They can provide constructive feedback about the beginning teacher in ways that promote growth rather than defensiveness.

Good urban teachers are equally concerned with the cognitive and affective domains. They care about students' self-esteem and actively strive to develop it and acknowledge how feelings can affect learning. They do not accept the conditions of students' lives as an excuse for their not learning, but seek to understand the effects of those conditions and overcome them. Effective mentors have high expectations of students. They do not refer to urban students as a distinct group of learners for whom expectations should be different but, rather, as learners whose potential can be tapped. Mentors can articulate their expectations and the rationales for them to beginning teachers. They can help novices develop realistic but challenging expectations for students and can be a source of support when expectations are not met.

Good urban teachers are able to respond to curricular mandates while teaching creatively and attending to the whole child, including cognitive and affective concerns. Teaching basic skills and raising test scores are preoccupations of urban schools. These teachers can meet the demand of a curriculum driven by those goals within a minimal timeframe or while teaching in a more holistic way. Mentors can work with beginners to help them meet the sometimes conflicting needs of children and demands of the curriculum.

## Mentor Interpersonal Skills

Successful urban teachers hear what others have to say and listen for meaning. They strive to understand rather than to judge. They hear what students, other teachers, administrators, and beginning teachers have to say. Their agendas have ample place for new ideas or questions. Mentors ask hard questions of beginning teachers and expect hard questions in return. Mentors also do not expect beginning teachers to accept unquestioningly what they do and say. They enjoy and seek out meaningful conversations about teaching.

Effective urban teachers enjoy working with young people and are able to articulate what they like about them and how they use that in their teaching. They transmit a sense of playfulness and joy in dealing with young people and can help beginners see the differences between playing and being playful in achieving academic goals.

# Conclusion

This article has described interactions between the teacher and the teaching context. It has posited that the urban school context is a unique teaching environment with significant and singular effects on beginning teachers and that mentors in urban schools must possess certain characteristics to nurture the professional growth of new teachers. Research on the urban school environment and urban teachers was used to identify at least some of these characteristics, but such characteristics, although necessary, are not sufficient. All good teachers are not good mentors. Good urban mentor teachers need to be able to articulate their beliefs and practices, and they need coaching skills to foster professional growth in beginning urban teachers. They need the characteristics of all good mentors as described in the literature, in addition to the attributes outlined above.

# References

Brookhart, S. M., & Rusnak, T. G. (1993). A pedagogy of enrichment, not poverty: Successful lessons of exemplary urban teachers. *Journal of Teacher Education*, 44(1), 17–26.

Campbell, K. P. (1990). Personal norms of experienced expert suburban high school teachers: Implications for selecting and retraining outstanding individuals. *Action in Teacher Education*, 12(4), 35–46.

Christensen, K. P., & Conway, D. F. (1990–1991). The use of self-selected mentors by beginning and new-to-district teachers. *Action in Teacher Education*, 12(4), 21–28.

Cochran-Smith, M. (1991). Learning to teach against the grain. *Harvard Educational Review*, 61, 279–310

Colbert, J. A., & Wolff, D. E. (1992). Surviving in urban schools: A collaborative model for a beginning teacher support system. *Journal of Teacher Education*, 43(3), 193–199.

Fagan, M. M., & Walter, O. (1982). Mentoring among teachers. *Journal of Educational Research*, 76(2), 113–118.

Gray, W. W., & Gray, M. M. (1985). Synthesis of research on mentoring beginning teachers. *Educational Leadership*, 43(3), 37–43.

Guyton, E., Dobbs, D., Gray, F., & Wright, J. (1987). The mentor teacher: A key component of induction programs. *Georgia Association of Teacher Educators Journal*, 2(1), 5–11.

Guyton, E. (1994, April). *First year teaching experiences of early childhood urban teachers*. Paper presented at the annual meeting of the American Educational Research Association, New Orleans.

Haberman, M. (1987). *Recruiting and selecting teachers for urban schools*. New York: ERIC Clearinghouse on Urban Education.

Haberman, M. (1991). The pedagogy of poverty versus good teaching. *Phi Delta Kappan*, 73, 290–291

Haberman, M. (1993). Contexts: Overview and framework. In M. J. O'Hair & S. J. Odell (Eds.), *Diversity and teaching: Teacher education yearbook* (pp. 1–8). Fort Worth, TX: Harcourt Brace Jovanovich.

Haberman, M. (1994). Gentle teaching in a violent society. *Educational Horizons*, 72(3), 1–13.

Hidalgo, F., & Huling-Austin, L. (1993). Emerging sources of Latino teacher candidates. In Y. Rodriguez-Ingle & R. DeCastro (Eds.), *Reshaping teacher education in the southwest: A response to the needs of Latino students and teachers,* 13–30. Claremont, CA: Tomas Rivera Center.

Huling-Austin, L. (1990). Teacher induction programs and internships. In W. R. Houston (Ed.), *Handbook of research on teacher education*. New York: Macmillan.

Knapp, M. S., & Shields, P. M. (1990). Reconceiving academic instruction for the children of poverty. *Phi Delta Kappan*, 71, 752–758.

Means, B., & Knapp, M. S. (1991). Cognitive approaches to teaching advanced skills to educationally disadvantaged students. *Phi Delta Kappan*, 73, 282–289.

Odell, S. J., & Ferraro, D. P (1992). Teacher mentoring and teacher retention. *Journal of Teacher Education*, 43(3), 200–204.

Schaffer, E., Stringfield, S., & Wolfe, D. (1992). An innovative beginning teacher induction program: A two-year analysis of classroom interactions. *Journal of Teacher Education*, 43(3). 181–192.

Weiner, L. (1993). *Preparing teachers for urban schools: Lessons from thirty years of school reform*. New York: Teachers College Press.

Zeichner, K. M. (1993). *Educating teachers for cultural diversity*. East Lansing, MI: National Center for Research on Teacher Learning.

# Modeling and Mentoring in Urban Teacher Preparation

GENEVA GAY
*University of Washington, Seattle*

Commonsense, wisdom and research findings suggest that when it comes to effectiveness in teacher preparation and classroom instruction, people make the most difference. For the most part, we know that teachers teach as they have been taught. This influence is apparent in philosophical beliefs about the role of teachers in the educational process, pedagogical techniques used in the classroom, relationships with students, and general professional development. One comes to the logical conclusion that the best way to ensure that individuals will learn to be good teachers is through practice under the tutelage of others who have established records of success. This explains why the idea of modeling and mentoring in teacher education is so enticing.

Although these general ideas warrant serious consideration in redesigning teacher preparation for urban settings, they are often obscured by some recurrent myths and fallacious notions associated with models and mentors. One of these is the assumption that anyone who is a successful teacher can guide and assist others through the process of becoming effective teachers. That is, good models are automatically good mentors. Although modeling and mentoring are closely related, they are not synonymous. By definition, a mentor is someone who facilitates and assists another's development. An essential element of that facilitation is being able to model the messages and suggestions being taught. By comparison, a model is a tangible embodiment of an idea or ideal. The former is a process and the latter is a product.

Simply because individuals have mastered a skill does not guarantee that they can teach it to others. They may know how to do good teaching but are unable to communicate the principles of their success to others. A case in point is writing and research abilities. Many individuals who are gifted writers cannot teach others how to write. Many college graduates have memories of professors who were stellar scholars in their disciplinary fields but horrendously poor instructors. Some teacher educators are cognitively insightful about the need to be creative in teaching but do not behave accordingly.

Embedded in many discussions about modeling and mentoring is the implication that these are monolithic, or one-dimensional, processes. Often the focus is on the technical skills of teaching—such as how to make lesson plans, evaluate student performance, deliver various techniques of teaching, and maintain effective classroom management and discipline. These "product-oriented" perceptions suggest that effective modeling and mentoring can be accomplished by simply showing the mentees what to do. No allowances are made for the fact that good teaching is contextual, situated, and developmental. A technique that works well for one

---

Geneva Gay, "Modeling and Mentoring in Urban Teacher Preparation," *Education and Urban Society,* November 1995, Vol. 28, No. 1, pp. 103–111. Reprinted by permission from Sage Publications, Inc. (US), Corwin Press.

individual in a particular time and place may not work equally well for someone else, or even for the same person when the context changes.

Although in social, informal, and interpersonal relationships, imitation may be the highest form of flattery, it is not a reasonable and responsible basis for the development of professional teaching skills. Teaching is a highly individualistic, personal, and personalized process. Because the personalities, abilities, and persona of teachers vary widely, as do who, what, why, and where they teach, professional preparation programs that emphasize imitating others are potentially catastrophic. Merely mimicking the actions of models and mentors creates a kind of dependency that constrains the potential for each teacher to become an autonomous, self-empowered professional.

Therefore, modeling and mentoring in teacher education, as well as in classroom instruction with K-12 students, should focus on processes instead of products. That is, what makes particular ways of being and behaving, as expressed by certain individuals in specific teaching situations, effective, and how, or if, these can be adapted for personal use. The benefit of having models and mentors, as opposed to relying on abstract ideas, is that they present an actual embodiment, a living example of the theoretical principles of good teaching. Teachers in training then see theory translated to practice and can better engage with the embodied example to determine how it can be modified for their own use.

The following suggestions for using modeling and mentoring in urban teacher education are offered in the context of beliefs that the power of models and mentors resides more in the processes of their being and behaving than in the finished products. They can be used to select professional educators for urban teaching and mentoring as criteria for teacher education students to assess their own progress toward developing similar skills, and as benchmarks for them to use in choosing role models and mentors for their own students once they begin working in actual classrooms. The suggestions are based on the assumption that good teaching involves two distinct yet complementary clusters of skills. One is technical and the other is personal. Although both have doing and being dimensions, the primary focus of each is somewhat different.

Technical skills emphasize the doing aspects of good teaching, and personal skills concentrate on the being elements of people who are teachers. Models and mentors may personify these skills in varying degrees. In actuality, it is probably rare that all of them are embodied in individuals. Most likely, models and mentors are constructs that incorporate a composite of attributes from many different individuals, some real, some imaginary, and some symbolic. Their particular configurations also vary according to purpose, context, time, and audience.

## Modeling Is Being

Teaching is very demanding physically, mentally, and emotionally. It requires tremendous energy and resilience. It is simultaneously exhausting and energizing, challenging, and exhilarating. Those who do it well, and have the potential to be models and mentors for others, have an inner strength that fuels them. This reservoir of strength has an uncanny ability to replenish itself. Teachers who are mentors and models for being are competent, passionate, caring, and compassionate. They bring the multidimensional levels of their humanness to the act of teaching and help students to do likewise. According to Noddings (1984), these individuals are more vested in pedagogy of caring than pedagogy of method, and caring relationships are the anchors of all their teaching. By caring, Noddings means stepping out of ones own personal frame of reference into those of others and considering others' points of view, objective needs, and expectations.

When we care, our attentions and mental engrossments are on the cared for, not on ourselves. Our reasons for acting are driven by both the wants and desires of others, and the objective elements of their problematic situations. Such teachers practice what Berman, Hultgren, Lee, Rivkin, and Roderick (1991) call "a curriculum for being." Within it, the significance of persons and continuous personal growth are paramount. As they explain, "Persons are considered as sacred, holy, whole, thinking, meaning-making, trusting, searching for collegiality and friendship, and questioning. . . . Knowing when to live with one's plans and when to build on the possibilities of the moment necessitates a skill and attunement to self and others" (Berman et al., 1991, p. 7). Therefore, at die heart of the effective teacher is a psychologically healthy human being.

How teachers engage with their own and others' humanity and how they see themselves in relation to others in general determines, to a large extent, their potential for being effective models and mentors. The process of determining who can fulfill these roles and functions begins with an assessment of their personal qualities and dispositions. These are individuals who are secure in their own sense of self and are very accepting and facilitative toward others. They are what Kelly (1962) describes as a "fully functioning self," and Maslow (1954) calls "self-actualizing" persons. In their caring relationships, they become attuned to and take others into themselves, show empathy, share power, and nourish their students' total being. They also develop personal meaning for themselves and others "through relationships of receiving and responding that occur around struggles to know" (Bolin, 1995, p. 37). Such people can give freely of themselves, share accolades of accomplishment, and celebrate others' success without feeling chagrined, jealous, competitive, compromised, or minimized. Their guiding principles cause them to think well of and be affiliated with self and others, to be interested in and feel some responsibility for the quality of the people around them, to live their lives in keeping with their own values, and to accept change as a desirable necessity.

Kelly (1962) concludes that because the fully functioning person is

> in a perpetual state of emerging and becoming . . . he realize[s] that the only thing he knows for sure about the future is that tomorrow will be different from today and that he can anticipate this difference with hopeful expectation. . . . Such a person is a doer, a mobile person, one who relates himself in an active way with others. . . . He has no need for subterfuge or deceit, because he is motivated by the value of facilitating self and others. . . . Life to him means discovery and adventure. (pp. 19–20)

In addition to having the psychological disposition of a fully functioning person that Kelly describes, mentors for teachers preparing to work in urban areas should have several other more specific personal attributes. Edwards and Polite (1992) describe a set of attributes for successful African Americans that is generalizable to teaching models and mentors. The nature of these traits led the authors to conclude that (a) success is the result of a kind of drama that gets played out first in the psyche; (b) successful individuals are affirmed and empowered by a positive sense of their racial, ethnic, and cultural identity; and (c) real success "lies in who you become on the journey to achievement" (p. 274). The specific traits in Edwards and Polite's (1992) model for Black success are modified slightly to apply to successful teachers in urban schools. Each one is discussed below.

## Cultural Consciousness and Positive Ethnic Identity

In addition to being psychologically healthy, models and mentors in urban schools should have a positive self-ethnic identity and high levels of cultural consciousness, and they should readily accept similar attributes in others. Because of the wide variety of ethnic and cultural diversity that characterizes our cities and urban schools, no teacher can function well in them without a genuine appreciation for diversity as embodied in values and behaviors. Such models and mentors must accept their own and others' ethnicity as a natural and significant part of their humanity, which should be understood, respected, and honored at all times. This attribute dispels the frequently held assumption that because most students in urban schools are children of color, models and mentors for them and their teachers must likewise be individuals of color. Quite the contrary is the case. Shared ethnic and racial identity is not an automatically dependable criterion for effective modeling and mentoring, at either the personal or technical level. A far more important issue is that teachers embrace their own and their students' ethnicity. Foster (1994) describes this as having cultural solidarity, affiliation, and connectedness.

Research (Bernal & Knight, 1993; Cross, 1991; Gay 1994b; Phinney, 1990; Tatum, 1992) indicates that ethnic identity development is a kind of psychological metamorphosis. As individuals move away from distorted, highly stylized conceptions of their ethnic identity as constructed by others, they become increasingly more accepting of and open about themselves; more willing and capable of engaging in a variety of intra- and interethnic group relations; respective of the inherent value of others' ethnicity and culture; and more fluid, flexible, and engaging in all dimensions of their being. As their sense of self improves, so do their feelings of personal efficacy and confidence, and the quality of their task performances. Good teachers in urban settings are at the top levels of this developmental process, and they are committed to helping their students reach similar states of being in relation to their own ethnicity.

## Personal Responsibility and Integrity

According to Edwards and Polite (1992), personal responsibility in successful individuals leads them to "make a way out of no way, to build where nothing was erected, to achieve where no success seem possible" (p. 241). Translated to the arena of urban education, this means that successful teachers succeed in achieving high levels of performance from students, despite the conception that city children cannot or do not learn. They do not see poverty, race, language differences, immigration, and family structures as reasons or excuses for not insisting that students perform nor for using any means necessary to meet these expectations. They assume, and act accordingly, that no choice is acceptable but success. They take student achievement as a personal responsibility, and when it is not forthcoming, this is considered a personal insult. What others might consider a failure, they see as merely a temporary level of knowing that has not yet been mastered; what to some is unconquerable, to them is merely a stimulating challenge to the imagination, a solvable puzzle. Neither failure nor success deters future efforts; failure means both teacher and student have to work harder, and success means striving to be even better.

These teachers also employ and promote multiple indicators of success. Ethics, morality, social responsibility, and respect for self and others are as important to learn and apply as academic skills. These commitments to develop the whole child were validated by Foster's (1994) review of research on effective African American teachers. Evidence from these studies showed that teachers' practices reflected their value commitments. For instance,

> They explicitly teach and model personal values—patience, persistence, responsibility to self and others—that can serve as a foundation to current as well as future learning. They foster the development of student attitudes and interests, motivation-aspiration, self-confidence and leadership skills... their practice evidences a "hidden curriculum" of self-determination designed to help students cope with the exigencies of a living society which perpetuates institutional racism while professing a rhetoric of equal opportunity. (Foster, 1994, p. 233)

## Facility To Take Considered Action

In psychological terms, these attributes of success are known as internal locus of control; sociologically, they might be called personal efficacy and empowerment. Teachers so inclined believe they can make a difference and they plan for making a difference. They are thoughtful and deliberate about what they do. They constantly engage in self-reflection and analysis about their teaching, seeking to determine why something works or not, and how it can be improved. These teachers are not necessarily perfectionists, but they are interested in continuous growth and development, and they believe that their thoughtful actions can have a direct effect on bringing this about. Edwards and Polite (1992, p. 245) describe this attribute of success as individuals organizing their lives around "a series of miniature planned action sequences that are followed by an evaluation of what is accomplished, and an adjustment of the action, if necessary, based on what the experience has taught them, and then another action."

Efficacious teachers also assume that they and their students will be successful. They build this into their teaching by using different kinds of styles and complexities of tasks to ensure all students will achieve at some level, and use success in one instance as a foundation for future efforts and even higher levels of attainment. These concerned individuals interject into their teaching a level of personal caring and conviction that commands the same kind of respect and high-quality performance from their students as they give to them (Foster, 1994). Therefore, good role models for culturally diverse students cultivate reciprocity in both interpersonal and task performance skills.

## Ability To Manage Others' Racial and Ethnic Perceptions

Successful teachers of urban students are very aware of the fact that their own and their students' ethnic and racial identities affect their educational experiences. But this does not deter them from their commitment to achievement. Although race and ethnicity are always in their consciousness as facts of life, they are not the points of primary focus in their instructional interactions. They understand why and how critics will use their race and ethnicity to question their credibility, integrity, and ability. But they do not allow these tactics to interfere with them conforming to their own standards of accountability. As Edwards and Polite (1992) point out, these individuals are both intellectually competent and socially skilled, and they have a finely tuned sense of self-direction and self-control. They can demonstrate proficiency and competence without appearing to be intellectually arrogant, chauvinistic, or arbitrary. Teachers with this facility can make others feel at ease and valued without necessarily capitulating to their position or compromising their own values. Their presentation of self is neither immune to the racial attitudes of others nor governed by them. They are able to transcend "a racial-victim perspective" (Edwards & Polite, 1992, p. 273) for themselves and their students. They understand how both may be victimized by racism, ethnocentrism, and cultural hegemony, but they refuse to be debilitated by them. These ways of being and behaving apply in all interactions—whether with students, peer colleagues, supervisors, or parents and other community members.

## Pioneers and Trailblazers

Pioneering teachers are neither intimidated nor ingratiated by being "the only one" or among "the very few" in schools where the student populations are from ethnic groups different from their own. They do not see that as something to bemoan or to celebrate; it simply is! Yet they are very aware that additional emotional energy may be needed for them to prevail in these environments, and they take actions to ensure that it is available when needed. Edwards and Polite (1992) suggest that at the root of pioneering is "the ability to adapt, to be flexible, and perhaps most importantly, to tap emotional and spiritual sources of satisfaction, reward, and power. This includes developing and nurturing a support system that helps sustain the pioneer in the new milieu" (p. 252).

This facility in successful urban teachers is manifested in the ability to find validation of and satisfaction in their own significance without having to depend on external, traditional signals of success, such as institutional awards, compliments from others, and recognition ceremonies.

They live by the rule that, "If I have done the very best that I could have in a given situation, then that is reward enough." Of course, doing the very best possible in all things, however minor or significant, is their routine modus operandi.

Trailblazing and pioneering teachers go where others fear to tread, both literally and figuratively. This may mean being the only White in an all-Black or Latino school. Or, it could be African American or Latino teachers who directly challenge the racism of their employing institutions and overtly advocate for pedagogical positions that they believe are helpful to students, but are clearly not approved by the school system. An illustration of this bravery and conviction is Jaime Escalante's daring proof that Latino students in a Los Angeles school, who had been declared hopeless, hands-off failures by other teachers, could master advanced placement calculus.

A trailblazing high school teacher in the Seattle public schools achieved similar success with at-risk Latino students in advanced placement Spanish language and literature (Ramirez, 1992). Another is Marva Collins's work at the Westside Preparatory School in Chicago where she has defied conventional assumptions and proved that poor inner city African American students on the verge of dropping out of school could learn high-level academics and the classics without sacrificing their cultural heritage. These pioneering individuals, and others like them, explore unknown terrains of teaching as welcome and promising new ventures of discovery instead of as directions to be actively resisted or reluctantly forced into. They are adventurers seeking new understandings, new strategies, new possibilities. They are motivated by novelty and newness. They are not intimidated by new ideas and possibilities, nor are they capricious or irresponsible in their explorations and actions.

Trailblazing teachers give careful and thoughtful consideration to new ideas and techniques, but they do not wait for "guarantees of success" before they are willing to try them. They believe anything that has the potential to improve what they are currently doing is worth trying. They are willing to be nonconformists and lone travelers if this is what it takes to improve educational quality and effectiveness for students. Simply because something is unpopular or popular with colleagues or is a trend in the profession is not sufficient reason for them to endorse it. Personal conviction, authenticity, and integrity are stronger motivators for them than popularity or group consensus.

## Self-Reliance and Self-Acceptance

Being self-reliant and self-accepting is a natural extension of some of the other traits of individuals who can be effective role models for urban teachers. These people are able to operate independently, with confidence and competence. They are risk takers and have a healthy element of nonconformity. They use their own standards of accountability and ethics to assess the quality of their performance. Although they are not insensitive to others, they do not depend on them for accolades of approval to decide what to do. On these points, Edwards and Polite (1992, p. 255) describe these individuals as ones who are able to "break with the pack, to take a chance," and they are least likely to "give a damn about what others may think" about their decisions and actions.

Another strong indication of just how self-reliant and self-accepting successful individuals are is the extent to which they integrate the multiple dimensions of their being into their professional lives. They embrace, enable, and express the totality of their humanness; they operate from a position of racial, ethnic, and gender strength. There are no abrupt separations of their personal and professional lives. These individuals do not try to escape from, contain, or camouflage any aspect of their composite identity. Instead, deliberate actions are used to maintain a balance in their lives. The spaces they inhabit radiate signs and signals of their self-embracement, such as artwork, books, and music from their own ethnic and cultural traditions. When entering this space, even in the absence of the owner, one is able to discern a great deal about who he or she is from these "artifacts and symbols of self-presentations."

This same kind of integration is evident in their cognitive and affective being as well. Self-accepting people are not embarrassed about being incorrect; or emotionally expressive, apologetic, or arrogant about their intelligence; or reluctant to learn from someone; or unwilling to incorporate their emotional states as a natural part of the instructional processes; or shy about celebrating others' successes and insights. They find joy and exhilaration in reaching deeper and broader levels of their own unique individual abilities, as well as those to be derived from being a member of the human collective and learning communities. They are at once playful and profound, certain and skeptical, serious and humorous, humble and confident. This kind of psychological grounding creates an aura, a sense of personal power that confounds, threatens, and simultaneously attracts others.

## Reaching and Giving Back

Achieving individuals who are genuine role models have a tendency to not take their success for granted, or to assume that it is the result of individual initiative. They know that wherever they are, "they did not get there alone." The African Americans who participated in the Edwards and Polite (1992) study wondered "by what dispensation they were singled out for the blessing of success over others who may have been smarter, more talented, more worthy" (p. 265). These individuals accept "lending a helping hand" to assist others toward their own maximum potential as a moral and ethical responsibility. Because they contribute their own success to a combination of will, skills, and the "divine intervention of grace, these blacks seek to honor, acknowledge, and reciprocate through the act of giving back, reaching back, and in so doing achieve both empowerment . . . and absolution" (p. 266). This attribute is particularly distinguishable because it is so much at odds with the prevailing mainstream values of individualism and "survival of the fittest." Yet it is one of the strongest bonds that effective urban teachers use to develop a sense of affiliation, kindredness, and community between themselves and their students.

Teachers with these dispositions go beyond the routine expectations to make learning of all kinds accessible to students. They try to empower students by teaching them how to learn, how to demystify different kinds of knowledge constructions, and how to analyze their own and others' thinking processes. They share off-the-record insights about what happens behind the scenes when educators come together in professional gatherings. They add whatever personal knowledge they have about the "experts" that learners are studying as a means of getting students into the habit of remembering that "these authorities are humans, too." They also share their own developmental matriculation toward success, as reminders that "I've not always been here," the "roads I traveled to get where I am may have parallel routes for you," and "my journey is not yet complete either."

## Spirituality

Good role models have a resounding faith in the possibility of all things. This faith is not necessarily anchored in any particular religious order in the traditional sense. Rather, it is driven by a belief in the inherent goodness, worth, dignity, and redeeming potentiality of humans and by a deep respect for the natural order of things. These individuals are positively inclined and refuse to be focused in negativism. They are guided as much by intuition and faith as reason and intellect. According to Edwards and Polite (1992), this belief in the possibility of things not yet seen or done gives these "Blacks their moral fiber, their psychic strength, their capacity for struggle and sacrifice and continuing success against the odds" (p. 270).

Most certainly, teachers who believe that anything is possible are able to persevere in the face of odds long after everyone else gives up. They feel that there is something good and capable in all students, to the point of being recalcitrant in this belief. This is not a missionary zeal, but a genuine trust in the inherent quality of human nature. The challenge is for teachers to find ways to help students unleash this potential. This belief feeds both their physical strength and creativity. They simply will not give up on students. What others see as failure and impenetrable obstacles in students, they see as the challenge yet unsolved of finding pedagogical solutions for how to access students' knowledge and abilities. And, they persevere in the struggle to do so with convictions that they will succeed. Part of this perseverance is encouraging, advising, and sharing with colleagues and others preparing to become teachers their stories and their "common wisdom" of successful being. Without necessarily planning to do so, without the formal rituals, these "experts in the art of victory" (Edwards & Polite, 1992) drift naturally into the role of mentoring.

## Mentoring Is Doing

The attributes exhibited by role models, discussed above, affect how they operate as mentors helping others to improve their own professional development. They reach out to others in the same tone and credence as they live within their own success. Yet they do not encourage the people they work with to merely imitate them. Instead, they are committed to helping them maximize their own strengths, potentials, and skills.

The behaviors of teachers who do this well in urban settings are embodied expressions of their ethos, ethics, and aesthetics of being. They strive to apply all of the attributes and expectations they have for themselves to those they teach. Because they integrate their whole being and strive for balance in their lives, the personal self is not divorced from how they perform their professional roles. Gloria Ladson-Billings (1994) personifies this integration in her preface to *Dreamkeepers*. She informs the reader that, "I have written this book in three voices: that of an African American scholar and researcher; that of an African American teacher; and that of an African American woman, parent, and community member. This book is a mixture of scholarship and story—of qualitative research and lived reality" (p. x). This projection of the whole self into the professional arena explains why others tend to describe these models and mentors in terms of the power of their personalities, the clarity of their convictions, the vigor of their vitality, the magnitude of their physical stamina, the positivism of their overall demeanor, the authenticity of their caring, and the rigor and range of their intellect.

## Cultural Context Teaching

Operationally, mentors and models in urban teaching evoke and employ their own cultural frames of reference and those of ethnically, racially, and socially diverse students to make teaching and learning intellectually challenging, emotionally stimulating, and task effective. Several leaders (Au, 1993; Boggs, Watson-Gegeo, & McMillen, 1986; Gay, 1994a; Hollins, King, & Hayman, 1994; Ladson-Billings, 1994; Shade, 1989; Trueba, Guthrie, & Au, 1981) in the field of cultural diversity in education have identified this style of operating as culturally contextual, culturally relevant, culturally sensitive, culturally responsive, or culturally congruent teaching. It will be referred to hereafter as cultural contextualized teaching. Being able to explain it conceptually, model it in one's own action, and teach others how to know when they are doing it themselves is the quintessential embodiment of pedagogical mentoring for teachers in urban schools.

Cultural contextualized teaching is grounded in the beliefs that teaching and learning are always cultural processes and that students learn better when classroom instruction builds on and is connected to their cultural references and experiences. Young (1990) offers a powerful explanation of this position:

> The culture to which one belongs . . . becomes the root of the individual's identity, because culture gives us a sense of power and confidence by giving us the basis for achieving our goals, determining what is desirable and

undesirable and developing the purposes of our life. Accordingly, to reject or demean a person's cultural heritage is to do psychological and moral violence to the dignity and worth of that individual . . . by knowing the cultural maps of our students, we can better facilitate the conditions for effective learning for them. (p. 24)

Students in urban schools represent many different ethnic and social groups, each with somewhat different cultural maps. Of necessity, then, urban teachers must be multicultural in their pedagogical styles. This approach to teaching transforms "the educational process to align it more closely with students' cultural knowledge and their indigenous ways of knowing, learning, and being" (King, 1994, p. 27). Its intent is to improve the quality of educational opportunities and outcomes for ethnically diverse students by using their own cultural frames of reference as conduits for teaching them academic, cross-cultural, interpersonal, socially conscious, and political activism skills.

At the heart of cultural contextualized teaching is an emphasis on personal caring, cultural bridging, shared responsibilities, and teachers and students working together in collaborative and cooperative relationships. These emphases, and the teaching techniques embedded in them, work well because their key features are compatible with the cultural orientations of many of the ethnic and social groups who live in inner cities. African Americans, Latinos, Asian Americans, recent immigrants, and the poor are accustomed to strong family ties, extended kinship networks, working together to solve problems of daily living, and living in physical and emotional proximity to each other. Therefore, teachers who are able to recreate the sense of cultural community in their classrooms and incorporate rituals and routines from diverse cultures into their teaching achieve more academic success for their students (Hollins, King, & Hayman, 1994; Ladson-Billings, 1994).

Ladson-Billings (1994) used the findings of her research on teachers who were effective with African American students to develop a set of descriptive features of cultural contextual teaching. They are included here as a set of indicators to select mentors for urban teachers and to pinpoint explanations of how and why their instructional personas model cultural context teaching. They also will serve as another benchmark for teachers in training to monitor their own professional development.

According to Ladson-Billings (1994), cultural contextual teaching includes three major aspects. They are teachers' conceptions of themselves and others, how social interactions are structured in the classroom, and teachers' conceptions of knowledge. Some specific descriptions of these aspects of the behaviors of individuals engaged in doing cultural contextual teaching, based on those provided by Ladson-Billings, are as follows:

- approaching teaching as an artistic process rather than a technical craft;
- working diligently to reduce the hierarchical power order between themselves and their students, and replacing it with more egalitarian relationships, including shared responsibilities, rights, and authority;
- providing many opportunities for students to succeed on a wide variety of tasks and challenges;
- building a sense of "community and camaraderie" in the classroom among students and employing consistent efforts to reveal connections among different kinds of communities locally, nationally, and globally;
- using a teaching style that is more facilitative than directive, in which students are invited and challenged to "get in touch" with and applaud their own knowledge;
- employing different methods to diffuse classroom climate factors that may interfere with the quality of student-teacher and student-student relationships;
- creating a classroom ambience and an ethos of informality, flexibility, excitement, exuberance, curiosity, discovery, and variety;

- extending teacher-student relationships beyond the usual one-dimensional relationships restricted to "instructional business only" and beyond the classroom door. Teachers place interest in the "person" before concern for the "student"; however, neither is neglected;
- constructing communities of learners where all students are equally enfranchised, and have full membership in all subsets of the "learning communities." They are able to give each of these "memberships" quality engagements and comparable nurturing, so that no one feels that they are not getting their "fair share" of the teacher's time, attention, and care;
- using cooperative groups and collaborative learning as the normative and routine procedures for teaching and learning;
- perpetually challenging and inviting students to "think about" what they are learning, to push the boundaries of their current levels of knowing, to remember that no existing knowledge need necessarily be "the final word"—that something yet not seen is always possible;
- conveying a sense of passion, energetic involvement, and enjoyment that pervades the entire teaching-learning process;
- routinely using examples and scenarios from a variety of cultural groups and experiences to illustrate instructional concepts, principles, and strategies;
- applying multiple criteria in constructing standards for and indicators of excellence in student performance;
- enabling students to emerge from their instructional encounters feeling personally embraced, affirmed, and empowered.

The overlap between these *doing* dimensions of cultural contextual teaching and the ways of *being* discussed earlier are obvious and unavoidable, because good role models live their ethics and beliefs, both personally and professionally. This explains why, when individuals are asked to identify those attributes that make these teachers good mentors, they are likely to key into what appear to be personality traits. It is as if their style of teaching takes on a personality of its own that is somewhat analogous to a person. It has many layers, expressions, nuances, and complexities, just as people do. Thus the teaching of mentors has a pulse beat, an energy, a feel, a tone, a flavor of its own. On occasion, individuals observing or experiencing it may be hard-pressed to give its attributes concrete description, but invariably they leave feeling, "I've been touched," and thinking, "That's the kind of teacher I want to be."

# References

Au, K. H. (1993). *Literacy instruction in multicultural settings*. New York: Harcourt Brace.

Berman, L. M., Hultgren, F. H., Lee, D., Rivkin, M. S., & Roderick, J. A. (1991). *Toward a curriculum for being: Voices of educators*. Albany: State University of New York Press.

Bernal, M. E., & Knight, G. P. (Eds.). (1993). *Ethnic identity: Formation and transmission among Hispanics and other minorities*. Albany: State University of New York Press.

Boggs, S. T., Watson-Gegeo, K., & McMillen, G. *(1986). Speaking, relating, and learning: A study of Hawaiian children at home and at school*. Norwood. NJ: Ablex.

Bolin, F. S. (1995). Teaching as a craft. in A. C. Ornstein (Ed.). *Teaching: Theory into practice* (pp. 26–40). Boston: Allyn & Bacon.

Cross, C. E., Jr. (1991). *Shades of Black: Diversity in African American identity*. Philadelphia: Temple University Press.

Edwards, A., & Polite, C. K. (1992). *Children of the dream: The psychology of Black success*. New York: Doubleday.

Foster, M. (1994). Effective Black teachers: A literature review. In E. R. Hollins, J. E. King, & W. C. Hayman (Eds.), *Teaching diverse populations: Formulating a knowledge base* (pp. 225–241). Albany: State University of New York Press.

Gay, G. (1994a). *At the essence of learning: Multicultural education.* West Lafayette, IN: Kappa Delta Pi.

Gay, G. (1994b). Coming of age ethnically: Teaching young adolescents of color. *Theory Into Practice,* 33, 148–155.

Hollins, E. R., King, J. E., & Hayman, W. C. (Eds.). (1994). *Teaching diverse populations: Formulating a knowledge base.* Albany: State University of New York Press.

Kelly, E. C. (1962). The fully functioning self. In A. W. Combs (Ed.), *Perceiving behaving becoming: A new focus for education* (pp. 9–20). Washington, DC: Association for Supervision and Curriculum Development.

King, J. E. (1994). The purpose of schooling for African American children: Including cultural knowledge. In E. R. Hollins, J. E. King, & W. C. Hayman (Eds.), *Teaching diverse populations: Formulating a knowledge base* (pp. 25–56). Albany: State University of New York Press.

Ladson-Billings, G. (1994). *Dreamkeepers: Successful teachers of African American children.* San Francisco: Jossey-Bass.

Maslow, A. H. (1954). *Motivation and personality.* New York: Harper & Row.

Noddings, N. (1984), *Caring: A feminine approach to ethics and moral education.* Berkeley: University of California Press.

Phinney, B. J. (1990). Ethnic identity in adolescence and adulthood: A review of research. *Psychological Bulletin,* 108, 499–514.

Ramirez, M. (1992, September 6). Time and reason: In this class, Spanish is an asset hurdle. *Pacific Magazine, Seattle Times,* pp. 12–20.

Shade, B. J. (Ed.). (1989). *Culture, style, and the educative process.* Springfield, IL: Charles C. Thomas.

Tatum, B. D. (1992). Talking about race, learning about racism: The application of racial identity development theory in the classroom. *Harvard Educational Review,* 62, 1–24.

Trueba, H. T., Guthrie, G. P., & Au, K. H. P. (Eds.). (1981). *Culture and the bilingual classroom: Studies in classroom ethnography.* Rowley, MA: Newbury House.

Young, P. (1990). *Cultural foundations of education.* New York: Macmillan.

# Teachers' Perspectives on School/University Collaboration in Global Education

TIMOTHY DOVE, JAMES NORRIS, and DAWN SHINEW

Relationships between many teachers and professors in central Ohio were strained at best when seven high school teachers and a university professor at Ohio State (OSU) sat down together in 1991 to discuss ways to improve the preparation of preservice teachers in global education. There was a history of professional teachers accusing university people of not knowing the *real world* of the public school classrooms. University professors had expressed frustrations over the schools' resistance to change and teachers' lack of interest in innovative teaching methods and current scholarship. Student teachers often found themselves in the middle of an ideological tug-of-war between what their cooperating teachers and their OSU supervisors perceived as *good teaching*. However, there was a group of teachers and professors who agreed that they needed to work together if the preparation of preservice teachers in global education was to be improved.

This chapter was developed from the shared experiences and reflections of teachers in nine schools in six school districts in Central Ohio who have worked with professors and supervisors over the past 5 years in the OSU Professional Development School (PDS) Network in Social Studies and Global Education. Written by three teachers, this chapter has three goals: First, we provide a background of the development of our collaboration in a globally oriented PDS Network. Second, we reflect upon what we have learned about the power of a school/university learning community in the preparation of preservice teachers in global education, and we provide some examples and illustrations of our approaches. Third, we make some suggestions for teacher educators who may be interested in developing their own partnerships in global education.

Let us introduce ourselves. Jim Norris, one of the founding members of our PDS Network, is department chair and social studies teacher at Linden McKinley High School, Columbus Public Schools, Columbus, Ohio. Linden McKinley provides considerable diversity for our preservice teachers because it is an English as a second language (ESL) magnet for high school students in Columbus. Jim began his international experiences as a American Field Service (AFS) high school exchange student to Finland in 1966. In other professional work, Jim has served as Central Ohio Representative and Legislative Liaison for the Ohio Council for the Social Studies and has developed courses of study in world area studies and K-12 global education for his district.

Tim Dove, who joined the PDS Network in 1993, teaches social studies as part of an interdisciplinary seventh-grade team at McCord Middle School in the Worthington School District, Worthington, Ohio. Interested in civil rights issues since he was 12, Tim continues to

---

Timothy Dove, James Norris, and Dawn Shinew, "Teachers' Perspectives on School/University Collaboration in Global Education," *Preparing Teachers to Teach Global Perspectives: A Handbook for Teacher Educators*, edited by Merry M. Merryfield, Elaine Harchow, and Sarah Pickert, 1997. Reprinted by permission from Sage Publications, Inc. (US), Corwin Press.

work for equity and justice through the development and teaching of curriculum that questions past and present practices. Tim was an AFS high school exchange student to Paraguay in 1975 and has worked with seventh- to ninth-grade students in providing international study programs.

Dawn Shinew is currently Graduate Research Assistant at the Mershon Center at The Ohio State University, where she works with the Education for Democracy Project, a cross-cultural project with Polish educators in civic education. Dawn taught social studies in Los Angeles for Duarte Unified School District for 6 years and provided in-service workshops for teachers on multicultural education. She has lived and worked in England and Sweden. In 1994–1995, Dawn supervised student teachers in our PDS and worked with Tim in a research project on school/university collaboration in global education. Together, the three of us bring 44 years of teaching and a wide variety of cross-cultural and international experiences to our work with preservice teachers.

## Developing School/University Collaboration in Global Education

A constantly changing world requires a continual reexamination of what we teach. Whether our students will live most of their lives within a 50-mile radius of Columbus or find themselves constantly moving and changing careers, their lives will be inexorably linked to other parts of the world through economic, political, technological, ecological, and cultural connections (Kniep, 1986). Today's students must understand these global connections and consider the interdependence of their own decision-making processes in the local community with the decisions and alternative futures of others around the world (Alger & Harf, 1986; Anderson, 1990). Young people must develop perspective consciousness and cross-cultural competence so that they are able to work effectively and manage conflict with people who are different from themselves (Hanvey, 1975; Merryfield & Remy, 1995). However, before students in the classroom can develop these understandings, their teachers must appreciate the challenges of an interdependent world and acquire global knowledge and skills. Whether they will teach in rural, urban, or suburban communities, preservice teachers must be able to prepare young people to participate effectively in the increasing cultural diversity and global interconnectedness of the late 20th century. Teacher educators need to work with professional teachers to reform teacher education so that it might more effectively prepare beginning teachers to become global educators (Merryfield, 1995). How does a teacher educator go about working with classroom teachers in global education?

## Getting Started

Our PDS Network developed from a professor's relationships with several globally oriented teachers during a time when she was growing increasingly concerned about her program's preparation of preservice teachers. In October 1991, Merry Merryfield, a professor in social studies and global education at Ohio State, invited Keith Bossard (Columbus Alternative High School, Columbus Schools), Sue Chase (Hilliard High School, Hilliard Schools), Shirley Hoover (Upper Arlington High School, Upper Arlington Schools), Jim Norris (described above), Bob Rayburn (Eastland Career Center, Eastland Vocational School District), Steve Shapiro (Reynoldsburg High School, Reynoldsburg Schools), and Barbara Wainer (Independence High School, Columbus Schools) to a meeting at her home to discuss how the preparation of preservice teachers in her program could be improved. Asking us to serve as an advisory board, Merry noted that she had come to believe she could not improve the program without the help of experienced teachers who were accomplished in global education. Because we knew her through her work in the schools, a research project on global teaching, and courses some of us had taken with her, most of us felt comfortable sharing our frustrations about what we perceived as the lack of preparation of preservice teachers for student teaching in our schools.

By the spring of 1992, we had moved beyond discussions and "advising." Deciding that drastic reform was needed in how teachers were prepared, we became full partners with Merry in bringing about real changes in the content of methods courses and field experiences. Working together, the eight of us set aside Merry's methods course and developed a new field-based course that we decided, with Merry's encouragement, to teach as a team. Through a series of brainstorming sessions and a work-to-consensus style of group decision making, we developed a list of competencies and overall goals that we agreed characterize good teaching in social studies and global education. (See Table 1 for the goals and outcomes we developed for our preservice teachers.) We discovered that our different school settings and styles of teaching did not stop us from having similar educational goals and beliefs about good teaching and global education. Although the focus of initial conversations centered around preservice teachers, we also found that we could not separate our vision of excellence in beginning teachers from the professional concerns we had for ourselves and our in-service colleagues.

## Integrating Field Experiences and Methods Seminars

From our goals, we began to think about areas of expertise that we would now take responsibility for teaching in the course: multicultural and global perspectives, higher-level thinking skills, cooperative learning, authentic assessment, technology, questioning techniques, independent research projects, and so forth. We decided that the preservice teachers needed to work with us in our classes and begin to teach our students as they were learning methods and developing lessons. One of the problems of Merry's course had been that the preservice teachers developed lesson plans to meet course requirements instead of teaching them to actual middle or high school students. We wanted authentic learning, where the preservice teachers would be applying all the content of methods seminars and readings to the real world of schools. We also wanted to ground them in the lives of our students and decided they should stay in the same schools for methods courses and student teaching. Most of OSU's preservice teachers with whom we work are white males from rural Ohio who have little knowledge or experience with cultural diversity or global education before they enter the OSU program. We tacitly knew that two academic quarters (over a period of 6 months) of working daily with us and our classes would help the preservice teachers learn much more about the cultures of our students, our school and community, administrative requirement, and expectations of the other professional teachers who would become their colleagues. We also assumed that these experiences would help build their classroom management skills and confidence.

As we continued to craft our methods course, a new language emerged. We decided to call the university students *preservice teachers* instead of students or interns. We wanted them to think of themselves as beginning teachers, not university students. We began to use the term *field professor* to describe our new roles as teacher educators who would be teaching the methods class on campus and supervising and mentoring the preservice teachers in our schools for Ohio State.

## Deciding on Readings in Global Education

We selected readings to complement the seminars and fieldwork. We chose Robert Hanvey's *An Attainable Global Perspective* (1975), Willard Kniep's "Defining Global Education by Its Content" (1986), the *NCSS Position Statement on Global Education* (National Council for the Social Studies, 1982), and Christine Bennett's (1995 is the latest edition) *Comprehensive Multicultural Education: Theory and Practice*. We decided to make available a wide variety of specialized publications on global education for secondary social studies published by the

Mershon Center, The Stanford Program on Intercultural Education (SPICE), the American Forum, and some other organizations. All of us were willing to share our professional libraries and instructional resources that support global education with our preservice teachers to help in their lesson planning and understanding of the world.

## Developing Assessments

As we taught methods and worked with the preservice teachers in our schools during autumn quarter 1992 and winter 1993, we purposefully developed congruence between the requirements of the methods course with immediate classroom responsibilities and the student teaching experiences that would follow. Global education was taught in seminars and readings, modeled by the field professors in our classroom teaching, and assessed in reflective journals, lesson plans, portfolios, and their teaching of our students. Assignments were developed based on the assumption that learning is most effective when the content is relevant to the learner. The preservice teachers wrote reflective journals to synthesize their readings, the methods seminars, and their experiences in the schools. They were also required to develop and present case studies based on their field experiences, as in *The Intern Teacher Casebook* (Shulman & Colbert, 1979). In addition, throughout the methods course and student teaching, the preservice teachers developed portfolios to demonstrate and document their evaluation of their own progress in relation to the goals we had developed. The portfolios also serve as a way for the preservice teachers to document their professional growth to potential employers.

One accomplishment in our work from 1993 through 1996 has been our development and field-testing of several assessment tools and rubrics so that the nine field professors are evaluating our preservice teachers in social studies and global education using the same criteria. For example, we assess all of our preservice teachers on their ability to teach perspective consciousness and multiple perspectives. We have developed some assessment tools to meet the needs of individual school cultures and contexts, because we have found such flexibility helps our preservice teachers better meet the needs of their students.

## Improving the Program Since 1992

Our Professional Development School Network in Social Studies and Global Education has grown and changed each year since our first team-taught methods course during autumn quarter 1992. When we added two middle schools to our network in 1993, Tim Dove became a field professor at McCord Middle School and Dave Fisher became one at Hilliard Weaver Middle School, Hilliard, Ohio. We now have 9 field professors and make decisions and teach as a team of 10. We have also moved all our PDS work off-campus and into our schools. Holding the seminars in our own schools (we take turns depending on who is leading the seminar) gives us opportunities to bring in colleagues or our interdisciplinary teams (5 of the field professors team-teach with colleagues in English, art, science or math) and share resources in our schools (computer labs, media). Occasionally, we even have our middle or high school students enriching the seminars with their perspectives, such as a session with Steve where his 10th graders compared their experiences with conventional and authentic assessment in World Connections, an interdisciplinary global education program that integrates math, science, language arts, and social studies (for more on World Connections, see Levak, Merryfield, & Wilson, 1993).

In 1993, we also brought together two 4-credit-hour methods courses and added additional field experience requirements so that we now teach a 10-credit-hour PDS methods block to preservice teachers as a prelude to their student teaching in our schools. In the PDS methods block, our preservice teachers work with their field professors' classes for 3 hours every day and attend our team-taught seminars on Tuesdays and Thursdays from 4:00 to 6:00. Unlike with the previous program that the PDS methods block replaced, our preservice teachers begin to teach

## Table 1

### *Goals and Expected Outcomes for Preservice Teachers*

At the end of the course, the preservice teachers will have demonstrated progress in:

A. Knowledge and skills in the basics of instructional planning, including

   1. planning for a specific course of study (e.g., Global History, World Cultures), through developing a plan for specific time periods including long-term (a term or semester) and short-term (a mini-unit or daily lesson plan)
   2. infusing global and multicultural perspectives into instruction
   3. dealing effectively with controversial issues in the curriculum and as they emerge in the classroom
   4. finding and using instructional materials and resources in the school, community, nation, and world
   5. using the K-12 scope and sequence of knowledge, skills, and attitudes/values

B. The use of a variety of instructional methods that encourage active learning, meet the different learning styles of students, and are congruent with content and educational goals

   The preservice teacher has demonstrated the ability to use a variety of teaching strategies in each of these areas:
   6. having student *actively find information* (from readings, A-V, statistics, library research, electronic databases such as CompuServe, interviews in the school or community, etc.)
   7. having students *process information* (categorize, chart, clarify, draw conclusions from, analyze, synthesize, evaluate, etc.)
   8. having students *use information* (to solve problems, make decisions, analyze values, teach others, relate/apply information to new situations, etc.)
   9. having students *examine global and multiple perspectives*
   10. *leading a discussion to get students to think* about and articulate what they are learning (from readings, guest speakers, videos, news, etc.)
   11. *having students evaluate the merit and worth of information* (its source/time-frame/geographic reference, point of view of author, possible bias, unstated assumptions, etc.

   The preservice teacher has demonstrated:
   12. selecting teaching strategies to fit social studies and global education content and student needs
   13. using a variety of assessment strategies to meet student and content needs
   14. being flexible in modifying lesson plans so that the structure and pacing of the lesson and transition from one activity to another meet the needs of the students
   15. keeping students actively involved in learning throughout each lesson

C. Awareness and support of their students as individuals and as learners

   The preservice teacher has demonstrated:
   16. building rapport with and respect from every student
   17. learning about each student's knowledge base, background, abilities, and interests
   18. making a concerted effort to see that every student learns

## Table 1 (Continued)

D. Questioning techniques that build higher-level thinking skills

    The preservice teacher has demonstrated:
19. asking questions that require students to go beyond recall of knowledge and comprehension to the application, analysis, synthesis, and evaluation of content, and include motivating (questions that create interest and connections) and divergent questions (open-ended questions that foster creative thinking)
20. leading a discussion effectively
21. being aware of problems of bias (gender, race, ethnicity, religion, language, national origin, physical appearance or seating, disability) in managing class discussions

E. Progress in reflecting on and improving their own teaching and learning as a professional educator

    The preservice teacher has demonstrated:
22. reflecting upon and continually improving his or her teaching and learning
23. articulating his or her own teaching style, including choices in such decisions as management and organization, discipline, expectations, extracurricular activities, and different roles with students, parents, colleagues, administrators
24. development as a professional teacher (for example, demonstrating integrity and ethics, taking all responsibilities seriously, demonstrating a positive attitude toward students and teaching, demonstrating continued learning, being involved in the community and professional organizations)

© Merry M. Merryfield, 1995

our middle or high school students during the first week of methods so that they can immediately put into practice what they are learning through the seminars and readings.

Because diversity and multiple perspectives are two of our major themes, each preservice teacher also spends an additional 6 clock hours each week with a field professor in a school that differs significantly from his or her primary site. For example, preservice teacher Joe[1] may have Tim's suburban middle school as his primary site, where he works 3 hours every day. On Mondays and Wednesdays, he also spends 3 hours with Bob at Eastland Career Center because he has an interest in the computer technologies and multimedia that characterize Bob's work. Tim's other preservice teachers (each field professor has three preservice teachers as his or her primary responsibility), Megan and Lou work with Jim for 6 hours each week because they want experiences with students who are new immigrants, and Jim's school has a large number of students from other parts of the world. One of the strengths of our PDS Network is that we have urban and suburban sites, middle and high schools, a regional vocational center, two high schools that are members of the Coalition of Essential Schools, and an urban alternative high school. Through our restructuring of PDS methods and student teaching, our preservice teachers now have an intensive 6-month experience in learning to teach global perspectives in our schools.

## Preparing Preservice Teachers in Global Education

What difference can it make for preservice teachers when classroom teachers and university professors collaborate in a PDS Network in global education? In our work together, we find that we have learned to (a) bring together theory and practice in global education; (b) appreciate the power of reflective practice in thinking about, talking about, and questioning what we do as teachers and learners in global perspectives; and (c) benefit from the multiplier effects of collaboration in global education that bring about wider networks and increased understanding of the nature of global education in our teaching and in our lives. We discuss each of these below with some illustrations from our work with preservice teachers.

## Bringing Together Theory and Practice

When we began our work together, conventional wisdom in many schools said that the university's program was "too theoretical," but some university professors perceived classroom teachers as "only interested in materials they can use on Monday." Our preservice teachers often remarked that the ideas they learned about in methods (particularly reflective inquiry and global education) were laughed at or ignored in the schools by teachers who saw them as controversial, impractical, or irrelevant. However, much of the theory of global education—particularly the constructs of perspective consciousness, multiple realities, local/global interconnectedness, global perspectives, cross-cultural understanding, and experiential learning—was very much valued by those of us who became field professors. (See Table 2 for our framework of global education.) Merry originally asked us to work with her *because* she knew we actually teach student-centered global education.

### Table 2

### *Elements in Infusing Global Perspectives in Education*

1. Human Values
    a. universal and diverse human values
    b. perspective consciousness/multiple perspectives
    c. recognition of the effect of one's own value, culture, and worldview in learning about and interacting with people different from oneself
2. Global Systems
    a. economic political, ecological, technological systems
    b. knowledge of global dynamics
    c. procedures and mechanisms in global systems
    d. transactions within and across peoples, nations, regions
    e. interconnections within different global systems
    f. state of the planet awareness
3. Global Issues and Problems
    a. development issues
    b. human rights issues
    c. environmental/natural resource issues
    d. issues related to distribution of wealth, technology and information resources
    e. issues related to dependency
    f. peace and security issues
    g. issues related to prejudice and discrimination (based on ethnicity, race, class, sex, religion, politics, etc.)
4. Global History
    a. acceleration of interdependence over time (J-curves)
    b. antecedents to current issues

Once we took charge of teaching the methods course, we integrated our global education expertise into all the seminars so there was congruence between the ideas of the assigned readings, the biweekly seminars, and the actual middle and high school teaching of global education that the preservice teachers became a part of in our classrooms. Whether we are teaching the preservice teachers approaches to cooperative learning, student research, or higher-level thinking skills, we demonstrate the theory and practice of global education. For example, Shirley and Keith teach a seminar on integrating community resources and service learning. They discuss experiential learning and share illustrations of their students' work with many international organizations and people within Ohio and the world community. During the same month, our preservice teachers are working with Steve as he brings South Africans into Reynoldsburg High School to act as consultants for his 10th graders' exhibition on conflict resolution. They also work with Bob as he develops local-global projects on the Internet at Eastland Career Center and with Tim, whose students are working on a research project in the community with parents from Japan, Eritrea, and India. Shirley and Keith's *theory* of cross-cultural experiential learning is supported by the *practice* of Steve, Bob, and Tim.

From their first week of the PDS methods, the preservice teachers are expected to apply the theory and practice of global education to their development of lesson plans and actual work with our students. Although their efforts are tentative at first, they are continuously assessed on their ability to teach with a global perspective. And they have us, their field professors, helping

### Table 2 (Continued)

- c. origins and development of cultures
- d. contact, borrowing, and diffusion among cultures
- e. evolution of global systems
- f. changes in global systems over time
5. Cross-Cultural Understanding
    - a. recognition of the complexity of cultural diversity
    - b. the role of one's own culture in the world system
    - c. skills and experiences in being one's own culture from others' perspectives
    - d. experiences in learning about another culture and the world from another culture's values and worldviews
    - e. extended experiences with other cultures
6. Awareness of Human Choices
    - a. by individuals, organizations, local communities, nations, regions, economic or political alliances
    - b. past and present actions and future alternatives
    - c. recognition of the complexity of human behavior
7. Development of Analytical and Evaluative Skills
    - a. abilities to collect, analyze, and use information
    - b. critical thinking skills (e.g., ability to detect bias, identify underlying assumptions, etc.)
    - c. recognition of the role of values in inquiry
8. Strategies for Participation and Involvement
    - a. opportunities for making and implementing decisions
    - b. experience with addressing real-life problems
    - c. attention to learning from experience

© Merry M. *Merryfield, 1995*

them every step of the way. We believe we have brought theory and practice together in such a way that our preservice teachers do actually become global educators through their 6 months in our schools.

## Appreciating the Power of Reflective Practice

One of the most fundamental tenets of global education is the importance of analyzing issues from a variety of perspectives, including those of people who may be marginalized culturally, historically, economically, or politically in local, national, or world communities (Hanvey, 1975). Multiple perspectives are essential to comprehending the complexity of an increasingly interconnected world (Merryfield & Remy, 1995). However, our preservice teachers often enter the classroom having had little or no experiences with diversity and, like many Americans, having stereotypical or biased information about people who are different from themselves (Zimpher & Ashburn, 1992). Consequently, our preservice teachers often exhibit ethnocentric, even xenophobic attitudes at the same time that they are learning to teach global perspectives. To move preservice teachers beyond stereotypical "knowledge" and prejudicial attitudes toward students, people in the school community, and the diverse cultures they will teach about in social studies courses, they must be challenged to examine their personal theories of people different from themselves. Left unexamined, preservice teachers' personal beliefs often bias their instruction and interaction with their students in very powerful ways. We have found that reflective practice within a learning community in global education can make a difference.

Two examples come to mind. First, a preservice teacher was presenting a lesson about Vietnam. During the lesson, Ed mentioned that some Vietnamese consider dog meat to be a delicacy. When the students groaned, he responded, "I know! Isn't that gross? I could never eat a dog!" In a discussion with his field professor and university supervisor that followed, we posed the following questions: Why did you choose to share that particular aspect of Vietnamese culture with the students? What aspects of American culture might seem equally "gross" to someone outside the culture? What impression does such a comment leave on your students? How do you think that aspect of the lesson offended the Vietnamese student in your class?

As Ed considered these questions, he was being asked to reflect upon perspectives of different cultural orientations from his own. During his next lesson, Ed and his students discussed elements of American culture that might seem unusual to others and also identified the many similarities between ways of life in the United States and Vietnam. In addition, Ed asked students why people always seem so interested in what is different, strange, or bizarre in other cultures. What followed was a sophisticated analysis of human nature and the "lure of the exotic" (as it was described by one of the seventh-grade participants). In a seminar that followed, Ed shared his thoughts about this teaching experience with those of other preservice teachers who were also struggling with the application of perspectives consciousness and with field professors who shared their experiences, ideas, and the materials they had developed.

Because our preservice teachers are learning to teach within the PDS Network in global education, they perceive reflective practice as a natural part of working with others to improve both teaching and learning. The long-term effects of our emphasis on thinking about global education as we teach has been noted by our graduates and the school systems that hire them. For example, Julie, who graduated from our PDS Network a year ago, was hired as a teacher in a high school in a neighboring district. When she was given her teaching assignment, Julie was also handed a graded course of study (the mandated curriculum) and a textbook. After looking at the text, she asked her department chair, "If this is to be a world history course, why is the American political view the only one presented in the texts?" She explained the need for multiple perspectives to develop the students' skills in perspective consciousness and critical thinking. She received permission to bring in other materials and went on to help her students research the advantages and disadvantages of different policies throughout history from many perspectives. The level of thinking skills jumped to a new level in her class when her students began to evaluate disparate and even conflicting sources of information. Explaining her ability to recognize the need to make changes in the curriculum, Julie spoke of her role in thinking through the effects of the materials upon the learning of her students. Her confidence in her knowledge of global education came from her months of reflective practice within the PDS Network.

## Benefiting From the Multiplier Effects of Collaboration

As we look back an our experiences over the past six years, we recognize that our efforts to bring about congruence between theory and practice in global education with a climate that supports reflective practice and collaboration across schools and districts have had many unforeseen consequences. The synergy of professionals interacting with one another on significant issues valued by the group leads to increased interest in new networks and opportunities for sharing and learning. We have experienced this multiplier effect as we have tackled concerns and issues in curriculum, collaboration, assessment, and competence of beginning teachers. Perhaps an example from our exploration of diversity will illustrate the multiplier effect.

Let's look at several not unrelated events and their impact on our professional growth and networks. While observing a class discussion in a "current issues" class, a field professor and a supervisor noticed that comments regarding the illegal immigration of Mexicans to the United States were quite hostile. Gary, the preservice teacher leading the discussion, seemed to agree with the class consensus that illegal immigrants were taking jobs from U.S. citizens and creating a drain on the social system. As is standard procedure after observing the teaching of a preservice teacher, the field professor and the supervisor met with Gary to discuss his lesson. During the conversation, they asked about his own experiences with Latinos. He described the rural area in which he was raised and the influx of migrant workers who were predominantly "Mexican" and "probably here illegally." Gary noted that many people in his home community routinely made negative, stereotypical comments about the migrant workers. When asked how this type of socialization might be affecting his presentation of materials in the classroom, Gary commented that he had never thought of himself as prejudiced but had to admit that he probably was. The supervisor suggested using two videotapes to present conflicting views of the illegal immigration issue, which she had seen used by another preservice teacher in a different school district. Having viewed both of the tapes, the preservice teacher later admitted that he had felt some cognitive dissonance that helped push his thinking along. He also decided to expand illegal immigration into a unit so he could explore the issue more fully with his students and tackle some of their stereotypes and assumptions.

This connection between personal experiences with diversity and learned prejudice in our own community has become a recurrent theme as other preservice teachers and their field professors have shared similar experiences during seminars and work in the schools. In more than one team meeting, field professors and Merry have discussed what we can do to help our preservice teachers examine their own beliefs about expectations for their African American, Appalachian white, and immigrant students as well as the middle-class white students whom our preservice teachers often perceive as being "like themselves."

In some discussions, we have found our own stereotypes of urban versus suburban schools and students coming forth as we talk. The more we have worked with and thought about the issues of diversity and equity in schools in Central Ohio, the more parallels we have found with global education, especially cross-cultural experiential learning, inequities of power and the role that privilege plays in education. We have learned that field professors and supervisors must play a critical role in countering parochial and ethnocentric attitudes and help the preservice teachers understand the intricate connections between global and multicultural education.

Our ongoing exploration of diversity and prejudice has led to some professional development and networking that probably would not have occurred without our collaboration and attention to theory and practice in global education. Let us share three examples as evidence of the multiplier effect on our ongoing attention to diversity. Recognizing that we have much to learn from one another and our different schools, last year we began to have our OSU supervisors substitute for the field professors so that we could be available to visit one another's schools without any costs to the districts or OSU. Most of the visits have been either suburban to urban or urban to suburban. As a group, we are coming to appreciate and learn from the diversity of other teachers' students and school cultures. Understanding different contexts and approaches to multicultural and global perspectives has led us to rethink some of our assumptions and try out alternatives in instruction and assessment

Second, field professor Barbara Wainer has just completed an article for publication on how to build on the student diversity of urban schools in the teaching of multiple perspectives in global education (Wainer, in process). Explaining why she decided to write the article, Barbara told us that through her work with preservice teachers and her colleagues in the PDS, she has developed approaches to teaching global perspectives that are especially relevant to her urban students. Because of the discussions in our PDS work and the interaction she experienced during a presentation of her ideas at a conference of the American Associate of Colleges for Teacher Education (AACTE), Barbara knows that others will be interested in and benefit from her writing.

Third, professors also grow professionally from such discussions and long-term grappling with significant issues. The PDS Network's dialogue and work on diversity and equity has influenced Merry's thinking and research about the relationships between multicultural and global education within teacher education programs. She is now on sabbatical, looking at a number of teacher education programs where purposeful connections are being made between these two fields (see also Merryfield, 1996). Undoubtedly, she will bring new ideas and materials back to our PDS Network. We know that as an ongoing learning community, we and our preservice teachers benefit from outgrowths of our work together.

Because of these and other experiences, we have developed a better understanding about preparing our own middle and high school students for the challenges and rewards of living in a diverse nation that is intricately connected to a rapidly changing global community. By raising our own consciousness about these issues, our shared experiences have helped us better model such perspectives in a positive manner and support our preservice teachers to move beyond the rhetoric of multicultural and global education. The result of our PDS collaboration is an increased capacity of all of us—preservice teachers, experienced teachers, and university professors and supervisors—to develop more complex understanding of ourselves, our students, and our world and to promote real and meaningful change in our teaching and professional work.

## Suggestions for Teacher Educators

Although the work we have started in our PDS Network is far from complete, the lessons we have learned along the way may provide insights for teacher educators interested in infusing global perspectives into their programs. First, we suggest that teacher educators identify and begin to work with experienced global educators in local schools. Classroom teachers who share your expertise and enthusiasm in global education are the best allies you can have in preparing preservice teachers to teach global perspectives. We believe that field professors must be carefully selected. In our case, the university professor took time to get to know the field professors and had collaborated with several of us on previous projects. She had spent time in our classrooms so she understood our teaching styles. We suggest you select teachers who are dedicated to global education, teach with instructional variety and student-centered approaches, actively seek out opportunities for professional growth, and want to improve the preparation of preservice teachers.

Second, we advise teacher educators to choose school districts and schools that offer diversity for your teacher education program. Diversity in K-12 student population is critical, because your preservice teachers must learn to interact with and teach people who are different from themselves in ethnicity, race, class, national origin, and other characteristics. Diversity in the school structure, mission, and curricula are also powerful factors if the preservice teachers are to learn the significance of context, school reforms, and the role of teachers in bringing about change.

Third, work with the teachers you select to develop a shared vision of excellence in global education and goals that you want to reach. All the people who are involved in the collaboration need to agree on what global education is and what it means in the practice of K-12 teaching. Although the details of the day-to-day teaching of global perspectives may depend on the grade level, the course, or other contextual factors, the field professors and university professors must have some agreements on expectations for preservice teachers and long-term goals of the school/university collaboration. Teachers and professors must work to consensus in developing instruments to assess the beginning teachers' abilities to teach global perspectives.

Fourth, school/university collaboration needs to go beyond preservice teachers to help practicing teachers grow professionally We coordinate full-day global education workshops for colleagues in our schools. We share strategies, materials, and resources across nine schools in six districts, work together on issues of concern, and examine ways in which we can improve our own teaching. Because they recognize the worth of these in-service days, our principals and school districts provide substitutes. We also work together to develop presentations at professional meetings and regularly present our work at meetings of the American Educational Research Association, the American Association of Colleges For Teacher Education, the National Council for the Social Studies, the Ohio Council for the Social Studies, and the American Forum on Global Education. Some of us have also written articles to share what we have learned (Levak, Merryfield, & Wilson, 1993; Merryfield & White, in press; Wainer, in process) and have contributed to textbooks for preservice teachers in global education (Shapiro & Merryfield, 1995).

Fifth, we suggest that there are attributes of professors who work well with teachers in long-term, intensive collaboration in global education. These professors respect and trust teachers and recognize that good global educators have specialized expertise, based on their work with K-12 students, to which professors and preservice teachers need access. They support a flexible vision of global education, where teachers may use different approaches and teaching styles to address shared goals. The professors model excellence in global education in their own teaching and are active in multicultural and global networks in the local community, the state, the United States, and the world. The professors bring both new knowledge and networks to teachers and help teachers share their expertise and become involved in networks that may be new to them. It is especially important that the professors can comfortably share power and decision making with teachers.

Finally, our last piece of advice is (borrowing from Nike), "just do it." We believe that too much planning time can stifle creativity and enthusiasm. School/university collaboration depends to a large extent on the individuals involved, because it is the distinctive mix of people's experiences, expertise, and beliefs about global education that shapes the educational plan and brings it to fruition. We learned about school/university collaboration by doing it. Our sense of ownership in our PDS Network and our shared pride in developing the next generation of global educators keep us excited and motivated. Come, visit us and our PDS Network in Social Studies and Global Education.

## Note

1. We have changed the names of the preservice teachers.

# References

Alger, C. F., & Harf, J. E. (1986). Global education: Why? For whom? About what? In R. E. Freeman (Ed.), *Promising practices in global education: A handbook with case studies* (pp. 1–13). NY: National Council on Foreign Language and International Studies.

Anderson, L. (1990). A rationale for global education. In K. A. Tye (Ed.), *Global education: From thought to action* (pp. 13–34). Alexandria, VA: The Association for Curriculum and Supervision Development.

Bennett, C. I. (1995). *Comprehensive multicultural education: Theory and practice.* Boston: Allyn & Bacon.

Hanvey, R. G. (1975). *An attainable global perspective.* New York: Center for War/Peace Studies.

Kniep, W. M. (1986). Defining a global education by its content. *Social Education, 50*(10), 437–466.

Levak, B., Merryfield, M., & Wilson, R. (1993). Global connections. *Educational Leadership, 51*(1), 73–75.

Merryfield, M. M. (1995) Institutionalizing cross-cultural experiences and international expertise in teacher education: The development and potential of a global education PDS network. *Journal of Teacher Education, 46*(1), 1–9.

Merryfield, M. M. (1996). *Making connections between multicultural and global education: Teacher educators and teacher education programs.* Washington, DC: The American Association of Colleges for Teacher Education.

Merryfield, M. M., & Remy, R. C. (1995). *Teaching about international conflict and peace.* Albany: SUNY Press.

Merryfield, M. M., & White, C. (in press). Teaching issues-centered global education. In R. Evans & D. W. Saxe (Eds.), *Handbook on teaching social issues.* Washington DC: National Council for the Social Studies.

National Council for the Social Studies. (1982). *The NCSS position statement on global education.* Washington, DC: Author.

Shapiro, S., & Merryfield, M. M. (1995). A case study of unit planning in the context of school reform. In M. M. Merryfield & R. C. Remy (Eds.), *Teaching about international conflict and peace* (pp. 41–123). Albany: SUNY Press.

Shulman, J., & Colbert, J. (1979) *The intern teacher casebook.* Washington, DC: The ERIC Center on Teacher Education and the Far West Lab.

Wainer, B. (in process). *Creating a program for social studies preservice teachers that moves them towards teaching from multiple perspectives and a global worldview.*

Zimpher, N. L., & Ashburn, E. A. (1992). Countering parochialism in teacher candidates. In M. E. Dilworth (Ed.), *Diversity in teacher education* (pp. 40–62). San Francisco: Jossey-Bass.

# School Uniforms and Safety

M. SUE STANLEY
*California State University, Long Beach*

School uniforms are one of several strategies being used by this nation's public schools to restore order in the classroom and safety in the school. Recently, it seems as if everyone—from the president to the national media to the local PTA—is talking about the utility of this approach. Why school uniforms? Principals believe that the use of school uniforms can have a positive effect on violence reduction and academic achievement, and can reduce the need for discipline (Portner, 1996). Popular press articles report that school uniforms control violence associated with attending school ("Dress," 1987; Harris, 1989b; Jolly, 1994; Leo, 1990), improve attendance rates (Harris, 1989b), modify behavior (Harris, 1989b; McManus, 1987), improve academic achievement (Pushkar, 1995), reduce the focus on fashion contests (Harris, 1989a; McManus, 1987), and promote ideas and achievement (McManus, 1987).

In his State of the Union message, President Clinton referred to school uniform policies, suggesting they be supported if they help deter violence (Howe, 1996). In a recent visit to Long Beach, California, President Clinton commended Long Beach Unified School District (LBUSD) parents, teachers, administrators, and students for taking the courageous step of implementing a mandatory school uniform policy in their quest to provide a safe and nurturing learning environment ("The Complete Text," 1996). The president's interactions with the LBUSD community culminated in the development and distribution of the *Manual on School Uniforms* (U.S. Department of Education, 1996) to provide guidance on implementing uniform policies within public schools (Howe, 1996).

The interest in encouraging or requiring uniforms for public school students is not a new phenomenon; that trend began in the late 1980s, particularly in inner-city locations. In the fall of 1987, Baltimore, Maryland, and Washington, D.C. experimented with public school uniforms, and 97% of the parents favored them (McManus, 1987). By 1989, 74% of the public schools in Baltimore had implemented policies on school uniforms; in addition, 32 public schools in Washington, D.C., 44 schools in Miami, Florida, and 30 schools in Detroit, Michigan had voluntary or mandatory uniforms at the elementary and middle/junior high school level (Harris, 1989b). Bridgeport, Connecticut also experimented with school uniforms that year ("What About," 1989). By 1990, use of public school uniforms had spread to Chicago, Illinois and New Haven, Connecticut ("Chilling," 1990). Currently, school districts in 10 states have implemented mandatory or voluntary school uniform policies, mostly at the elementary and middle/ junior high school level (U.S. Department of Education, 1996).

In many communities, implementation of policies on school uniforms is just one part of a larger program of school reform. Such school-specific efforts might include introducing conflict resolution skills training in the curriculum, strictly enforcing rules, increasing police surveil-

---

M. Sue Stanley, "School Uniforms and Safety," *Education and Urban Society,* August 1996, Vol. 28, No. 4, pp. 424–435. Reprinted by permission from Sage Publications, Inc. (US), Corwin Press.

lance, locking school doors during school hours, installing metal detectors, or canceling after-school activities and athletic events (McCarthy, 1996; "Regulating," 1994; "Restricting Gang Clothing," 1994).

## Uniforms in Perspective

Uniforms have been worn historically under a variety of circumstances (e.g., nurses, athletic teams, military personnel) and may be associated with positive or negative roles. Experts have pointed out that uniforms can serve several functions, both for the wearer and for those outside the organization (Joseph, 1986; Kaiser, 1985):

- Uniforms act as a symbol of group membership. Group members can identify with each other, and nongroup members can identify group participants.
- Uniforms can reveal roles. These clothing symbols tell observers that the wearer can be categorized as a student, priest, security guard, or a Boy/Girl Scout.
- Uniforms legitimize roles in given situations by certifying membership and role. Persons dressed in fire uniforms would be expected to extinguish fires; those in medical garb would be expected to give assistance to someone injured in an accident.
- Uniforms define group boundaries, promote group goals, and reduce role conflict. Football team members do not compete against each other, but work cooperatively to achieve the goal of winning the game, using behavior appropriate to the football field.

Communities across the United States are considering school uniforms for several reasons. They appear to provide ready solutions to some of the headline-grabbing aspects of school safety, gang violence, weapons in school, and assaults associated with theft of expensive clothing. Gang attire causes rival gangs to be openly hostile to each other and creates an atmosphere of intimidation or disruption; an accidental wearing of the wrong color can put a child's life at risk. Also, in recent years, some youth fashions have encouraged wearing "baggy" clothes; weapons can be, and have been, hidden in oversized pants or overalls, for example. In addition, some children have been robbed, injured, and even killed for their expensive clothing on the way to and from school ("Dress," 1987; Harris, 1989b; Jolly, 1994; Kennedy & Riccardi, 1994; Leo, 1990; Polacheck, 1994; "Regulating," 1994; "Should Public Schools," 1994). These kinds of offenses would probably be favorably affected by encouraging or requiring uniformity in student school wear.

Uniforms are used, or are under consideration, for reasons other than the more obvious links to safety noted above. Kaiser (1985) suggests seven benefits historically cited by proponents of school uniforms; namely, the belief that (a) discipline and (b) respect for the teacher are increased; (c) group spirit is promoted; (d) academic standards are maintained through uniformity; (e) strain on parental budgets is eased and (f) there is a decrease in the race for social status, accompanied by an ability to de-emphasize socioeconomic differences by limiting "fashion statements"; and (g) intruders on the school campus can be more easily identified.

The introduction of uniforms can reduce the emphasis on fashion wars and reinforce the acceptability of more practical, less costly school clothing. Uniforms can ease the strain on parental budgets, a particular advantage in low-income families or during economic times when adults are struggling to do more with less. Uniforms also reduce the use of clothing as an indicator of status and wealth. No logos, no labels, no pressure (Balter, 1994; Harris, 1989a; Polacheck, 1994; "Regulating," 1994; "Should Public Schools," 1994).

In addition to encouraging students to concentrate on learning, rather than on what to wear, uniforms can be social equalizers that promote peer acceptance, as well as school spirit and school pride. By wearing uniforms, students can take ownership in school membership, bridge differences among widely diverse ethnicities and economic levels, and become more unified through the reduction in cultural and ethnic tensions (Kennedy & Riccardi, 1994; McManus, 1987; Polacheck, 1994).

Further, uniforms can support the connection between school, learning, and future success. Appearance is an important part of the nonverbal communication individuals use to establish their credibility in roles. It is suggested that students who come to school "dressed for success" and ready to learn have a higher probability of achieving their goals (Joseph, 1986; Kaiser, 1985; Kennedy & Riccardi, 1994; Polacheck, 1994; "Regulating," 1994; "Should Public Schools," 1994).

However, not everyone is enthusiastic about the use of school uniforms. Some are not convinced that wearing a uniform to school will change the way a child learns (McCarthy, 1996). Some groups say that uniforms are just a fad and will, in the long run, have little effect on education reform ("Goal Seven," 1996). Others maintain that gangs will not go away just because kids wear uniforms to school ("School Uniforms," 1994). It has also been suggested that uniforms may undermine efforts under way in many schools (as well as other institutions to understand and appreciate diversity, Howe, 1996). Also, some sectors of society may be adversely affected by the introduction of school uniform policies. These policies have contributed to reduced sales of standard children's clothing, but a boom in uniform purchases, for example. Nevertheless, some retailers feel profits are a worthwhile sacrifice if resulting actions make kids safer at school (Kennedy & Riccardi, 1994).

## Policy Issues Related To Uniforms

The use of school uniforms by many of the nation's school districts raises a number of policy issues. These include (a) legal concerns about requiring uniforms, (b) the implications for parents who oppose uniform policies, and (c) how to provide uniforms for children in low-income or indigent families.

Local concerns about school uniforms focus on constitutional rights of self-expression and freedom of expression. Private schools have long required their students to wear uniforms. Mandatory use of uniforms in public school settings is a relatively new concept, and one that is largely untested from a legal standpoint; however, the courts have affirmed the legitimacy of stricter dress codes prohibiting student attire that:

- is insulting, vulgar, indecent, or obscene;
- displays messages that are contrary to educational objectives, such as clothing that seemingly endorses underage drinking or drug use (Majestic & Smith, 1995).

Legal opinion varies about how courts might decide on the constitutionality of mandatory public school uniforms. Some suggest that courts that have historically permitted school officials to determine student dress and grooming codes (e.g., hair length) also will uphold school district policies that mandate uniforms. Others believe that courts may give more weight to students' liberty interests before validating such restrictive requirements, unless school systems can document that uniforms actually do improve the learning environment. In general, it appears that mandatory school uniform programs are likely to be permitted to the extent that these are tied to valid educational purposes.

School districts that have introduced uniforms typically have implemented such policies on a voluntary rather than mandatory basis, and some have still garnered more than 90% compliance with voluntary uniform codes (Majestic & Smith, 1995). School districts most commonly have introduced uniforms for elementary or middle-school students, who tend to be more compliant and less concerned about their individuality than adolescent youth in higher grades. Majestic and Smith (1995) suggest this may well be the path of least resistance, because the students are less likely to demand the right to self-expression.

What about parents who are opposed to school uniforms? Some parents who feel uniforms are not appropriate for their children transfer their children to a school where uniforms are not worn ("School Uniforms," 1994). Others take advantage of clauses in district policies and obtain permission for their children to wear standard clothes to school ("School Uniforms," 1994).

One frequently raised concern associated with school uniforms is the financial burden their purchase imposes on low-income families. The U.S. Department of Education *Manual on School Uniforms* (1996) discusses strategies employed by several school districts for making uniforms available to every child who wishes to wear one. For example, local community businesses and charitable organizations have taken leadership roles in providing funds to purchase uniforms. Grants from foundations have also been used to buy uniforms. In some school districts, education funds are allocated for uniform purchases. Uniform recycling centers also keep costs down ("School Uniforms," 1994).

Based on the experiences of the Long Beach Unified School District (LBUSD), Carl Cohn (1996), LBUSD's superintendent of schools, suggests that school districts need to meet five criteria to be ready to implement a school uniform policy. First, a stable school board is needed. The LBUSD board members were enthusiastic supporters of a uniform policy. Second, there cannot be a successful uniform policy without collaborative efforts among parents, schools, and community members. Eighty percent of the Long Beach community favored school uniforms. Third, the district must have the resources to defend the policy against possible legal challenge. Fourth, the proactive, enthusiastic support of principals is necessary for implementation of a uniform policy, as it is for any other systemwide change. And fifth, a mechanism to provide uniforms for children who cannot afford them should be developed. Long Beach schools had the support of the business community and 100 local charities and individuals for this purpose.

To avoid problems that might arise due to adoption of a uniform policy, districts might consider the following:

1. Is everyone involved in the process (i.e., parents, community leaders, businesses, teachers, administrators, and students)?
2. Are students' rights to public safety, religious expression, and freedom of expression protected?
3. Has the ethnic, religious, and socioeconomic composition of the school district been considered when developing the policy?
4. Will uniforms be mandatory or voluntary?
5. Will there be any consequences for not wearing a uniform when parents approve of the uniform policy?
6. Are the apparel items chosen as part of the uniform easy to obtain in the community? Are a wide range of sizes available? Do they have growth features? Are they easy to care for? Are they affordable for most of the community's families?

Communities that decide to implement a uniform policy also need a good, year-round public relations plan to keep the community informed. Posters, flyers, newspaper and newsletter articles, and television ads should include pictures of children in uniforms as well as written uniform criteria. A telephone hotline for parents has been successful in Long Beach. A uniform information table at school registration sites is also helpful.

## Mandatory School Uniforms in Long Beach, California

On July 5, 1994, school uniforms became mandatory in all 70 Long Beach Unified School District elementary and middle schools. This bold step was taken as part of a comprehensive strategy that included, in addition to mandatory uniforms, increased emphasis on basic skills, student accountability, and respect for others (Polacheck, 1994). The rationale developed by the district

Ad-Hoc Advisory Committee to Recommend Procedures to Assure Full Implementation of the Mandatory Uniform Requirement included countering the influence of gangs, minimizing disruption, and improving the learning environment (Eveland, 1994).

Long Beach Unified School District elementary and middle schools chose their uniforms through a school site advisory board consisting of parents, teachers, and principals. The typical school uniform consists of a white shirt or blouse with a collar; dark (navy, red, green, or black) pants, shorts, skirt, or jumper; dark shoes; and a dark jacket, sweatshirt, or sweater. Plaid skirts, shorts, and jumpers are acceptable at some schools. Jeans, T-shirts, and clothes many sizes too large are unacceptable. Typical uniforms cost $70–$90 per year for a set of three (Cohn, 1996).

As part of the program's implementation, each school in the district communicated information to parents about the composition of the school uniform, where to purchase uniforms, how much they cost, how to receive financial assistance to buy uniforms, procedures for exemption to the uniform policy, and compliance incentives. Parents or guardians who preferred not to participate in the uniform policy were asked to complete an Application for Exemption From Uniform Program, submit the application to the child's school administrator, and complete an interview with the school administrator in which objections to participation were discussed (Office of the Superintendent, 1994).

School staff and volunteers were instructed to work closely with families who needed financial assistance with purchasing school uniforms (Office of the Superintendent, 1994). The district compiled a list of businesses, agencies, and individuals in the community who wished to provide uniforms for needy students, and linked them with participating schools,

## The Longitudinal Study of Mandatory School Uniforms

In the spring of 1995, the Long Beach Unified School District began a longitudinal study on mandatory school uniforms. The purpose of the study is to collect empirical data on the impact of school uniforms.

Questions assessing student, teacher, administrator, and parent perceptions of mandatory school uniforms were developed and included in the annual all-district survey. Questions were structured around district goals and the rationale for mandatory school uniforms developed by the district Ad-Hoc Advisory Committee to Recommend Procedures to Assure Full Implementation of the Mandatory Uniform Requirement. Most of the questions were directly linked to the uniform policy. Topics included the subjects' perception of school safety, which are reported and discussed below. In addition, the surveys addressed academic issues (e.g., desire to learn, homework and assignment completion) and other attitudes and behaviors, including perceptions about uniforms (e.g., uniforms making it easier to get dressed in the morning).

All of the elementary and middle-school teachers (2,050), administrators (65), and school counselors (97) present on the day of the all-school evaluation in June 1995 participated in the study. The study also included all middle-school students (12,051) and elementary school students in the 4th and 5th grades (10,325). High school students (9,845) were also surveyed (although they do not wear uniforms) to obtain baseline data regarding perceptions of school uniforms. A telephone survey of parents was also conducted during July and August. This involved a purposive random sample consisting of 966 computer-selected adults who were legally responsible for a child/children attending an elementary and/or middle school within the district. The sample was representative of the ethnic and economic makeup of the school district.

In addition to the survey results, the author was given district counts on the following: (a) reported class disruptions, (b) reported playground acts of violence, (c) suspensions, and (d) reported violations of the dress code. The school district also provided data on selected reported crimes occurring in Grades K–8 in Long Beach public schools during the 1993–1994 and 1995 school years (Cohn, n.d.).

## First-Year Results

Early research findings indicate that Long Beach schools are remarkably safer, although it is not clear that these results are entirely attributable to the uniform policy. District data indicated suspensions decreased when comparing the 1993–1994 and the 1994–1995 academic years. Elementary school suspensions declined from 3,183 in 1993–1994 to 2,278 in 1994–1995, a reduction of 28%. Middle-school suspensions were reduced even more substantially, from 2,813 in 1993–1994 to 1,814 in 1994–1995, a reduction of 36%. Decreases in reported crimes in Grades K-8 over the same 2 school years were even more encouraging:

- Assault/battery decreased 34% (from 319 to 212)
- Assault with a deadly weapon decreased 50% (from 6 to 3)
- Fighting decreased 51% (from 1,135 to 554)
- Sex offenses decreased 74% (from 57 to 15)
- Robbery decreased 65% (from 29 to 10)
- Extortion decreased 60% (from 5 to 2)
- Possession of chemical substances decreased 69% (from 71 to 22)
- Possession of weapons or look-alikes decreased 52% (from 165 to 78); possession of dangerous devices decreased 50% (from 46 to 23)
- Vandalism decreased 18% (from 1,409 to 1,155) (Cohn, n.d.).

Survey results indicate that adults, particularly school administrators, perceived that uniforms had a positive influence on student behavior. In addition, all of the administrators, 85.6% of the counselors, and 66.1% of the elementary and middle-school teachers indicated that there was a safer environment at school. Administrators perceived less class disruption (81.5%), improved student behavior (95.4%), increased student cooperation (78.4%), improved student attitude (90.7%), increased work ethic (78.4%), fewer playground fights (73.9%), fewer suspensions (57%), fewer dress code violations (83%), increased cooperation among students (53.8%), and a small increase in student courtesy (47.7%).

Other adult in the school system also reported positive effects. Counselors perceived students as more cooperative (83.5%). Elementary and middleschool teachers perceived uniforms as influencing fewer student disruptions (49.6%) and improved student behavior (56.9%).

Parents similarly reported positive perceptions regarding uniforms. Nearly 67% of the parents surveyed indicated that they perceived the school environment to be improved. Parents also felt that uniforms influenced increased citizenship grades (67%), increased ease in children getting along (82.4%), and reminded children they were going to school to learn (81.7%).

Students responding to the survey, however, did not perceive uniforms as positively as adults did. The majority of middle-school children indicated uniforms did not reduce fights (80.9%), did not help them fit in at school (76.4%), and did not make them feel a part of the school (68.7%). In addition, 71.2% of the middle-school students reported they did not feel safer going to and from school. Elementary school students similarly reported that uniforms did not make students fight less (77%), but that they did make them feel a part of the school (61%). Half of the elementary school students reported they felt safer going to and from school. High school students did not perceive that uniforms would help them be a part of the school family (62.5%).

First-year results from LBUSD suggest that school uniforms may affect school safety. Adult perceptions and school district data indicated uniforms had a positive influence on student behavior; it was also perceived that schools were safer. Administrators perceived less class disruption, improved student behavior, fewer playground fights, fewer suspensions, and fewer dress code violations. Counselors perceived students as more cooperative. Elementary and middle-school teachers perceived uniforms as influencing fewer student disruptions and improved student behavior. Parents perceived uniforms as influencing increased citizenship grades and an increased ease in children getting along, and indicated schools had an improved environment.

Despite the positive perceptions of adults, students indicated that they did not feel safer going to and from school, nor did most students perceive that uniforms reduce behaviors such as fighting, or that they promoted feelings of belonging to the school.

A variety of factors may have contributed to the discrepancies between adult and student perceptions. In oral responses, some students indicated they were bored with uniforms. By the end of the school year the uniforms, like other apparel items, may have been too small, were discolored and faded, or just worn out. Two independent studies are being developed to look at the differing views of adults and students. Second- and third-year data should give further insights into student perceptions of uniforms.

Data on classroom disruptions, playground violence, suspensions, and dress code violations clearly support the link between school uniforms and school safety. Decreases in crime data also are encouraging. However, caution should be used in interpreting such data. As Paliokas and Rist (1996) suggest, other factors may affect the decrease in crime. These include other changes in the schools or community occurring at the time the uniform policy was implemented (such as changes in community policing practices and introduction of other school security measures). Alternatively, the trend of violence may have been at its peak and ready to decline, regardless of any interventions implemented. Finally, early data may reflect the "Hawthorne effect," with changes resulting from the short-term visibility and attention associated with the new policy.

It is also possible that adult perceptions regarding student behavior, and even some adult responses to student behavior such as imposing school suspensions or other forms of discipline, may reflect adult responses to the wearing of uniforms, rather than actual changes in behavior. If youths are wearing uniforms instead of fashions that may cause adults to perceive them as potentially dangerous—such as styles commonly associated with youths involved in gangs or crime—their behaviors may also be perceived as less threatening. This may explain some of the positive perceptions regarding behavior reported by adults. Adults may also refrain from imposing stringent disciplinary actions, such as suspensions, because they are interpreting behavior, or the intent underlying behavior, differently due to the more socially acceptable appearance of youths wearing uniforms.

Exciting as these results are, it is important to remember that they are *preliminary,* first-year results in a longitudinal study. Nonetheless, the results suggest that uniforms may have a positive impact on school safety. Because they are a low-cost intervention that is unlikely to do harm, it appears they are well worth considering.

# References

Balter, J. (1994, August 14). School-uniform policy in Long Beach, Calif., is one worth copying. *The Seattle Times.*

Chilling the fashion rage. (1990, January 22). *Time.* p. 27.

Cohn, C. A. (1996, February). Mandatory school uniforms: Long Beach's pioneering experience finds safety and economic benefits. *The School Administrator,* pp. 22–25.

Cohn, C. A. (n.d.). *Implementing a mandatory uniform policy: Community and legal implications.* Long Beach, CA: Long Beach Unified School District.

The complete text of Clinton's speech. (1996, February 25). *Long Beach Press-Telegram,* p. A 14.

Dress, right, dress. (1987, September 14). *Time,* p. 76.

Eveland, E. (1994, January 18). *Mandatory uniforms for all elementary and middle schools beginning with the 1994–95 school year and action to obtain legislative authority for such requirement* (Motion passed by the Long Beach Unified School District Board of Education). Long Beach, CA: Long Beach Unified School District.

Goal seven: Safe schools. (1996, March 6). *Daily Report Card,* 6(23),4.

Harris, R. (1989a, November 12). Children who dress for excess. *Los Angeles Times* [Orange County Edition], pp. A1, A26–A28.

Harris, R. (1989b. November 14). Following suit. *Los Angeles Times* [Orange County Edition], pp. E1, E12.

Howe, H. II. (1996, April 3). School uniforms: Leaning toward the Spartans and away from the Athenians. *Education Week,* 15(28), 36, 52.

Jolly, V.(1994, February 10). Garden Grove district ushers in uniforms. *Orange County Register,* p. E1.

Joseph, N. (1986). *Uniforms and nonuniforms: Communication through clothing.* New York: Greenwood.

Kaiser, S. B. (1985). *The social psychology of clothing and personal adornment.* New York: Macmillan.

Kennedy, J. M., & Riccardi, N. (1994). Clothes make the student, schools decide. *Los Angeles Times,* pp. A1, A23.

Leo, J. (1990, April 30). The well-heeled drug runner. *U.S. News & World Report,* p. 20.

Majestic, A. L., & Smith, T. (1995, March). Student dress codes in the 1990s. In A. A. Blumberg, R. T. Dowling, J. L. Horton, M. E Howie, A. L. Majestic, R. A. Schwartz, B. C. Shaw, & B. W. Smith (Eds.), *Legal guidelines for curbing school violence* (pp. 55–62). Alexandria, VA: National School Boards Association.

McCarthy, C. (1996, March 16). Uniforms aren't the answer. *The Washington Post,* p. A17.

McManus, K. (1987, October 5). Uniform idea for student bodies. *Insight,* pp. 56–57.

Office of the Superintendent, Long Beach Unified School District. (1994, August 11). *Guidelines and regulations for implementing the mandatory uniform policy in grades kindergarten through eight.* Long Beach, CA: Long Beach Unified School District.

Paliokas, K. L., & Rist. R. C. (1996, April 3). School uniforms: Do they reduce violence—or just make us feel better? *Education Week,* 15(28), 36, 52.

Polacheck, K. (1994, July 5). Uniforms help solve many school problems. *Long Beach Press-- Telegram,* p. D5.

Portner, J. (1996, March 6). Department to issue guidelines on school uniforms. *Education Week,* 15(24), 27.

Pushkar, L. (1995, January 17). Dressed for success: Uniform solution to crimes of fashion. *Village Voice,* p. 12.

Regulating student appearance: A new trend. (1994, March). *School Safety Update,* p. 6.

Restricting gang clothing in the public schools. (1994, March). *School Safety Update,* pp. 1–4.

School uniforms growing in favor in California. (1994, September 3). *New York Times,* p. 8.

Should public schools use uniforms to deter violence? (1994, March). *School Safety Update,* p. 8.

U.S. Department of Education. (1996). *Manual on school uniforms.* Washington, DC: Author.

What about school dress codes? (1989, November 13). *Long Beach Press-Telegram,* p. A5.

# Service Learning

## An Introduction to Its Theory, Practice, and Effects

RICHARD J. KRAFT
*University of Colorado at Boulder*

After a decade characterized as narcissistic and individualistic (Bellah, Madsen, Sullivan, Swidler, & Tipton, 1985), the 1990s appeared to be destined as a time with strong communication and service overtones. President Bush's Points of Light campaign, the congressional passage of the National and Community Service Act in 1990, and President Clinton's National Service Trust Act of 1993, were preceded by the foundation of the national Campus Compact by college and university presidents in 1985. Throughout the 1980s, state and local boards of education and hundreds of schools across the country began service-learning programs or required community service for graduation. The decline in volunteerism on college campuses was halted as growing numbers of young adults again found meaning in giving back to their communities.

Although volunteerism has a long and honorable history in American society, community service has often come to mean a court-ordered sentence for misdemeanors. Civic, or citizenship, education has theoretically been part of the school social studies curriculum for a century, but with increasing youth violence and other social pathologies, it has received increased attention in recent years. Until the Republican congressional victories in the 1994 off-year election, there had been a growing acceptance of and coalescence around the concepts of service learning. It remains to be seen how much of the current movement is dependent upon federal funding, and how much will remain as part of a broader movement toward the rebuilding of community and the reform of public education.

In this brief introduction to service learning, an attempt is made to lay out some of the historical and sociological antecedents to the current movement, to define some of the related terms, and to report on some of the findings of research and evaluation on the effects that it has on participants and society.

Although service learning and voluntary service in general are not uniquely American concepts, they have taken root in our schools and the broader society in new and powerful ways. It is important to understand some of the historical antecedents of the current movement and the sociological explanations for them as we approach the turn of the century. Bellah et al. (1985) documented in *Habits of the Heart* the constant competing pressures in American society, throughout our history, between what they term *individualism* and *commitment*. They express this dilemma in the following way:

---

Richard J. Kraft, "Service Learning: An Introduction to Its Theory, Practice, and Effects," *Education and Urban Society,* February 1996, Vol. 28, No. 2, pp. 131–159. Reprinted by permission from Sage Publications, Inc. (US), Corwin Press.

> We found the classic polarities of American individualism still operating: the deep desire for autonomy and self-reliance combined with an equally deep conviction that life has no meaning unless shared with others in the context of community. (p. 150)

The oft-quoted and seldom-read de Tocqueville (1835/1969) held in the 1830s that in traditional European societies one's status and role was carefully delineated and people knew where they stood in relationship to others, whereas in the United States, ties between individuals were more casual and transient, in part because of their "restlessness in the midst of prosperity," and because Americans "never stop thinking of the good things they have not got" (p. 565). This restlessness, according to de Tocqueville, is intensified by the "competition of all" (p. 536). Whereas Americans could be characterized as perhaps the most individualistic of all peoples, de Tocqueville went on to show the near equal importance that we as a people place on "being with" others in social relationship (p. 538). Whereas our ancestors may have felt oppressed by the civic, religious, and moral cultures from which they fled, they almost immediately formed similar associations on landing in the New World.

Bellah et al. (1985) conclude that implicit in this penchant for getting involved is the peculiarly American notion of the relationship between self and society. Individuals are expected to get involved—to choose for themselves to join social groups. Barber (1992) and other observers of contemporary society conclude that in the last half century, individualism has triumphed over commitment, citizenship demands, and civic responsibility, and that only as we rebuild a sense of community will we be able to rebalance the two poles of our national dilemma. He and others have concluded that community service, citizenship education, and service learning are crucial to the survival of American society.

## Historical Background

It is difficult to know where to trace the beginnings of service learning, but if one accepts its antecedents in voluntary service to community, then it can be easily traced to its Judeo-Christian roots, though certainly not exclusively to these two great world religions. Dass and Bush (1992) have connected the roots of helping to the Hindu tradition, and others have presented evidence from other world religions.

Whereas philosophers before him certainly confronted questions of the "good" and "living in community," most scholars trace the tying of service to schooling to the writings of Dewey (1902). His concept of "associated living" as the basis for both education and democracy was a precursor of much later writing about rebuilding the connections between the school and community. In his classic works, *Experience and Education* (1938/1963) and *Democracy and Education* (1916), he provided the intellectual undergirdings for such critical service-learning components as student involvement in the construction of learning objectives; working together rather than in isolation on learning tasks; using "educative" and minimizing "miseducative" experiences; the organic relation between what is learned and personal experience; the importance of social and not just intellectual development; and the value of actions directed toward the welfare of others.

James (1910) stated:

> What we need to discover in the social realm is the moral equivalent of war; something heroic that will speak to man as universally as war does, and yet will be as compatible with their spiritual selves as war has proved to be incompatible. (p. 17)

This "moral equivalent of war" theme has been struck by progressive and experiential educators for most of this century and most recently by service-learning advocates, who believe that through service learning, education can again be a moral force, something too often lacking in traditional schooling.

More directly related to service learning, Kilpatrick (1918), a Dewey disciple and leader of the Progressive movement, advocated the adoption of the "project method" as a major curricular and pedagogical tool of education. Social reform, education outside the school setting, and real-life problems became the focus for many progressive schools between the First and Second World Wars. Also during the interim between the wars, the Civilian Conservation Corps, although primarily a youth unemployment program, became a forerunner for countless youth service programs and corps in the 1980s and 1990s. Classic works such as Count's (1932) *Dare the Schools Build a New Social Order?* and Hanna's (1937) *Youth Serves the Community* provided additional underpinnings to the service-learning movement over a half century later.

Although the 1950s are generally not known or remembered as a period of reform or progressivism, the Citizenship Education Project (CEP) at Teacher's College set the framework for a wide variety of "active learning," "community studies," and social and political action programs that came to renewed popularity in the 1970s. Many of the ideas developed and updated by Newman and Rutter (1986) and by Barber (1992) could be found in the CEP materials from the quiescent 1950s.

Despite a few such curricular efforts, little in the way of gains was made during the 1950s and 1960s, but with the 1970s came a host of state and national reports on educational reform and the need to escape the passivity of schooling and the "irrelevance" of school to either students or the broader society. The Panel on Youth of the President's Science Advisory Committee (1972), the National Committee on Secondary Education's (1972) *American Youth in the Mid-Seventies*, the Carnegie Commission on Higher Education (1973), Coleman's (1974) *Youth: Transition to Adulthood,* the National Manpower Institute (1975), the National Panel on High School and Adolescent Education (1976), and Martin's (1976) *The Education of Adolescents* made a host of proposals on a range of topics: service programs; experience-based learning; job preparation; service graduation requirements; real and meaningful tasks; interaction with a greater range of people; reintegration of the young into the community. Little in the way of broad reform, however, was started until the publication of *A Nation at Risk* (National Commission on Excellence in Education, 1983), and by then the pendulum had swung away from the "progressive" aspects of the 1970s reports and had returned to a focus on the basics.

In spite of the emphasis on basics in many of the reform documents in the 1980s, there still were a considerable number of commissions and influential individuals calling for one form or another of community service. Goodlad (1984), in *A Place Called School,* included community service, as did Boyer (1983) in *High School,* which called for a service requirement for graduation from high school. Two Carnegie reports (Harrison, 1987; Carnegie Task Force on Education of Young Adolescents, 1989) focused on the needs of middle school youth and also called for service opportunities.

Suffice it to say that service learning is the most recent manifestation of what is now almost a 100-year history of American educational reform attempt to bring the school and community back together, to build or rebuild a citizenship ethic in our young people, and to bring more active forms of learning to our schools. Service-learning advocates are generally careful to not claim the movement as a panacea for all that ails American schools. With strong evidence that classroom pedagogy and curriculum has not changed significantly in the 100 years since Dewey began to call for reform, there are those who see the service-learning component of citizenship education, caring, community building, and active pedagogies as a possible "Sleeping Giant of School Reform" (Nathan and Kielsmeier, 1991). Whereas national and state commissions have provided political support for service learning, it has been primarily a grassroots movement, with thousands of teachers and professors discovering the power of the methodology and using it with their classes, or developing programs, often with little or no money or external support. This grassroots nature would appear to indicate a longer life than the more typical autocratic, top-down reforms that have generally failed in recent decades.

## Definition of Terms

*Volunteerism* has a long and honorable tradition in our society and generally refers to the millions of citizens who perform some service or good work of their own free will and without pay. Scouting, coaching, church work, community food share, Meals on Wheels, crisis lines, and thousands of other voluntary organizations and opportunities make up the voluntary sector.

Whereas service learning has obvious roots in traditional volunteerism, it is the traditional volunteer ethic that has proved troublesome for those seeking to bring service learning into the schools. When states, schools, or colleges mandate service for graduation or as part of course requirements, there has been a public outcry about the oxymoronic nature of "mandatory volunteerism." Only as service-learning advocates have more carefully defined it as a pedagogical tool, rather than as voluntary activity, has the negative rhetoric lessened, though it certainly has not yet disappeared.

*Community Service,* as indicated earlier, has several meanings. Those familiar with the criminal justice system recognize the punitive aspects of its current meaning, whereby thousands of adolescent and adult offenders are sentenced each year to picking up trash or doing other menial tasks in the community in exchange for jail time. Because of the negative and punitive connotations of the words, those within the service-learning movement have generally abandoned the use of the term community service. In the school definition of the word, it has generally meant volunteering in the community, although that might also include a range of tutoring or other programs on the school campus. Volunteering alone generally is differentiated from service learning by having an emphasis on service without a formal, structured learning component. *Community-based learning* also involves learning that occurs out in the community through outdoor education, field trips, internships, or apprenticeships, but it generally does not involve any service component.

Many schools have a wide range of *peer-helping* programs. These are generally cocurricular and involve students in peer or cross-age helping services. Although such programs are good examples of many aspects of service learning, they tend to be separated from the regular curricular subject areas. Among the many such programs begun or continued in recent years with state or federal service-learning funding have been conflict mediation, peer and cross-age tutoring, health and drug education programs, and counseling programs

## Service Learning: Definition and Principles

Whereas there is still much discussion in the field about what actually constitutes service learning, the Commission on National and Community Service (CNCS, 1993) provides perhaps the most widely accepted definition:

A service learning program provides educational experiences:

a. under which students learn and develop through active participation in thoughtfully organized service experiences that meet actual community needs and that are coordinated in collaboration with school and community;
b. that are integrated into the students' academic curriculum or provides structured time for a student to think, talk, or write about what the student did and saw during the actual service activity;

c. that provide a student with opportunities to use newly-acquired skills and knowledge in real-life situations in their own communities; and
d. that enhance what is taught in school by extending student learning beyond the classroom and into the community and helps to foster the development of a sense of caring for others. (p. 15)

Other definitions of service learning speak to the blending of both service and learning goals in such a way that both occur and are enriched by the other. Most service-learning advocates also emphasize the importance of a reflective component where students use higher order thinking skills to better understand and extend the formal learning from the service experience.

The principles that often guide the creation of service-learning programs are those created by the Johnson Foundation (1989) in the *Wingspread Special Report.* In the report are 10 principles preceded by the following preamble:

> We are a nation founded upon active citizenship and participation in community life. We have always believed that individuals can and should serve. It is crucial that service toward the common good be combined with reflective learning to assure that service programs of high quality can be created and sustained over time, and to help individuals appreciate how service can be a significant and ongoing part of life. Service, combined with learning, adds value to each and transforms both. Those who serve and those who are served are thus able to develop the informed judgment, imagination, and skills that lead to a greater capacity to contribute to the common good. (p. 1)

The principles that follow claim to provide criteria for effective programming. The resulting model is one that:

1. Engages people in responsible and challenging actions for the common good.
2. Provides structured opportunities for people to reflect critically on their service experience.
3. Articulates clear service and learning goals for everyone involved.
4. Allows for those with needs to define those needs.
5. Clarifies the responsibilities of each person and organization involved.
6. Matches service providers and service needs through a process that recognizes changing circumstance.
7. Expects genuine, active, and sustained organizational commitment.
8. Includes training, supervision, monitoring, support, recognition, and evaluation to meet service and learning goals.
9. Insures that the time commitment for service and learning is flexible, appropriate, and in the best interests of all involved.
10. Is committed to program participation by and with diverse populations. (Johnson Foundation, 1989, pp. 2–3)

## Critical Questions about Service and Service Learning

Although these principles have been widely disseminated and accepted, it is important to raise some critical questions about them. Principle 1 states that effective service learning "engages people in responsible and challenging actions for the common good." How is this common good determined? Are agencies identifying and representing individual strengths and weaknesses of their clients in service-learning projects with the same rigor that students are being assessed and represented by the schools? Is the common good a service ethic that responds to the equally important needs of both partners in service or is it simply a chance to cure the ills of society by one group serving people whose needs are collectively predetermined? The common good must reflect an empowering benefit for both partners in the service relationship. Both partners should be provided an opportunity to feel responsible and challenged as the service project is designed to enhance the individual growth of all partners: *That* is common good.

Principle 2 states that effective service learning "provides structured opportunities for people to reflect critically on their service experience." Based on the service-learning models in literature, "people" in this statement refers exclusively to the student: Students are required to keep journals that allow them opportunities to reflect upon their experiences in the service project. The *Wingspread Special Report* (Johnson Foundation, 1989) elaborates:

> This reflective component allows for intellectual growth and the development of skills in critical thinking. It is most useful when it is intentional and continuous throughout the experience, and when opportunity for feedback is provided. Ideally, feedback will come from those persons being served, as well as from peers and program leaders. (p. 25)

Besides being a blatant exclusion of the partner in service, this principle merely suggests rather than requires a discourse between the service partners. It assumes that the student engages in service, thinks about what he or she has done, writes down reflections, then hopefully receives feedback on these observations. Without a foundation grounded in the quest for shared understanding, only the student is encouraged to reflect, and he or she may do so in a vacuum.

In research on service learning conducted as a result of CNCS venture grants in Colorado (Maybach, 1994), reflection for or with service recipients was reported in only 1% of the grant recipients' projects, and in only 4% of the projects was discourse encouraged among students and recipients regarding the effects and/or design of the service. Discourse throughout the project between service partners should be the hallmark of the service experience. Freire (1970) makes the point:

> For us, however, the requirement is seen not in terms of explaining to, but rather dialoguing with the people about their actions. The pedagogy of the oppressed, which is a pedagogy of the people engaged in the fight for their own liberation, has its roots here. (p. 35)

> The important thing, from the view of libertarian education, is for people to come to feel like masters of their thinking by discussing the thinking and views of the world explicitly or implicitly manifest in their own suggestions and those of their comrades. Because this view of education starts with the conviction that it cannot present its own program but must search for this program dialogically with people, it serves to introduce the pedagogy of the oppressed, in the elaboration of which the oppressed must participate. (p. 105)

The opportunities for cross-cultural learning are greatly enhanced if the service partners are engaged in written and verbal reflection that is shared with each other throughout the service experience. In this interactive, dialogical form of reflection, individuals can explore each other's opinions, thoughts, desires, and perspectives. Noddings (1992) writes that "dialogue is a common search for understanding, empathy, or appreciation. . . . It is always a genuine quest for something undetermined at the beginning" (p. 23). Without this emphasis on dialogue between individuals, service learning again becomes one-sided, focusing on the isolated views and perceptions of the student without true understanding of each individual's perspective. Misunderstandings and missed opportunities for learning can occur in isolated reflection.

Principles 3 through 10 are closer to acknowledging more or all voices in the service relationship. However, as Cruz (1994) points out, a perspective of diversity must be the lens from which these principles should be conceptualized. Emphasized in these principles is the need to include all participants in the goals, outcomes, process, evaluation, and publicity of service-learning ventures. As the report expands upon these principles, it continues to frame the involvement of service partners as service providers and recipients. As previously stated, this focus does not go far enough in acknowledging the strengths of the served, and it does not adequately address the barriers keeping these individuals from fully participating in society. The concept of "partners in service" needs to be embraced in the principles themselves for programs to emulate a paragon of equality.

Which vision of service are we supporting through service learning? Do we only go so far to embrace a vision of service in which people are serving others in need? Or do we expand upon that vision to include a more empowering model of service that acknowledges the strengths and weaknesses we all bring to a relationship, a vision that moves people away from the margins of society through partnerships based on equal concern, equal voice, equal opportunity to serve and explore new perspectives, a vision that allows each partner to learn from the other, an opportunity for every participant to grow in an environment that nurtures social, cognitive, emotional, physical, spiritual, and occupational growth. The latter should be our service ethic. It is only when equal consideration and voice is acknowledged from all members of a community that we begin to move toward a truly democratic society. In an effective service-learning model, no voice is silenced, no role is invisible.

The questions above must be revisited if we are to equally address the needs of all individuals in the service relationship. It is only through this equal consideration that service can truly be mutually beneficial: to allow for growth in both the student and his or her partner in service. What is needed is a new paradigm of service learning in which the service ethic involves students engaged in projects that do not focus solely on the learning and growth of the student but that focus also on the voice and empowerment of the individual involved with the student in service.

## The Practice of Service Learning

Whereas some agreement on the definition of service learning has been achieved in recent years, its practice in schools and colleges varies widely. Many of the practices do not strictly fit the definition and guidelines described above but still are listed under the general rubric. Cocurricular service activities with special clubs or through a volunteer clearinghouse are generally not directly connected to the curriculum of the school, although students can often receive academic credit for their involvement in the community. One of the most dominant forms is that of individual or group class projects. When they are carefully tied to curricular objectives, contain academic content, involve the student in reflection, and contain an evaluative component, they can be considered service learning. If these components are missing, they fit more comfortably into community-based learning or volunteerism. The following list, prepared by the National Youth Leadership Council (Cairns & Kielsmeier, 1991), is exemplary of the types of projects carried out by literally hundreds of thousands of school and college students today:

> bicycle shop, Big Buddies, blood drive, board membership, building projects, clothes collection, community education classes, community history, cooking meals, crisis centers, day care, emergency services, environmental research, environmental cleanup, fund-raising for charities, gardens, helping the homebound, home chores, hot lines, Meals on Wheels, overseas volunteers, paint-a-thons, peer helpers and tutors, performing arts, planting trees, public awareness, public media, reading for the blind, recreation programs, recycling, research, special equipment, Special Olympics, tax preparation, tutoring, victim aid, visiting institutionalized people, visual art, voter education, youth agencies, youth leadership, and youth sports.

Service within the school is one of the largest forms of service learning, with numerous opportunities for students to tutor, counsel, mediate conflicts, mentor, and address other needs within the school community. Often these activities tend to be internships rather than service learning, or to involve minimal skills development with little or no connection to the curriculum. Many teachers use service as an extension of their regular classroom. Service-learning activities become a means for completing course requirements and going into greater depth into a topic. Community service classes and programs can be found in many schools, with colleges even developing complete interdisciplinary majors with service as the focus.

The ultimate goal of many in the service-learning movement is to have service integrated into the curriculum of all subjects and at all levels. It thus becomes an ongoing part of the curriculum, pedagogy, evaluation, and ethic of the class or school, and not an add-on, dependent on external funding or the particular philosophical whims of the teacher, administration, or board. This goal has been reached in a minority of colleges, particularly many religiously based institutions that have a long history of service as part of their educational philosophy. It is also becoming increasingly common in the elementary and middle schools, where greater philosophical compatibility exists than is generally true at the high school level. Active, interdisciplinary, community-based, cooperative, and other forms of learning have long been part of primary education and are now deeply embedded in the more recent "middle school philosophy." The pressures for college admission, academic rigor, and success on standardized examinations make the high school perhaps the weakest link in the current service-learning movement.

## The Extent of Service Learning in American Schools and Colleges

It is difficult, if not impossible, to give an exact number of schools and colleges that have developed service-learning programs of one form or another. The following statistics are taken from the 1994 report of Abt Associates (Melchior, Jastrzab, Bailis, & Frees, 1994), the national evaluator for CNCS, which awarded $64 million in grants to 150 states, colleges, community-based organizations, Indian tribes, and other institutions. These organizations, in turn, distributed the funds to literally thousands of school, college, and community organizations. Among the many findings of this first-year report to Congress are those enumerated in the list that follows. The statistics do not include the thousands of programs that involve students in service to their communities without external funding, but the data do give a snapshot of the extent to which service learning is now going on in our schools and society.

1. Approximately 200,000 young people and adults took part in ongoing community service programs. An additional 45,000 took part in short-term or one-time event.
2. CNCS-funded programs generated nearly 6 million hours of community service.
3. CNCS-funded programs also involved participants in another 4.2 million hours of nonservice activities. These activities were generally basic education and/or service learning.
4. The average hours of direct service per participant varied widely among the programs. Serve-America (K-12 programs) participants provided an average of 16 hours of direct service, compared to 39 hours in Higher Education programs and averages of 344 and 507 hours in Service Corps and National Service Demonstration programs, respectively.
5. Community service programs engaged a broad range of individuals, from kindergarten students to senior volunteers, representing a diverse array of racial, ethnic, educational, and economic backgrounds. The vast majority of participants were school-aged youth and young adults. Of the total, 56% were women, 36% were non-White, and 19% were economically disadvantaged.

6. Community service programs also provided a broad range of services. Approximately 40% of service hours were focused on conservation and environmental projects and 22% on education and human needs.
7. Federal CNCS funds were matched significantly from other sources. Local programs contributed an additional $1.38 from other government and nongovernment sources for every $ 1.00 of federal CNCS funds.

## Research and Evaluation of Service Learning

Whereas service learning, community service, and volunteer programs have been a part of schools and colleges in the United States for decades, and there have been a range of research and evaluation studies, there is a general lack of solid evidence on the effect of these programs. One of the major difficulties in evaluating or researching service-learning programs is the lack of agreement on what is meant by the term service learning and exactly what it is meant to accomplish. Whereas some programs emphasize social growth, character development, or civic responsibility, others attempt to study psychological development and effects of programs on self-concept. Moral judgment studies have sought to evaluate the effects of service on moral and ego development, and other studies have attempted to measure the effects of service on the broader community. Perhaps the most difficult arena has been in the area of intellectual, cognitive, and academic effects. It has been difficult to design tight experiments to isolate the effects of service on specific academic achievements. A recent experimental study (Markus, Howard, & King, 1993) of students in a university political science course provides some of the first evidence of the positive academic effects of service learning.

A challenge for evaluators and researchers in the field is the dramatically different nature and duration of the programs that go under the guise of service learning. It is difficult to compare one-term service events for a group of 8-year-olds in an elementary classroom with full-time, paid programs for young adults in conservation corps. In-depth, semester-long academic courses in international settings for college students differ greatly from once-a-week volunteer visits to a senior center. Yet all of these can and do meet basic criteria for service learning.

## General Surveys

Krug's (1991) research was on the effects of service learning on four groups of high school young people: at-risk youth in a special program, student assistants (primarily minority) within the school, nature guides, and tutors at a primary school. Preliminary results indicate that although all the experimental groups gained on measures of potency, activity involvement in the community, self-concept, and other factors, the statistically significant growth at the .01 and .05 levels was found almost exclusively with the at-risk and minority young people. The control group, as predicted, did not change on the pre- and post-instruments.

Newman and Rutter (1986) estimated that in 1984, approximately 27% of all high schools offered some forms of service program, involving 900,000 students in 5,400 schools. Service took on the form of (a) school clubs or cocurricular organizations; (b) service-learning credit or requirement; (c) a laboratory for an existing course; (d) a service-learning class; or (e) a schoolwide focus. Nonpublic schools were more likely to offer service, and suburban and large schools did so more often than urban, rural, or small schools. Alternative public and Catholic schools were more likely to offer service than were regular public or non-Catholic private schools. An estimated 6.6% of all high school students were involved in 1984, with 2.3% of their activities tied to the curriculum. This compared with 52% of seniors involved in team sports and 34% in the performing arts. Time spent was an average of 4 hours per week across all programs, and 6 hours in elective programs. Those with service as a high school graduation requirement spent 1 hour per week. Schools where a majority of students were non-White were more likely to offer programs than White majority schools; they also were three times as likely to offer community service as an elective course and to award academic credit. Programs involved students in near-equal proportion from the college prep, general, and vocational tracks. At-risk students and those with behavioral problems were found to be nonexistent in service programs. Thirty-four percent of programs were in schools, not in the community.

Harrison (1987) reported that among voluntary programs, most (61%) involved less than 10% of the student body. Ninety percent of the students put in less than 200 hours, about half the time required by one season of high school football. Sixty-five percent of service programs were within the schools themselves.

Whatever the actual numbers of students involved in service learning might be, the surest conclusion that can be drawn is that school-based service learning is an educational concept that has endured throughout this century but has not yet become an integral part of the high school experience for more than a small group of students. In addition, few programs involve participation by at-risk and minority youth, and a majority of school-sponsored programs are focused on college-bound White students.

## Social Growth

Riecken (1952) studied college students involved in 2 months of intensive, full-time summer experience, designed to strengthen humanitarian ideals by having youths participate in physically useful labor in an economically deprived community. Using a questionnaire, he discovered that participants became less prejudiced, more democratic, less authoritarian, and more service oriented, and they developed greater ego strength.

Smith (1966), in a study of 44 Peace Corps volunteers who taught in Ghana during a period of 2 years, discovered that after the first year in which the volunteers displayed initial and perhaps naive optimism, a more reasoned but no less committed moralistic philosophy emerged. They demonstrated more realism, autonomy, and independence, and significantly increased levels of self-worth and insight. In addition, they became more service oriented in terms of their own career aspirations.

Hunt and Hardt (1969) found that both White and Black groups in a Project Upward Bound, precollege enrichment program for high school students achieved nearly identical increases in motivation, self-esteem, and academic achievement. Other researchers have indicated positive results in social growth from less intensive school service programs. Marsh (1973) concluded that participation in community affairs as part of a high school experimental course increased, as did interest in political activities and a desire to support political issues.

Using a model based on Mosher's moral education, Newman's citizen education, and Hampden-Turner's psychosocial development, Bourgeois (1978) concluded that democratic values were accepted by young teenagers, that an urgency for personal competence existed, and that community activities helped to develop civic competence.

Wilson (1974) examined open-mindedness and a sense of political efficacy in a community-based alternative education program. Wilson concluded that because the learning environment became one of openness, changed authority relationships between students and teachers, and student self-selection of the subject matter and process of curriculum, greater open-mindedness and political efficacy on the part of participants were able to occur.

Corbett (1977) studied the effects of high school students' participation in a yearlong community program that aimed to develop student commitment to the solution of social problems. He found that during the first year when the program was teacher centered and teacher directed, gains in student moral and psychosocial development were nonsignificant, but in the second year, when it became student centered and reflective in nature, significant gains on personality measures and emotional and task competence were found. He concluded that students who worked with individuals in providing service developed more commitment to the solution of social problems than did the students whose volunteer work was focused upon group situations.

Stockhaus (1976) sought to determine if 20 hours of helping in social service agencies would positively affect self-esteem, political efficacy, social responsibility, and community responsibility in high school seniors. Stockhaus found that participants in one school developed a greater sense of social responsibility, community responsibility, and altruism than did nonparticipants and controls, but that strong support for community involvement programs to bring about positive changes in citizenship attitudes was lacking. Changes were too small to be of practical significance.

Broudy (1977) delineated problems that limited the effective development of moral/citizenship, experiential, and service-learning programs in the public schools. They included heterogeneity of values and lifestyles, discrepancies between educational objectives and community behaviors, discrepancies between structured classroom teaching and students' informal community learning, and community experiences of differing intensity and quality. Conrad and Hedin (1982) found that students in service and other experiential programs developed more favorable attitudes toward adults and also toward the type of organizations and people with whom they were involved.

Luchs (1981) reported that high school students involved in community service gained a more positive attitude toward others, a greater sense of efficacy, and higher self-esteem than nonparticipating comparison students. Calabrese and Schumer (1986) reported lower levels of alienation and isolation, and fewer disciplinary problems among junior high school youth involved in service as part of a program for students with behavioral difficulties.

In summary, the research findings on social outcomes as a result of students' involvement in experiential and service-learning programs are mixed. Intensive, full-time, communal living programs have generally proven to be more successful in changing attitudes; these programs also have usually included older students who may already have committed themselves to achieving program objectives, primarily because they entered the programs as volunteers. Too many of the studies suffer from small sample size, lack of strict controls, the effect of previous volunteer experiences on the part of students, and the uneven quality of students' experiences in the program.

## The Impact on Psychological Development

A number of research studies have concentrated on the student's psychological development as a result of participation in experiential education and service-learning programs. Taking full responsibility for one's own actions, developing a sense of self-esteem and ego strength, reaching a high level of moral reasoning, and becoming psychologically mature were seen to be key determinants for success in school and for active involvement in positive citizenship (Stockhaus, 1976). Unfortunately, traditional school curricula frequently not only do not promote these aims but, conversely, appear to affect them negatively (Bidwell, 1965; Coleman, 1961; Cusick, 1973; Goodlad & Klein, 1990; Jackson, 1990; Martin, 1975; Silberman, 1990; Sturges, 1979).

Advocates of experiential education and service-learning programs believe that development of psychological strength will occur more strongly in such programs than in traditional school programs (Coleman, 1974; Conrad & Hedin, 1982; Dewey, 1938/1963; Erikson, 1968; Frankena, 1965; Kohlberg, 1970; Piagét, 1970; Rich, 1962; Rogers, 1969; Schwebel & Ralph, 1973).

Bontempo (1979) conducted field interviews with students and coordinators, and studied program documents from the various schools. Her conclusions were that this type of learning was clearly grounded in consistent philosophies of learning and was making valuable and extensive use of community resources in students' education. Students who were enrolled demonstrated positive self-concepts and increased feelings of self-worth.

Kazunga (1978) concluded that voluntary youth helping experiences promoted a more positive self-concept among youth and significantly helped to improve the community.

Sager (1973) studied 22 high school seniors who volunteered for 9 weeks during their summer vacations at state hospitals. Young people increased their self-esteem and self-confidence significantly on 30 of 34 subscales on seven personality inventories. In addition, they were more self-accepting and felt more adequate and worthwhile in interactions with their peers and with the persons they were helping.

Kelly (1989) found that therapeutic helping behavior generated positive changes in self-concept and other self-perceptive dimensions on the part of the helper. He found that students who helped on a one-to-one personal level underwent significantly greater positive changes in self-concept and other related measures than did those in more general types of service activities.

To determine whether self-concept of students who had experienced school behavioral problems of apathy, vandalism, and delinquency would be improved by enrollment in a voluntary curriculum with a traditional school setting, Martin (1977) employed a case study approach to a yearlong study of 30 male and female high school students. By the end of the year, student behavior had positively changed as measured by teacher interviews and by students' own self-reflections as reported to the researcher. Both teachers and students believed that students had also developed more positive self-concepts as they changed their former negative behaviors.

Exum (1978), in addition to investigating interpersonal behaviors and ego development, also studied the results of systematic reflective discussions of college students' helping experiences upon the development of self-concept. Conclusions indicated that a combination of actual experiences and systematic reflective discussions were the most important components in the curriculum and that participants showed significant growth in self-concept and ego development.

Rutter and Newman (1989) found that the potential for service enhancing social responsibility was dependent on the presence of a reflection seminar. The opportunity to discuss their experiences with teachers in small peer-group setting greatly affected whether students reported a positive interaction with the community. Saunders (1976) investigated whether junior and senior high school student tutors would demonstrate a positive attitude change in self-concept, in reading, and toward school when compared to student nontutors. Although no significant difference was found, Saunders concluded that the program had an effect on maintaining positive attitudes.

Soat (1974) examined college students in an introductory psychology course as to whether one's cognitive style and self-concept were related to expressed willingness to help others. He found no significant relationships.

In summary, the research evidence does give some indication that experiential and service-learning programs may have a positive effect upon the development of a positive self-concept in those students involved in such a program. More research must be done for that evidence to be definitive.

## Service Learning and Moral Judgment

Alexander (1977) investigated whether moral thinking, ego development, and the presence of prejudice in youth could be changed by an alternative education curriculum. Significant changes were discovered in moral reasoning, ego development, and level of prejudice. Edwards (1974) studied experiential education as it relates to moral development and explored the influence of environment upon moral reasoning development. Studying 103 high school and university students in Kenya, she confirmed the following hypotheses relating to the effects of intellectual and social experiences:

1. Students who attended multicultural secondary schools displayed higher levels of moral judgment than did students who attended ethnically homogeneous schools.

2. An atmosphere of mutual trust and cooperation stimulated students in preconventional (Stages 1 and 2) reasoning postures to develop toward more adult postures (Stages 3 and 4).
3. Students who resided at boarding schools displayed more Stage 3 and 4 moral reasoning than did students living at home.
4. Students who studied law and social sciences displayed more Stage 3, 4, and 5 moral reasoning development than did students who studied primarily science and engineering.

Reck (1978) attempted to determine whether participation in a school service-learning program was positively related to moral development, whether the amount of time given to service was related to students' positive moral development, and whether students with little experience in service activities experienced more moral development than did students with more prior experience. On only 2 of 16 variables were there significant differences between experimental and control groups: (a) Students who pretested low in moral development demonstrated greatest gains in the posttest, and (b) students who served only during the program in their assigned tasks showed significant growth.

Mosher (1977) concluded that moral and ego development can be enhanced by service-learning programs, with the most powerful being those that combine discussion of moral issues with the experiences.

Although the research results in the area of moral judgment are mixed, they do tend to indicate that experiential and service-learning programs may have an impact upon the development of moral judgment. What has not been answered is whether there are consistently effective ways in which moral judgment may be developed, what types of students will benefit from what programs, and what formats will be most successful.

## The Impact on Academic Learning

Houser (1974) recorded significant gains in an experimental group versus a control group in the development of both reading skills and self-concept at the seventh- and eighth-grade level for students participating in a student-aide program involving elementary school students. Lewis (1977) recorded significant gains in his investigation of whether learning by doing (experiential learning) was as effective a method of teaching subject matter concepts to adolescents and adults as was expository learning. Although expository learning was effective in a number of situations, learning by doing coupled with receipt of procedural knowledge learned both by declarative and procedural knowledge was more effective.

Hedin (1987), in a comprehensive meta-analysis on peer tutoring by high school students involved in service, found increases in reading and math achievement scores both on the part of the tutor and tutee. Although the achievement score increases in reading and math were modest, the author defends the analysis on the basis that small increases are evident with most learning and growth in general.

Hamilton and Zeldin (1987) found that when the measuring instrument is a general test of knowledge, there is usually no difference between students in service programs and those in conventional classrooms who do not participate. Consistent gains in factual knowledge have been found, however, when researchers have used tests designed to measure the kinds of information students were more likely to encounter in their field experiences.

Braza (1974) studied 15 experimental and 8 control group students in an attempt to discover significant improvements in knowledge, behavior, and attitudes recorded as a result of a community-based service-learning procedure. Control group students received traditional classroom instruction in health problems of disadvantaged groups, whereas the experimental group students were given intensive community experiences. Posttest results demonstrated that both methods were equally effective in promoting knowledge gains; in addition, both groups expressed essentially identical increased commitment to the study of health problems of disadvantaged persons.

Markus et al. (1993) reported results of an experiment in integrating service learning into a large undergraduate political science course. Students in service-learning sections of the course were significantly more likely than those in the traditional discussion sections to report that they had performed up to their potential in the course, had learned to apply principles from the course to new situations, and had developed a greater awareness of societal problems. Classroom learning and course grades also improved significantly as a result of students' participation in course-relevant community service. Finally, pre- and post-survey data revealed significant effects of participation in community service upon students' personal values and orientations. The experiential learning acquired through service appears to compensate for some pedagogical weaknesses of classroom instruction.

Thus the findings on intellectual learning and participation in experiential and service-learning programs are mixed. It may be that positive intellectual outcomes are found most frequently for tutoring because it is the form of service learning that is most school-like, and the knowledge and skills examined are most like those the tutors have been using. In the instances when student in other forms of experiential and service learning have been tested for gains in factual knowledge, the results have been less conclusive. In most cases, the test instruments used to measure intellectual gain were developed by the same individual responsible for the service-clearing program, therefore raising questions of researcher bias and lack of test validity.

## Community Impact and Effects on Those Served

Ellington (1978) studied the effects of contact with and education about the elderly in three experimental classes of high school seniors. Although no differences were discovered between students who received only contact with the seniors and the control group, and none were discovered between the attitudes of the two groups receiving inductive and deductive teaching, the study did find that a combination of contact with the seniors and learning about their problems appeared to positively change young people's attitudes.

Glass and Trent (1979) concluded that adolescents' attitudes toward the elderly can be changed through classroom experiences. Owens (1979) sought to determine whether student attitudes toward academic and vocational goals would change in a positive direction after involvement in a yearlong service-learning program. He concluded that students in the experimental group experienced significantly larger attitudinal changes than did the control group in the areas of more positive self-confidence and more clarity in educational direction and career paths.

Shoup (1978) saw service learning as a viable alternative to the set secondary curriculum, and as a valuable method for expanding the traditional classroom experiences to promote citizenship attitudes. Clayman (1968), in a study of a program to train preservice teachers to become familiar with community resources, discovered that although student Watchers were committed to using the community as a resource, supervision of their activities was complex and difficult.

Conrad (1979) chose 11 experiential and service-learning programs from various cities for intensive study. The 11 programs from nine schools involved more than 600 students in nine experimental and four control groups; foci included community service, outdoor adventure, career exploration, and community action. The overall conclusions of the study were that experiential education and service-learning programs can promote social, psychological, and intellectual development; that they appear to do so more effectively than classroom-based programs; and that the key factors in promoting growth are (a) that the experiences be significant and provide for the exercise of autonomy and (b) that there be opportunity for active reflection on the experience.

Keene (1975) examined whether students involved in an elective sociology high school course where classroom instruction was coupled with 5 hours of volunteer direct experience per week for one semester at various social agencies would have a more positive attitude change toward poverty and minority problems than would students who took only a required political science and economic course. She found no significant difference in the groups, but the experience was perceived as positive by parents, students, and the community, and so was continued.

Newman (1978) found a negative impact on attitudes toward the disabled held by elementary students placed in contact with severely emotionally disturbed children, as compared to attitudes held by students who received classroom instruction about handicapped children. Tobler (1986) conducted a meta-analysis of 143 studies on drug prevention programs and found that peer-helping programs were identified as the most effective on all outcome measures. Sprinthall and Sprinthall (1977), reporting on a series of studies of high school students engaged as teachers, tutors, and peer counselors, observed that in addition to showing other gains, many students had developed higher-level counseling skills than those achieved by graduate students in counseling.

The findings on community impact and the effects on those served are primarily positive, indicating that young people enrolled in experiential education and service-learning programs that focus upon making a difference in terms of community do, in fact, positively affect community members. In addition, the attitudes of young people frequently are significantly changed in the process of helping others.

The evaluation of Service-Learning Colorado was conducted by a team of researchers from the University of Colorado at Boulder (Kraft, Goldwasser, Swadener, & Timmons, 1993). It looked at all K-12 Serve-America, Youth and Conservation Corps, and Higher Education programs funded from grants made by the Commission on National and Community Service to the Colorado State Commission. To give a sense of the wide range of possible outcomes of service learning, the following list indicates the impact domains, participant and teacher attitudes, participant behaviors, and institutional and community impacts that were looked at in the Colorado research:

> civic/social responsibility, self esteem, leadership, poverty, career aspirations, moral development, empowerment, service/community, gender, alienation, social justice, race, efficacy, environment, peers, elderly, younger children, handicapped, family, reflection, and cross-cultural experience.

More than 2,000 students and staff from middle school through higher education responded to the pre- and post-attitude survey. The survey instrument was developed by the researchers and was based on previous research on the effects of service learning. Among the results gleaned from the pre- and post-attitude survey were the following.

1. There were few items on which the students made statistically significant gains in positive attitudes toward service, possibly due to the short time frame of most of the programs, often only once a week for 6 to 8 weeks.
2. Teachers, all of whom had received grants to administer service-learning programs, were significantly more committed on almost all items to the goals of service learning than were their student participants.
3. There were few statistically significant differences between middle school service-learning participants and those in high school as far as their attitudes toward items on the service-learning instrument. This could be seen as a surprising finding, because research by the Search Institute (Benson, 1993) found that high school students were significantly less committed to serving others than were younger students in grades six through eight.
4. Students in higher education tended to be more positive in their attitudes toward service learning than were students at the younger grades.

5. Short-term service-learning experiences did not have a statistically significant effect either way on attitudes of students.
6. On almost all attitudinal items, girls were significantly more positive in their attitudes to service and related values than were boys.

## Conclusions

It is too early to predict the long-term impact of service learning on educational reform, citizenship education, community building, or pedagogical and curricular change. There were great hopes during the Bush and early Clinton administrations that a consensus had been reached on the value and importance of involving the young in their communities and of reforming the public schools through the use of service learning and community service. Service has again become a political football between liberals and conservatives, and this may well end much, if not most, of the federal funding for K-12, Higher Education, and Americorps programs. If the movement is a genuinely grassroots one, as many of its advocates claim, and if the effects are as positive as some of the research and personal testimony indicates, it is likely that service learning will continue as one of the educational reform mechanisms into the next century.

## References

Alexander, R. (1977). *A moral education curriculum on prejudice.* Unpublished doctoral dissertation, Boston University.
Barber, B. (1992). *An aristocracy of everyone.* New York: Ballantine.
Bellah, R., Madsen, R., Sullivan, W., Swidler, A., & Tipton, S. (1985). *Habits of the heart.* Berkeley: University of California Press.
Benson, P. (1993). *The troubled journey.* Minneapolis, MN: Search Institute.
Bidwell, C. (1965). Schools as organizations. In J. March (Ed.), *Handbook of organizations.* Chicago: Rand McNally.
Bontempo, B. (1979). *A study of experience based learning in alternative public high schools: Implications for a new role for educators.* Unpublished doctoral dissertation, Indiana University.
Bourgeois, M. (1978). *Experiential citizen education for early adolescents: A model.* Unpublished doctoral dissertation, University of North Carolina, Greensboro.
Boyer, E. (1983). *High school.* New York: Harper & Row.
Braza, G. (1974). *A comparison of experiential and classroom learning models in teaching health problems of the poor.* Unpublished doctoral dissertation, University of Utah.
Broudy, H. (1977). *Moral citizenship education: Potentials and limitations* (Occasional Paper No. 3). Washington, DC: National Institute of Education.
Cairns, R., & Kielsmeier, J. (1991). *Growing hope: A sourcebook on integrating youth service into the school curriculum.* Roseville, MN: National Youth Leadership Council.
Calabrese, R., & Schumer, H. (1986). The effects of service activities on adolescent alienation. *Adolescence, 21,* 675–687.
Carnegie Task Force on Education of Young Adolescents. (1989). *Turning points: Preparing American youth for the 21st century.* New York: Carnegie Council on Adolescent Development of the Carnegie Corporation
Clayman, C. (1968). *Experiencing the role of teacher as liaison: Case study of the training of pre-service teachers through direct community experiences.* Unpublished doctoral dissertation, Boston University.
Coleman, J. (1961). *The adolescent society.* New York: Free Press.
Coleman, J. (1974). *Youth: Transition to adulthood.* Chicago: University of Chicago Press.

Commission on National and Community Service. (1993). *What you can do for your country.* Washington, DC: Government Printing Office.

Conrad, D. (1979). *The differential impact of experiential learning programs on secondary school students.* Unpublished doctoral dissertation, University of Minnesota.

Conrad, D., & Hedin, D. (1982). Youth participation and experiencial education. *Child and Youth Services, 4*(3/4), 57–76.

Corbett, R. (1977). *The community involvement program: Social service as a factor in adolescent moral and psychological development.* Unpublished doctoral dissertation, University of Toronto.

Counts, G. (1932). *Dare the schools build a new social order?* New York: John Day.

Cruz, N. (1994). *Diversity principles of good practice in combining service and learning.* Manuscript in preparation.

Cusick, P. (1973). *Inside high school.* New York: Holt, Rinehart & Winston.

Dass, R., & Bush, M. (1992). *Compassion in action: Setting out on the path of service.* New York: Crown.

de Tocqueville, A. (1969). *Democracy in America* (J. P. Mayer, Ed., G. Lawrence, Trans.). New York: Doubleday. (Original work published in 1835)

Dewey, J. (1902). *The school and society.* Chicago: University of Chicago Press.

Dewey, J. (1916). *Democracy and education.* New York: Macmillan.

Dewey, J. (1963). *Experience and education.* New York: Collier. (Original work published in 1938).

Edwards, C. (1974). *The effect of experiences on moral development: Results from Kenya.* Unpublished doctoral dissertation, Howard University.

Ellington, J. (1978). *The effects of three curriculum strategies upon high school seniors' attitudes toward the elderly.* Unpublished doctoral dissertation, University of California.

Erikson, E. (1968). *Identity, youth, and crisis.* New York: Norton.

Exum, H. (1978). *Cross age and peer teaching: A deliberate psychological education program for junior college students.* Unpublished doctoral dissertation, University of Minnesota.

Frankena, W. (1965). *Three historical philosophies of education.* Glenview, IL: Scott, Foresman.

Freire, P. (1970). *Pedagogy of the oppressed.* New York: Continuum.

Glass, J., & Trent, C. (1979). *The impact of a series of learning experiences on ninth grade students' attitudes toward the aged: Summary and lesson plans.* Raleigh: North Carolina State University.

Goodlad, J. (1984). *A place called school.* New York: McGraw-Hill.

Goodlad, J., & Klein, F. (1990). *Behind the classroom door.* Worthington, OH: Charles Jones.

Hamilton, S., & Zeldin, R. (1987). Learning civics in community. *Curriculum Inquiry, 17,* 407–420.

Hanna, P. (1937). *Youth serves the community.* New York: D. Appleton Century.

Harrison, C. (1987). *Student service: The new Carnegie unit.* Princeton, NJ: Carnegie Foundation for the Advancement of Teaching.

Hedin, D. (1987). Students as teachers: A tool for improving school climate and productivity. *Social Policy, 17*(3), 42–47.

Houser, V. (1974). *Effects of student-aide experience on tutors' self-concept and reading skills.* Unpublished doctoral dissertation, Brigham Young University.

Hunt, D., & Hardt, R. (1969). The effect of Upward Bound programs on the attitudes, motivation and academic achievement of Negro students. *Journal of Social Issues, 25,* 117–129.

Jackson, P. (1990). *Life in the classrooms.* New York: Holt, Rinehart & Winston.

James, W. (1910). *The moral equivalent of war* (International Conciliation No. 27). New York: Carnegie Endowment for International Peace.

Johnson Foundation. (1989). *Wingspread special report: Principles of good practice for combining service and learning.* Racine, WI: Author.

Kazunga, D. (1978). *A youth program in the local community's context: The case of the young farmers of Uganda program.* Unpublished doctoral dissertation, University of Wisconsin.

Keene, P. (1975). *Social problems analysis through community experience: An experiential study of high school seniors' attitudes toward race and poverty.* Unpublished doctoral dissertation, Pennsylvania State University.

Kelly, H. (1989). *The effect of the helping experience upon the self-concept of the helper.* Unpublished doctoral dissertation, University of Pittsburgh.

Kilpatrick, W. H. (1918). The project methods. *Teachers College Record,* 19, 319–335.

Kohlberg, L. (1970). Education for justice: A modern statement of the Platonic view. In F. F. Sizer (Ed.), *Moral education.* Cambridge, MA: Harvard University Press.

Kraft, R. J., Goldwasser, M., Swadener, M., & Timmons, M. (1993). *First annual report: Preliminary evaluation: Service-learning Colorado.* Denver: Colorado Department of Education.

Krug, J. (1991). *Select changes in high school students self-esteem and attitudes toward their school and community by their participation in service learning activities at a Rocky Mountain high school.* Unpublished doctoral dissertation, University of Colorado at Boulder.

Lewis, J. (1977). *What is learned in expository learning by doing?* Unpublished doctoral dissertation. University of Minnesota.

Luchs, K. (1981). *Selected changes in urban high school students after participation in community based learning and service activities.* Unpublished doctoral dissertation, University of Maryland.

Markus, G., Howard, J., & King D. (1993). Integrating community service and classroom instruction enhances learning: Results from an experiment. *Education Evaluation and Policy Analysis,* 15(4), 410–419.

Marsh, D. (1973). *Education for political involvement: A pilot study of twelfth graders.* Unpublished doctoral dissertation, University of Wisconsin.

Martin, J. (1975). *Report of the national panel on high schools and adolescent education.* Washington, DC: Government Printing Office.

Martin, J. (1976). *The education of adolescents.* Washington, DC: National Panel on High School and Adolescent Education.

Martin, W. (1977). *A participant observation study of an outdoor education experiential curriculum experiment in a public secondary school.* Unpublished doctoral dissertation, Michigan State University.

Maybach, C. W. (1994). *Second year evaluation of three components of Colorado Campus Compact.* Denver: Colorado Campus Compact.

Melchior, A., Jastrzab, J., Bailis, L., & Frees, J. (1994). *Serving America: The first year of programs funded by the commission on national and community service.* Boston: Abt Associates.

Mosher, R. (1977). Theory and practice: A new E.R.A.? *Theory Into Practice,* 16(2), 166–184.

Nathan, J., & Kielsmeier, J. (1991) The sleeping giant of school reform. *Phi Delta Kappan,* 72, 738–742.

National Commission on Excellence in Education. (1983). *A nation at risk: The imperative of educational reform.* Washington, DC: Government Printing Office.

National Committee on Secondary Education. (1972). *American youth in the mid-seventies.* Reston, VA: National Association of Secondary School Principals.

Newman, F., & Rutter, R. (1986). A profile of high school community service programs. *Educational Leadership,* 43(4), 65–71.

Newman, R. (1978). *The effects of informational experiential activities on the attitudes of regular classroom students toward severely handicapped children and youth.* Unpublished doctoral dissertation, University of Kansas.

Noddings, N. (1992). *The challenge to care in schools: An alternative approach to education.* New York: Teachers College Press.

Owens, A. (1979). *The effects of experimental learning on student attitudes.* Unpublished doctoral dissertation, Temple University.

Piagét, J. (1970). *Science of education and the psychology of the child.* New York: Viking.

Reck, C. (1978). *A study of the relationship between participation in school service programs and moral development.* Unpublished doctoral dissertation, Saint Louis University.

Rich, J. (1962). Learning by doing: A re-appraisal. *The High School Journal, 45,* 338–341.

Riecken, H. (1952). *The volunteer work camp: A psychological evaluation.* Cambridge, MA: Addison-Wesley.

Rogers, C. (1969). *Freedom to learn.* Columbus, OH: C. W. Merrill.

Rutter, R., & Newman, F. (1989). The potential of community service to enhance civic responsibility. *Social Education, 53*(6), 371–374.

Sager, W. (1973). *A study of changes in attitudes, values, and self-concepts of senior high youth while working as full-time volunteers with institutionally mentally retarded people.* Unpublished doctoral dissertation, United States International University.

Saunders, L. (1976). *An analysis of attitude changes toward school, school attendance, self-concept and reading of secondary school students involved in a cross age tutoring experience.* Unpublished doctoral dissertation, United States International University

Schwebel, M., & Ralph, I. (1973). *Piaget in the classroom.* New York: Basic Books.

Shoup, B. (1978). *Living and learning for credit.* Bloomington, IN: Phi Delta Kappa.

Silberman, C. (1990). *Crisis in the classroom.* New York: Random House.

Smith, M. (1966). Explanation in competence: A study of peace corps teachers in Ghana. *American Psychology, 21,* 555–566

Soat, D. (1974). *Cognitive style, self-concept, and expressed willingness to help others.* Unpublished doctoral dissertation, Marquette University.

Sprinthall, R., & Sprinthall, N. (1977). *Educational Psychology: A developmental approach* (2nd ed.). Reading, MA: Addison-Wesley.

Stockhaus, S. (1976). *The effects of a community involvement program on adolescent student's citizenship attitudes.* Unpublished doctoral dissertation, University of Minnesota.

Sturges, A. (1979). High school graduates of the stormy 60s: What happened to them? *Educational Leadership, 36,* 502–505.

Tobler, N. (1986). Meta-analysis of 143 adolescent drug prevention programs: Quantitative outcome results of program participants compared to a control or comparison group. *Journal of Drug Issues, 16,* 537–567.

Wilson, T. (1974). *An alternative community based secondary school program and student political development.* Unpublished doctoral dissertation, University of Southern California.

# The Teacher's Place in the Formation of Students' Character

CAROLYN GECAN
*Thomas Jefferson High School*

BERNADETTE MULHOLLAND-GLAZE
*Mt. Vernon High School*

## Introduction

For a long time now, perhaps the last 20 years, many of us in teaching have been pretending that we don't, can't, or shouldn't teach moral education in the public high school. We don't teach moral education because there is no room in the curriculum. We can't teach moral education because by the time kids get to high school, their characters have already been shaped by their families—along with the advertising and entertainment industries. And we shouldn't teach moral education because that is the responsibility of parents and churches; moreover, because we live in a multicultural society there is no way to determine what is good and bad character.

We teach social studies at Thomas Jefferson High School for Science and Technology, a Governor's Magnet School serving several counties in northern Virginia. Five years ago the emptiness of the thinking described above was brought home to us in the weeks after a particularly vicious senior prank at our school impugned the integrity of two male faculty members and compelled our faculty to examine its role in the moral lives of our students. Two seniors had stolen negatives from the yearbook office, cropped them to give the illusion that the two men were involved in a homosexual activity, and printed the resulting photograph with logo on T-shirts which they were planning to sell for $18.00 to other students. When administrators learned about the T-shirts, they immediately confiscated the shirts and suspended the two boys.

Responses to the prank varied enormously: One camp held that it was just a silly prank and excused the pranksters by saying, "boys will be boys." Another group focused on First Amendment rights and supported the boys in their contention that what they had done constituted protected speech. A third group acknowledged that the prank bordered on the obscene and had devastated the two adults involved but excused the seniors because they hadn't meant to hurt anyone. The administration and many other faculty members were outraged by both the incident itself and more particularly by the lack of moral imagination demonstrated by the students and their defenders. Conversation among faculty about what had happened was painful and difficult.

---

Carolyn Gecan and Bernadette Mulholland-Glaze, "The Teacher's Place in the Formation of Students' Character," *Journal of Education*, 1993, Vol. 175, No. 2, pp. 44–57. Reprinted by permission from School of Education, Boston University.

There was considerable confusion. What was the incident about? Pranks, feelings, rights, responsibilities?

Talking with our fellow teachers about the incident, however, was academic compared to the discussion we had with the three seniors who drafted a petition in support of the pranksters. These intelligent students focused their argument solely on the issue of freedom of speech. We asked them what they thought about the fact that the two maligned faculty members were devastated by the prank, and they replied that the feelings of the people involved had nothing to do with freedom of speech. The students were not willing to admit that the incident had anything to do with right or wrong. It was at this point that we realized that the words "right" and "wrong" had years ago been stripped from our school vocabulary. The gap between the students' understanding of the T-shirt incident and our own moral outrage seemed unbridgeable. High schools in our experience had been values-neutral for so long that our ability to engage students in conversations about moral issues had become rusty. We were not even sure what our role was in the moral education of our students.

As we talked about the incident and its aftermath with each other and our colleagues, we began to question some of our assumptions about our school. We knew that as a group we were successful at producing academically accomplished students who consistently scored high on standardized tests and were admitted to the finest colleges and universities in the country. But was this enough? In the year preceding the T-shirt incident, a School Climate Committee had been formed. This group addressed topics such as cheating, illegal software copying, stealing of personal and school property, and student lack of respect for peers and adults. However, none of us made the connection between these behaviors and the role of schools in moral education. One solution put forward was to develop new and more extensive rules for each type of transgression. But we sensed that this response would not create a school climate that would foster the personal and intellectual growth that we wanted for our students.

As the two of us talked with our social studies colleagues and our principal, we began to realize that we did not want a school bound by even more rules and regulations., But what did we want instead? Did we have a role in the moral education of our students? If so, what was it? Both of us had observed that our school curriculum largely omitted consideration of ethical questions. Except for the "lifeboat" discussions that occasionally arose, we did not see much evidence that our school was establishing a climate for our students to develop what Dr. Edwin Delattre calls their moral imaginations. We knew that the omission was not the result of deliberate choice on anyone's part, but rather the result of our own training and experience as teachers. We came to realize that our faculty community needed to reflect upon and to discuss some of the great works of philosophy that serve as the foundations of ethics as a way to help us begin answering the questions we had. We wanted to undertake a shared intellectual journey, but we knew that we needed help and guidance.

There are many ways that a concerned faculty can begin to address the issues which perplexed us. Our approach was to design a program to bring outside scholars in the field of ethics and moral education to our school community. Our principal, Geoff Jones, encouraged and supported us from the very beginning. We wrote a grant proposal to the National Endowment for the Humanities [NEH] to develop what we eventually called the Jefferson Institute on the Foundations of Ethics in Western Society. We followed some guiding principles as we wrote; they were based on our experiences as veteran classroom teachers:

- One-day in-service programs frequently fail.
- Teachers need time away from school to read, write and reflect.
- Teachers enjoy intellectual challenges.
- Teacher institutes serve as pedagogical models for the participants.
- A serious program such as we were considering would take a lot of time and a lot of hard work.

From the very beginning of our project, we were fortunate to have the guidance and active participation of Dr. Steven Tigner, professor *emeritus* from the University of Toledo, a philosopher with experience in working with teachers in similar institutes. We told him what we wanted the focus of our proposed institute to be and asked for his help in shaping the basic program, selecting the core texts, and finding scholars to work with us. All these things he willingly did. After consulting with Dr. Tigner, we designed the original Jefferson Institute, which ran from June 1990 through August 1991, specifically for a group of 24 of our colleagues: teachers of English, social studies, computer science, biology, physics, plus two counsellors and our principal. This institute was successful beyond our imaginings, and drew praise from participants, scholars, and Fairfax County public school administrators. With the encouragement of an area school superintendent, we wrote another proposal modeled on the first that has also been funded by NEH. This second Jefferson Institute began in the summer of 1993 and includes 28 teachers, counselors, and an administrator. These educators represent 11 of the 23 Fairfax County high schools. The following is an overview of the two institutes and what we learned along the way.

## Intellectual Focus and Format of the Institute

The overall purpose of our institute was to provide teachers with a surer grasp of the nature of ethics in order to help them foster a school climate in which students consider issues of right behavior. Some key questions we addressed were the following:

1. What is a good life?
2. What constitutes good character?
3. What is virtue?
4. What is a good society?
5. What should the relationship be between the individual and society, and between communities and the state?
6. What kind of human beings should we be?

The core of the Institute was comprised of two main phases. Phase one consisted of a three-week summer program, preceded by a one-day preparation session in late May when participants received their materials. Phase two was actually a series of seven one-day follow-up sessions spread throughout the ensuing school year. The summer program was guided by Dr. Tigner who was joined each week of the three weeks by a different professor: Dr. Jon N. Moline of St. Olaf College; Dr. Richard D. Parry from Agnes Scott College; and Dr. William Wians from Boston University.

A typical Jefferson Institute day included a combination of directed presentation by the scholars, individual journal writing by the participants, and small-group discussions based on either a focus question or issues that emerged from the journals. Writing in the journal enabled participants to understand the texts by expressing basic ideas in their own words and by reflecting on the power of the arguments in the texts. We believe that this combination of reading, writing, and discussing provided an important pedagogical model for the teacher participants. Because many teachers teach the way they are taught, we wanted to ensure that the Institute modeled the intrinsic relationship between language and learning. We very consciously worked with the scholars to establish an atmosphere of trust so that the participants were willing to take intellectual risks in their conversations, their presentations, and their writing. All of us felt comfortable in asking questions of Dr. Tigner and the other scholars, the texts, and each other. None of us was afraid to admit to confusion, or to expose a different interpretation or point of view to the scrutiny of scholars and peers. As two participants wrote in one of their periodic evaluations:

> [This experience] has caused me to think so much that I am literally exhausted. I appreciate the learning process that has been so participatory and your gentle support that has guided us through. It is amazing to me that I had never been guided to or allowed to expose my soul to Plato previously. The gods have agreed that I am finally worthy of contemplating the philosophical and virtuous life.
>
> It seems that whenever men have sought the counsel and guidance of other men, a creative minority of helpers, healers, and guides have arisen to meet the need. Though their words and dress have differed in the many times and places in which they have appeared, these gurus have come forward as spiritual guides, as these special sorts of teachers who help men to make their passage from one stage of their lives to another. I feel that's what Steve [Tigner] has been for us. . . .

During the summer phase of the Institute, we investigated the foundations of ethics in Western society through reading selected works of Plato, Aristotle and the Bible. The central themes and issues of these works served as the focus for the entire program. Morning sessions of the first week, led by Dr. Moline, focused on Plato's *Euthyphro, Apology, Crito, Meno,* and *Phaedo.* These works contain accounts of the life and death of Socrates, his mission and methods, and an inquiry into the nature of right conduct. Dr. Tigner's afternoon sessions were devoted to an examination of Genesis 1–11, Exodus 20, Samuel I & II, and Matthew 5–7. Our discussions of these readings centered on the Judeo-Christian view of the nature of human beings and the right rules of living. The issues and questions raised by these fundamental texts provided for lively writing and discussion. Participants all agreed that this was the first time in years that they had spent time considering what it means to live an ethical life, defining virtue and the good life, and considering what constitutes good character. Through their writings and discussions we saw them begin to make connections, in some cases profound ones, between the ideas brought out in the Institute and their roles as parents, spouses and friends. These personal connections led to the forging of a close intellectual community which has continued in the ensuing years.

The second week's sessions were devoted entirely to study of the complete text of Plato's *Republic*. With Drs. Parry and Tigner, we explored the nature of justice and goodness: Is it better to be just or unjust? Is the just person wiser, stronger and happier than an unjust person? If one could live a life in which one "gets it all" (even if that means elbowing grandma in the face), or a life in which one is just but does *not* get it all, which would you choose? One of the most fruitful afternoons of this week stemmed from a discussion of Gyges' Ring from Book II. The issues raised by Gyges' Ring brought participants to some of the core concerns about the role of schools in character education:

- How do we get students to take themselves and their souls seriously?
- How do we make behavioral controls intrinsic rather than extrinsic?
- How do we get students to like doing the right thing?
- What means can schools use to habituate students to doing the right thing?

The discussions of Plato's ideas provided us with more questions than answers. We were assured throughout the first two weeks that Aristotle had answers for just about everything,

We read the most challenging of the summer texts, Aristotle's *Nicomachean Ethics,* during the third week. Aristotle, unlike Plato, deals directly with the central question of moral education: Is virtue teachable? Drs. Tigner and Wians led participant teams through an analysis of the intricate arguments presented by Aristotle on how virtue is learned. One of the most valuable dimensions of this week was the application of Aristotle's teachings on the nature of happiness, the acquisition of virtue through habituation, and the importance of practical wisdom to our personal and professional lives. The following are two examples of Dr. Tigner's typical team discussion assignments based on Aristotle. After assignments such as these, participants made informal presentations to the group.

1. Pool your own experience and thought for examples that would appeal to your students in illustrating not only the point that character affects moral perceptiveness, but also that the perceptions of a person of good character and judgment warrant the most respect. Why is this important for students to grasp?
2. Collectively write a dialogue with a student who is asked for the *point* (the end, the good) of engaging in some classroom or school activity, assignment, or project (*"Why do we have to learn fractions, study the partition of India, do a community service project, etc.?"*). Try to lead the student to see how this activity, assignment, or project ultimately contributes to the student's own "happiness." Draw on your own experience with students in being as realistic as you can in constructing the student's side of the dialogue and be as Aristotelian as you can in writing your own side. Select two of you to "perform" this dialogue when we reconvene.

These team topics, and others like them, along with the subsequent presentations evoked intense discussions about how to apply what we were learning to our classrooms and our school. Some of the most powerful answers included the following:

- Teaching by examples as they naturally, and sometimes unexpectedly, arise in current events and everyday student experiences in the classroom, cafeteria, and hallways.
- Providing students with opportunities to examine ethical issues through literature and personal writing.
- Tapping into students' natural idealism by giving them a chance to examine issues important both to the school and to the extended community, and to decide what actions they might take as responsible citizens.

Phase two of the Jefferson Institute, the seven one-day sessions spread throughout the school year, extended the ideas from the foundation texts through further reading and discussion. The sessions were conducted by other scholars in the field of ethics: Dr. William Kilpatrick, from the School of Education at Boston College; Dr. Kevin Ryan and Dr. Edwin J. Delattre, both from the School of Education at Boston University; and Dr. Leon Kass, from the University of Chicago. Dr. Tigner also returned for one of the early sessions. Participants worked with these scholars who led them in an examination of how ethical concerns arise in more recent history, education, literature, science and technology, law, and government. We discussed ways in which the curriculum can assist in the development of a school climate fostering more thoughtfulness about good conduct, the components of a good life, and the habits of mind and heart grounded in an understanding of ethical behavior.

The major thesis of Dr. Kilpatrick was that the most beneficial approach to moral education is the use of stories to enable students to identify character traits that they ought to know and need to practice until these traits are internalized. One teaching tool that he modeled for us was the use of movies in the classroom. He showed us selected scenes from *Captains Courageous* as an illustration of how to teach about the ways young people are initiated into moral maturity. He emphasized that rather than merely discussing the topic, students would be able to see the behavior in a "real" situation. He referred us to a variety of short stories, novels, and films that would lend themselves to discussion of character development. On this day we began to realize that incorporating moral education into our curriculum was not going to require an infusion of new funds for materials nor a major rewriting of the curriculum. Rather it was a matter of choosing different titles in the short story and novel collections already present in our classrooms, and shifting the emphasis of our classroom discussions to include issues such as fictional characters making the right decisions and doing the right thing in the stories.

Our foundation work during the summer had prepared us very well for Dr. Kevin Ryan's presentation. After presenting a convincing argument that "what schools do consciously and unconsciously to help students think and feel and behave in a morally effective way" constitutes moral education, he defined other crucial terms: "moral," "morality," "values," and "ethics." "Moral" he defined as an adjective describing right and wrong. "Morality" is a system of dealing with right and wrong, a world view about right and wrong, and the ground out of which individual moral decisions are made. "Values" are the things we want and desire, some of which are moral and some of which are material. The values clarification movement mixed these up. "Ethics" is a system that enables you to know how to deal with problems when you face them.

One of the most significant aspects of Dr. Ryan's presentation addressed the conflict between some scholars and public educators regarding moral education in schools as being a violation of religious freedom. He provided specific arguments to support the view that teachers *do* have a significant role in the moral education of their students. For example, both Socrates and John Dewey insisted that the passing on of moral traditions is the duty of schools. The Founding Fathers also saw schools as the places where knowledge for the exercise of responsible citizenship must be learned; they knew that a high level of moral sensibility was needed in a democracy. Dr. Ryan quoted from state laws and recent public opinion polls which indicated a high degree of support for moral education in the schools. He suggested several ways schools could develop an ethical climate: personal and historical example; explanation; exhortation; experiences; and expectations. Finally, he articulated something that we had become convinced of in the summer, namely that it is impossible not to affect kids when they are in school for twelve years. In a way, Kevin Ryan validated the impulses behind our entire Jefferson Institute.

Dr. Leon Kass focused the group's attention on two of his writings which we had read prior to his visit. Participants discussed the potentially dehumanizing effects stemming from scientific and technological advances. Improvements in science and technology have enabled us to have better and healthier lives, but there is always a price to pay. Following the pattern set by earlier presenters, Dr. Kass emphasized the importance of stories in shaping moral vision and the formation of character. He urged us to use appropriate questions to "get to the things that are indeed on our students' minds and that do matter . . . to get into their own thinking . . . [and to] get kids to connect stories to their own lives."

For the three visits from Dr. Delattre, participants read a variety of materials including, *Anne Frank: The Diary of a Young Girl,* Dr. Martin Luther King, Jr.'s "Letters from a Birmingham City Jail," selections from William James, and Ibsen's *Enemy of the People,* plus several short essays on character, some written by himself and others by W. J. Bennett. In our discussion of these works, we considered the definition of integrity as being the same person in public and in private. Dr. Delattre discussed with us the "fundamental questions: What is morality? Can morality be taught? Can morality be learned? Can adults possibly avoid influencing the moral habits and attitudes of the children and youths who keep company with them?" He elicited a discussion of heroes and heroines such as Anne Frank, Helen Keller, and Sojourner Truth as models of courage and good character in real life. These people serve as inspiration because they are made of the same stuff as the rest of us, or else they could not be effective models. In one memorable session, we returned to a discussion of the importance of habit first brought out during the summer when we read Aristotle. As one participant wrote, Dr. Delattre brought us "full circle from Aristotle on habit to James on habit." Through the powerful words of William James we came to see that any worthwhile character-education program must provide students with opportunities to practice good habits; for example, high school students engaging in community service projects or third-grade students tutoring nursery school children. The development of good habits is vital to the character education of the young because that is the only way good behavior becomes instinctive.

One special follow-up session held during the first institute in 1990–91 was scheduled for just the two of us as co-directors to conduct. The focus of this day was self-examination: Where were we? What had we learned? What did all of this mean for our school? Where should we go from here? Participants had the opportunity to share their answers to these and other questions.

One particular issue, dishonesty, which had historically been a problem at the school since it opened, engaged a great deal of our time. We brainstormed how we and our colleagues might help to promote a school atmosphere which supports the development of integrity and trust among faculty and students. Unfortunately, we did not find the definitive solution to that problem and are to this day still struggling with the issue. However, one significant outcome of this session was the beginning of many curriculum adjustments we started making in order to incorporate what we were learning in the institute into our various programs. A second valuable outcome was the decision that we needed to go beyond our group of 24 to engage the entire faculty of our high school in this worthwhile endeavor. The result was a two-day faculty retreat held at the end of the summer on the weekend before we had to report back to work. It is significant that 66 out of our faculty of 115 participated at such a time of great pressure,

Our purpose was to discuss our school's ethical climate, and we charged our colleagues with these questions: What are the qualities of a school that promotes right behavior? How can one create an atmosphere in which students can grow? How do we build an atmosphere that promotes the formation of good character? With Dr. Delattre as keynote speaker and major contributor, we began by identifying what was right and good about our school. Having read several articles by James and by Delattre prior to the retreat, our colleagues with great seriousness of purpose deliberated these very hard questions. We agreed on several things, especially on the fact that we all shared the responsibility for continuing to work as a group to change the school atmosphere. We also agreed that a weekend retreat could help us to focus but could not provide all the answers, that it is important to deliberately take on the task of changing the school climate through positive not negative actions, and that these changes must come from the entire range of the curriculum and from our active attention throughout the entire school day, in class and out.

## Conclusion

The success of the Jefferson Institute went beyond the expectations of all involved, including the visiting scholars, the co-directors, and the project evaluator. However, the most powerful accolades come from the evaluations of the participants themselves:

> Aristotle says that the contemplative life is the best. Our world at school leaves us pulled in so many directions that the contemplative part seems trampled. This institute brought the importance of the contemplative part back in focus. I have *never* enjoyed nor felt as challenged by any intellectual pursuit as much as this.

> Probably the greatest strength of [the Institute] has been the ability to make me think. The combination of teaching techniques-lecture, large-group discussion, small-group discussion, writing, and reading has been outstanding.

> This has really been a lovely and affirming experience. Not only have I learned a lot of philosophy, but I recaptured the love of learning that was a large part of my original motive for going into teaching ten years ago. The complications and drudgery of teaching can overwhelm a teacher so that the goal becomes surviving the week rather than learning. I guess I'm tying to say I'm happy to be reminded how joyful learning is.

> This has been an excellent experience. What a learning luxury this Institute has been! I think it's interesting that everyone I've told about this Institute thinks it's a great idea, too. Not just scholarly types, but say, my dentist and

my stepdaughter (who is busily planning a wedding). People say "Wow—that sounds neat," and they are being sincere. [The Institute] fosters greater community . . . and counteracts the isolating tendencies of [the] daily grind. The institute also encouraged me to think about the ethical side to subject matter. I taught government this year, 160 or whatever number of days, but rarely, if ever, did I address such fundamental questions as: "What is a good society?"

Throughout the evaluations, participants emphasized personal growth, renewed love of learning, and enriched collegiality. During the course of the ensuing year, they also discussed ways in which the Institute was enriching their teaching. A ninth-grade English teacher began teaching Harper Lee's *To Kill a Mockingbird* along with George Eliot's *Silas Marner* as examples of moral courage and moral failure. The work with Aristotle during the first summer, and William James during one of the follow-up sessions, provided the framework for teaching these works as ethical novels. An interdisciplinary team of eleventh-grade American studies teachers restructured the framework of their course. They began the year with their students in a discussion of the central questions posed in the story of Gyges' Ring: What does it mean to be a good person and what does it mean to lead a "good life"? They revisited these questions in each of the major units of their course. For example, when studying the civil rights movement of the 1950s and 1960s, they examined Dr. Martin Luther King, Jr.'s "I Have a Dream" speech. In discussing his words, students focused on how the "good life" has been denied to a segment of the American population. In another unit, through the works of Emerson, Thoreau, and Twain, students examined the extent to which an individual's vision of the "good life" is compatible with that of society at large. Throughout the year, when reading the works of Native Americans, African-Americans, and immigrant Americans, their goal was to habituate students toward an attitude which accepts the basic elements of humanness that cannot be betrayed or compromised. A counselor in the second institute wrote that "When students come to me with problems and decisions to make, I feel confident now in asking them what the 'right thing to do' is. We make the decisions based more on what is 'right' instead of what is 'wanted.'" An English teacher, also from the second institute stated:

> . . . I . . . notice a difference in the way I view my school, classroom, peers, students, curriculum. It's like observing something familiar through a new filter. The discussions about literature that I've been having with my students, for example, have gravitated towards ethical issues and choices and away from so much literary "stuff." We still *do* talk about the literary stuff, but the quality is different.

And a highly regarded veteran biology teacher in the second institute wrote:

> Every time I hear about the moral decline and the lack of ethics in the schools, I gloat a little as I think I possess one of the solutions for improvements of the ethical tone in students and staff. I am convinced talking over ideas with others refreshes and focuses thoughts, intensifies my grasp of what I am about and brings a number of the staff to consider ethics. You should ask me in a year or so about results I see from incorporation of ethics into the curriculum and the biennial plan. I have found my own in this group of teachers. . . . I take delight with those who love philosophy.

When we started this enterprise five years ago, we had little idea what to expect. Moving from our concern about our school community, we entered what for us was uncharted territory. Some teachers questioned whether or not it was possible to have an "ethics institute" in today's climate. But we were encouraged from the very first by our principal, by several supportive colleagues, by wise and generous scholars, and by our own sense of purpose. Our feeling from the very beginning was that what we were doing was important. We believed then and still do that teachers, counselors and administrators share in the responsibility for the moral education of their students, and that the classic texts have much to offer educators struggling with this crucial aspect of education.

## Jefferson Institute Texts

Aristophanes, *Three Comedies: The Birds, The Clouds, The Wasps,* William Arrowsmith, ed. (Ann Arbor: University of Michigan Press, 1969).

Aristotle, *Nicomachean Ethics,* W. E. Ross, trans., revised by J. L. Ackrill & J. O. Urmson (New York: Oxford University Press, 1980).

Frank, Anne, *Anne Frank, Diary of a Young Girl,* Enriched Classic edition (New York: Oxford University Press, 1980).

Ibsen, Henrik, *An Enemy of the People,* Arthur Miller, adapter (New York: Penguin Books, 1977).

James, William, "On a Certain Blindness," "Habit," "What Makes a Life Significant," in John J. McDermott, ed., *The Writings of William James: A Comprehensive Edition* (Chicago: The University of Chicago Press, (1977).

King, Martin Luther Jr., "Letter from a Birmingham City Jail," in James Melvin Washington, ed., *A Testament of Hope: The Essential Writings of Martin Luther King, Jr.* (San Francisco: Harper and Row, 1986).

Plato, *Five Dialogues (Euthyphro, Apology, Crito, Meno, Phaedo),* G. M. A. Grube, trans. (Indianapolis: Hackett Publishing Co., 1966).

Plato, *Gorgias,* Donald J. Zeyl, trans. and ed. (Indianapolis: Hackett Publishing Co., 1992).

Plato, *Republic,* G. M. A. Grube, trans., revised by C. D. C. Reeve (Indianapolis: Hackett Publishing Co., 1992).

Sophocles, *Three Tragedies: Antigone, Oedipus the King, and Electra,* H. D. Kitto, ed. and trans. (New York: Oxford University Press, 1962).

Urmson, J. O., *Aristotle's Ethics* (New York: Basil Blackwell, 1988).

# Family Group Conferencing

## BRUCE R. TAYLOR and GLENN KUMMERY

> While typical disciplinary procedures can often make matters worse, an approach in which student offenders face their victims and others affected by their misbehavior can lead to real healing.

An intense caustic odor permeated the school. Hundreds of students had to be evacuated from their classrooms, and many individuals suffered allergic reactions and respiratory problems. What was the cause? A few students had sprayed pepper mace on classroom doorknobs.

Students, teachers, and administrators were furious and felt victimized by this hurtful behavior. After identifying the responsible students, who acknowledged their involvement, we suspended them, contacted the police, and decided to try our district's new alternative option: a Family Group Conference.

In the Central Bucks Schools, a rapidly growing Suburban district near Philadelphia, we are increasingly using this approach to address incidents of misconduct and violence. The Community Service Foundation, which provided our training, imported the conferencing process from Australia to use in its own alternative schools and group homes serving four Pennsylvania counties. The foundation now provides training throughout North America under the name *Real Justice* (McDonald et al. 1993).[1]

Family group conferences provide a forum for those most affected by an incident to sit down together and honestly address the incident and related concerns. They are helping us create a safer, more supportive environment where teachers can teach and students can learn.

## Good Shame and Bad Shame

In *Crime, Shame, and Reintegration* (1989), Australian criminologist John Braithwaite talks about the importance of shame in social control. Shame helps us develop a sense of right and wrong. Braithwaite distinguishes between two types of shaming: *stigmatizing* shame, which rejects and permanently labels offenders, and *reintegrative* shame, which rejects the deed but not the doer.

Family group conferences use reintegrative shaming. By allowing offenders to move beyond their inappropriate behavior, discard the offender label, and return to their community, conferences foster positive personal change. Reintegrative shaming is what occurs in healthy families. When children do something wrong, their parents reprimand them, but ultimately forgive them and bring them back into the family. Societies in which reintegrative shaming has a strong cultural dimension have lower crime rates.

---

Bruce R. Taylor and Glenn Kummery, "Family Group Conferencing," *Educational Leadership,* September 1996, Vol. 54, No. 1, pp. 44–46. Reprinted by permission from Association for Supervision and Curriculum.

Stigmatizing shame, on the other hand, tends to encourage the development of negative subcultures. For example, excessive use of punishment often results in further labeling and alienation of offenders. Sometimes those punished will band together to seek status and approval, and their behavior may become more and more offensive. The small group makes them feel accepted for what they do, even though it is hurtful to others around them.

Nathanson, in *Shame and Pride* (1992), points out that the "compass of shame" has shifted since World War II. Whereas people used to "attack themselves" and show deference when experiencing shame, they now typically "attack others." For example, parents frequently criticize school administrators when they are confronted with their child's inappropriate behavior. Conferencing, however, provides a setting that can overcome these inappropriate responses.

## A Scripted Process

In order to hold a family group conference, the offender must first acknowledge committing the inappropriate behavior and be willing to participate in the process. Members of the conference include the victim and the victim's support group (family and friends), the offender and the offender's support group, and any school staff or others who have been affected by the behavior. Then, under the direction of a trained family group conference coordinator, the group moves through a scripted process, which includes two phases.

*1. First, the group explores the depth and variety of ways people have been hurt by the specific victimizing behavior.* In this phase, the coordinator encourages all present to express their experiences, reactions to the incident, and feelings and concerns in a nonblaming way.

It soon becomes apparent that victimizing behavior has a negative ripple effect. Often these negative results are unintended, but the harm caused to others is real and must be acknowledged. The process allows the victims the chance to be heard and to begin to move beyond the harm done by the incident.

The goal for the offender during this phase is to begin to take ownership for the victimizing behavior, as well as to gain a greater understanding of the variety of ways people have been harmed by it. In short, the young offender gains empathy.

*2. During the second phase, the coordinator seeks to engage everyone in finding specific ways to heal the harm.* In discussing concerns, the group seeks to reach specific agreements, which the coordinator writes down. This process empowers everyone to be a part of the resolution. The list of agreements might include an acknowledgment of responsibility for the incident, an expression of apology or regret for any harm that was done, an assurance that the victimizing behavior will never occur again, financial reparation for destroyed property, community service work, the support of everyone to help rebuild the sense of trust and community, an effort to improve relationships, and a review of progress at a future date.

The extent of the agreements reached depends on the concerns and creativity of the group. At the end of this phase, all the participants sign a written agreement, everyone receives a copy, and the coordinator brings the conference to a close.

Afterward, the participants remain to "break bread" with some light refreshments. In this relaxed setting, people can continue to express their feelings. In fact, this time can be one of healing and reintegration.

## A Powerful Experience

Unlike our usual disciplinary strategies, the family group conference provides the offending student with an opportunity to move past shame and begin to make amends. There is nothing magical about a family group conference, but it has the potential to be a powerful learning experience.

An effective family group conference brings into focus for the offender the disappointment of family, friends, victims, and school staff, and causes more serious thinking about the consequences of inappropriate behavior. While shame and disapproval are part of the process, the family group conference provides the opportunity for integration back into the family or school community. The agreements reached clarify specific ways to right the wrong or heal the harm and to reduce feelings of alienation and hurt. The goal is to enhance feelings of connectedness, care, and social consciousness to reduce the possibility of future victimizing behavior.

Here are some of the lessons learned through family group conferencing:

1. Honestly acknowledging issues, feelings, and responsibility is a much more effective approach to solving problems than blame, denial, minimization, or retribution.
2. There is power in a sincere apology.
3. Building up the sense of caring and community can be much more gratifying than tearing it down.
4. Thinking through the consequences of one's actions is much better than impulsive action.
5. Inviting everyone who was affected by the victimizing incident to be part of the process of healing the harm can empower everyone involved.
6. Providing people with the opportunity to express their feelings and concerns and to collaborate in righting the wrong empowers everyone to move on in a healthy way and begin to put the incident to rest.

## Figure 1

| **Current Disciplinary Procedure** vs. | ***REAL* JUSTICE Family Group Conferencing** |
|---|---|
| • Offense defined as violation against school rules | • Offense defined as harm to a person or the school community |
| • Focus on establishing blame, on guilt, on the past | • Focus on problem solving, on repairing the harm |
| • Victim ignored | • Victim's rights and needs recognized |
| • Offending student passive | • Offending student encouraged to take responsibility |
| • Accountability defined as punishment | • Accountability defined as demonstrating empathy and helping repair the harm |
| • Response focused on offender's past behavior | • Response focused on harmful consequences of offender's behavior |
| • Stigma of offense unremovable | • Stigma of offense removable through appropriate actions |
| • Nominal encouragement for repentance and forgiveness | • Repentance actively encouraged and forgiveness possible |
| • Dependence upon school professionals | • Direct involvement by those affected |
| • Strictly a rational process | • Allows for expression of emotion |

—Adapted with permission from H. Zehr, (1990), *Changing Lenses*, (Scottsdale, Pa.: Herald Press).

## Healing and Closure

In the pepper mace incident, unlike the typical disciplinary procedure, the offenders actually had to face some of the people they had hurt (including their parents) and experience their emotions. They also had an opportunity to make amends and take actions that would move them beyond their shame and back into the school community. Conferencing allows offenders to begin to shed their stigmatizing "offender" label, and they are often amazed at the generosity of others once they show remorse.

In another elementary school incident, two boys hurt another boy. Simple punishment in this case would have only caused resentment by the offenders toward the victim. A family group conference, however, revealed a pattern of persistent bullying by the two boys. The bullies, the victim, and their respective parents all had a chance to speak honestly and emotionally. As a result, the two boys sincerely apologized and promised to help stop other students from persecuting the victim, the victim forgave the boys, their parents felt that the school had responded effectively, and harmony was restored among the children.

We have also begun using conferencing in the elementary schools as a violence prevention technique. Guidance counselors and classroom teachers involve the children in role plays of conferences to deal with common incidents of bullying, teasing, or violence. The children play the roles of offenders, victims, parents, and school staff. Not only are the role plays surprisingly realistic, but the subsequent discussions hit on very adult issues, Such as questioning the effectiveness of punishment.

Abraham Maslow said, "If the only tool I have is a hammer, I tend to treat everything as a nail." Family group conferencing is a different tool to use when school staff are seeking to respond meaningfully to victimizing behavior. It is proactive and seeks to minimize the negative impact of hurt, anger, helplessness, labeling, stigmatization, and alienation. Conferencing is making our school community a safer, more supportive place where students can learn and grow.

## Note

1. For more information, contact *REAL* JUSTICE, P.O. Box 229, Bethlehem, PA 18016-0229; (610) 807-9221.

## References

Braithwaite, J. (1989). *Crime, Shame, and Reintegration.* Cambridge: Cambridge University Press.
McDonald, J., D. B. Moore, T. O'Connell, and M. Thorsborne. (1995). Real *Justice Training Manual: Coordinating Family Group Conferences.* Pipersville, Pa.: The Piper's Press.
Nathanson, D. L. (1992). *Shame and Pride: Affect, Sex, and the Birth of the Self.* New York: W. W. Norton.
Bruce R. Taylor is the Supervisor of Special Education and Glenn Kummery is a Social Worker in Central Bucks School District, 16 Weldon Dr., Doylestown, PA 18901-2359.